METHODS IN MOLECULAR BIOLOGY

For further volumes:
http://www.springer.com/series/7651

Calpain

Methods and Protocols

Edited by

Jeannette S. Messer

Department of Inflammation and Immunity, Lerner Research Institute, Cleveland Clinic, Cleveland, OH, USA

Humana Press

Editor
Jeannette S. Messer
Department of Inflammation and Immunity
Lerner Research Institute, Cleveland Clinic
Cleveland, OH, USA

ISSN 1064-3745 ISSN 1940-6029 (electronic)
Methods in Molecular Biology
ISBN 978-1-4939-8987-4 ISBN 978-1-4939-8988-1 (eBook)
https://doi.org/10.1007/978-1-4939-8988-1

Library of Congress Control Number: 2018965755

This Springer imprint is published by the registered company Springer Science+Business Media, LLC, part of Springer Nature.
The registered company address is: 233 Spring Street, New York, NY 10013, U.S.A.

Dedication

This volume is dedicated to Dr. Hiro Sorimachi:

Dedication to Calpain research and determination to enjoy science have been messages that we have learned from Dr. Hiro Sorimachi. He has been a Calpain expert and, therefore, was a very tough as well as fair judge. Getting him convinced with our data always took some rounds of discussion, but when he cast off his doubt with a hearty laugh, all the hard work was enormously rewarded. It has been our privilege to have known him as our mentor. He will be dearly missed.

Yasuko Ono and Shoji Hata

Preface

The calpains have been something of an enigma in the world of proteases. They are a large family of cysteine proteases with limited proteolytic activity. This means that calpain-mediated cleavage events create protein fragments, often with novel functions, rather than degrade proteins. The two best-characterized calpains, calpain 1 and calpain 2, are ubiquitously expressed. However, there are now 15 characterized calpains, many of which are expressed exclusively or predominantly in specific tissue types. To add to the complexity of this system, calpains also have an endogenous inhibitor (calpastatin), and calpain cleavage events can be regulated by unrelated proteins. Understanding calpains and their complexity is crucial since calpains are involved in many fundamental developmental and homeostatic processes and have been implicated in a number of human diseases through disease-associated single nucleotide polymorphisms in calpains or calpain substrates.

The large number and wide reach of calpains has led to a strong need for many investigators to pursue calpain-related studies but also a high level of trepidation when approaching these studies. In this volume, we have attempted to provide a library of protocols useful both for scientists dedicated to calpain research and those whose work has led them to calpain in other ways. We have included experimental methods to study calpains, their inhibitors, and their substrates along with bioinformatic approaches to calpain substrate identification. We have also paid special attention to tricky or challenging aspects of these methods and included tips on how to overcome these issues in the Notes section of each protocol.

I would like to thank all of the authors who lent their experience and expertise to developing and writing these protocols. These types of endeavors are increasingly important as the complexity of techniques continues to increase while the space allotted to describe methods in journals has decreased. It is therefore my hope that these efforts will open up calpain investigations to new investigators and expand the experimental repertoire of seasoned calpain scientists. Perhaps most importantly, I hope that this volume will provide information and inspiration that leads to the new discoveries needed to diagnose and treat calpain-related diseases.

Cleveland, OH, USA *Jeannette S. Messer*

Contents

Contributors

IOANNA ANTONIADES • *Department of Biological Sciences, University of Cyprus, Nicosia, Cyprus*

HELENA ARAUJO • *Institute of Biomedical Sciences, Federal University of Rio de Janeiro, Rio de Janeiro, Brazil; Institute of Molecular Entomology, Rio de Janeiro, Brazil*

MONICA AVERNA • *Department of Experimental Medicine (DIMES)—Biochemistry Section, University of Genova, Genova, Italy*

ASHIM KUMAR BISWAS • *Division of Post-Harvest Technology, ICAR-Central Avian Research Institute, Bareilly, Uttar Pradesh, India*

CLAIRE BRUN • *University of Lyon, UMRS 449: General Biology—Reproduction and Comparative Development, Lyon Catholic University (UCLy), Ecole Pratique des Hautes Etudes (EPHE, PSL), Lyon, France*

MAIRA CARDOSO • *Institute of Biomedical Sciences, Federal University of Rio de Janeiro, Rio de Janeiro, Brazil*

ANNA CHARALAMBOUS • *Department of Biological Sciences, University of Cyprus, Nicosia, Cyprus*

NEOPHYTOS CHRISTODOULOU • *Department of Biological Sciences, University of Cyprus, Nicosia, Cyprus*

MARALICE CONACCI-SORRELL • *Department of Cell Biology, University of Texas Southwestern Medical Center, Dallas, TX, USA; Harold C. Simmons Comprehensive Cancer Center, University of Texas Southwestern Medical Center, Dallas, TX, USA*

PETER L. DAVIES • *Department of Biomedical and Molecular Sciences, Queen's University, Kingston, ON, Canada*

ROBERTA DE TULLIO • *Department of Experimental Medicine (DIMES)—Biochemistry Section, University of Genova, Genova, Italy; Centre of Excellence for Biomedical Research (CEBR), University of Genova, Genova, Italy*

MARIA DEL CARMEN LAFITA-NAVARRO • *Department of Cell Biology, University of Texas Southwestern Medical Center, Dallas, TX, USA; Harold C. Simmons Comprehensive Cancer Center, University of Texas Southwestern Medical Center, Dallas, TX, USA*

DAVID A. DUVERLE • *Graduate School of Frontier Science, The University of Tokyo, Kashiwa, Chiba, Japan; Artificial Intelligence Research Center, AIST, Koto-ku, Tokyo, Japan*

JEAN-MARIE EXBRAYAT • *University of Lyon, UMRS 449: General Biology—Reproduction and Comparative Development, Lyon Catholic University (UCLy), Ecole Pratique des Hautes Etudes (EPHE, PSL), Lyon, France*

YONG-XIAN FAN • *Guangxi Key Laboratory of Trusted Software, Guangxi Colleges and Universities Key Laboratory of Intelligent Processing of Computer Images and Graphics, Guilin University of Electronic Technology, Guilin, China*

YAN GAO • *Department of Pathology and Molecular Medicine, Queen's University, Kingston, ON, Canada; Division of Cancer Biology and Genetics, Cancer Research Institute, Kingston, ON, Canada*

PETER A. GREER • *Department of Pathology and Molecular Medicine, Queen's University, Kingston, ON, Canada; Division of Cancer Biology and Genetics, Cancer Research Institute, Kingston, ON, Canada*

CHRISTINE HALL • *Department of Pathology and Molecular Medicine, Queen's University, Kingston, ON, Canada; Division of Cancer Biology and Genetics, Cancer Research Institute, Kingston, ON, Canada*
GUOCHANG HU • *Department of Anesthesiology, University of Illinois College of Medicine, Chicago, IL, USA; Department of Pharmacology, University of Illinois College of Medicine, Chicago, IL, USA*
SEIICHI KITAGAWA • *Department of Physiology, Graduate School of Medicine, Osaka City University, Osaka, Japan; Shitennoji Tamatsukurien Clinic, Osaka, Japan*
ZHE LIANG • *Department of Biological Sciences, National University of Singapore, Singapore, Singapore*
XIAOQIAN LIU • *Department of Anesthesiology, University of Illinois College of Medicine, Chicago, IL, USA*
WILLIAM LOSTAL • *Généthon INSERM, U951, INTEGRARE Research Unit, University Paris–Saclay, Evry, France*
JAMES MACLEOD • *Department of Pathology and Molecular Medicine, Queen's University, Kingston, ON, Canada; Division of Cancer Biology and Genetics, Cancer Research Institute, Kingston, ON, Canada*
VINIT B. MAHAJAN • *Department of Ophthalmology, Byers Eye Institute, Stanford University School of Medicine, Palo Alto, CA, USA*
HIROSHI MAMITSUKA • *Bioinformatics Center, Kyoto University, Uji, Kyoto, Japan; Department of Computer Science, Aalto University, Espoo, Finland*
CHRISTIAN-SCOTT E. MCCARTNEY • *Department of Biomedical and Molecular Sciences, Queen's University, Kingston, ON, Canada*
ELARA N. MOUDILOU • *University of Lyon, UMRS 449: General Biology—Reproduction and Comparative Development, Lyon Catholic University (UCLy), Ecole Pratique des Hautes Etudes (EPHE, PSL), Lyon, France*
DANIELLE OLIVEIRA • *Department of Biochemistry, Institute of Chemistry, Federal University of Rio de Janeiro, Rio de Janeiro, Brazil*
HILDE-GUNN OPSAHL-SORTEBERG • *Department of Plant Sciences, BIOVIT, Norwegian University of Life Sciences, Ås, Norway*
XIAOYONG PAN • *Department of Medical Informatics, Erasmus MC, Rotterdam, The Netherlands*
ISABELLE RICHARD • *Généthon INSERM, U951, INTEGRARE Research Unit, University Paris–Saclay, Evry, France*
HONG-BIN SHEN • *Institute of Image Processing and Pattern Recognition, Shanghai Jiao Tong University, Shanghai, China; Key Laboratory of System Control and Information Processing, Ministry of Education of China, Shanghai, China*
PARIS A. SKOURIDES • *Department of Biological Sciences, University of Cyprus, Nicosia, Cyprus*
DANIEL STOCKHOLM • *Généthon INSERM, U951, INTEGRARE Research Unit, University Paris–Saclay, Evry, France*
SAMARTH TANDON • *Division of Post-Harvest Technology, ICAR-Central Avian Research Institute, Bareilly, Uttar Pradesh, India*
KATHERINE J. WERT • *Department of Ophthalmology, Byers Eye Institute, Stanford University School of Medicine, Palo Alto, CA, USA*
SARA ZANARDELLI • *Department of Biological Sciences, University of Cyprus, Nicosia, Cyprus*
YANG ZHANG • *Department of Computational Medicine and Bioinformatics, University of Michigan, Ann Arbor, MI, USA*

Part I

Production and Purification of Calpains

Chapter 1

Single-Step Purification of Calpain-1, Calpain-2, and Calpastatin Using Anion-Exchange Chromatography

Ashim Kumar Biswas and Samarth Tandon

Abstract

Purification and separation of calpains and calpastatin are used to determine the individual activities of calpain-1 and calpain-2 and their inhibitor calpastatin. We discuss here a method to purify these enzymes using dialysis followed by separation using anion-exchange chromatography coupled with gradient elution. Swollen DEAE Sephacel is used as the column matrix in this method. Calpastatin and both domains of calpain are weakly basic molecules that effectively bind with the DEAE Sephacel and separate well using a stepwise, increasing gradient of NaCl to elute the proteins. Calpastatin binds most weakly with the column matrix, so it elutes first, followed by calpain-1 and, finally, calpain-2.

Key words Purification, Chicken, Muscle tissues, Calpain, Calpastatin, DEAE Sephacel, Anion-exchange chromatography

1 Introduction

Purification of calpains and their inhibitory agent calpastatin is important to better understand activities of these enzymes in tissues. Calpain-1 and calpain-2 are heterodimers composed of distinct 80 kDa catalytic subunits and an identical 28 kDa regulatory subunit, are calcium dependent, and autolyze in the presence of calcium [1, 2]. Calpastatin contains a variety of subunits, but the 66 kDa subunit has the greatest inhibitory activity [3]. Similar to calpains, calpastatin also undergoes degradation in the presence of calcium [1]. But degradation of calpastatin does not lead to complete loss of inhibitory activity, and even in extensive proteolysis, some inhibitory activity remains [4].

Earlier work suggests that four different fractionation steps are required to purify calpain-1, calpain-2, and calpastatin [5]. Later, in many studies, a single chromatographic step is used to purify these proteins [6–8]. In some studies, it has been reported that purification of calpains and calpastatin cannot be achieved by stepwise elution, while other researchers reported that long elution times

Jeannette S. Messer (ed.), *Calpain: Methods and Protocols*, Methods in Molecular Biology, vol. 1915,
https://doi.org/10.1007/978-1-4939-8988-1_1, © Springer Science+Business Media, LLC, part of Springer Nature 2019

are required using low-pressure chromatographic techniques and the number of samples that can be processed in a given time is therefore limited. An alternative method using hydrophobic interaction chromatography followed by ion-exchange chromatography is also reported [6]. In comparison to this, application of anion-exchange chromatography using DEAE Sephacel as column matrix has greater efficiency in separation of calpain-1, calpain-2, and calpastatin [9, 10], and further involvement of an additional purification step improves the purity. In many instances, a dialysis step is introduced before column chromatography separation to remove low molecular weight (MW) reactive molecules and also to reduce ion suppression effects of sample matrix [11].

2 Materials

All the solvents and reagents required for this experiment should be of analytical quality. Protease inhibitors should be highly purified forms. Prepare all solutions using double-distilled water (ddH_2O), and store at 4 °C until further use, unless otherwise indicated.

2.1 Tissue Lysis Buffer

50 mM Tris–HCl, (pH 8.3), 10 mM Ethylenediaminetetraacetic acid (EDTA), and 0.05% (v/v) 2-mercaptoethanol (2-MCE, assay 99%). Add a cocktail mixture of protease inhibitors from the following stock solutions (phenylmethanesulfonyl fluoride (PMSF), leupeptin, and ovomucoid) to the tissue lysis buffer immediately before use.

1. *PMSF* (100 mM stock solution): Weigh 174 mg of PMSF, and transfer to a 15 mL conical plastic tube. Add 10 mL of isopropanol, dissolve well, and store at −20 °C. Add 195 μL of stock solution to 10 mL of tissue lysis buffer immediately before use.
2. *Leupeptin* (5 mg/mL stock solution): Weigh 10 mg of leupeptin dry powder, transfer to a micro-centrifuge tube, and dissolve with 2 mL of ddH_2O. Store the solution at −20 °C. Add 12 μL of stock solution to 10 mL of tissue lysis buffer immediately before use.
3. *Ovomucoid* (300 mg/mL stock solution): Weigh 150 mg of ovomucoid dry powder, transfer to a micro-centrifuge tube, and dissolve with 500 μL of ddH_2O. The solution is stable for up to 2 months at −20 °C. Add 66 μL of stock solution to 10 mL of tissue lysis buffer immediately before use.

2.2 SDS-PAGE

1. *Resolving gel buffer*: 1.5 M Tris–HCl, pH 8.8. Store at 4 °C.
2. *Stacking gel buffer*: 1.32 M Tris–HCl, pH 6.8. Store at 4 °C.
3. *Polyacrylamide stock solution*: (40:1.6) Acrylamide to bis-acrylamide. Place 4 g of acrylamide and 0.16 g

bis-acrylamide into a 50 mL Falcon tube, and add 5 mL of ddH_2O to dissolve the chemicals. After the chemicals are dissolved, increase the volume to 10 mL using ddH_2O. Filter the solution through Whatman filter paper 1, and then use it immediately, or store it at 4 °C, protected from light. The stock solution can be used for up to 1 month.

4. *SDS buffer*: 10% SDS (w/v) in ddH_2O.
5. *Ammonium persulfate solution (APS)*: 10% ammonium persulfate (w/v) in ddH_2O. Prepare fresh whenever required.
6. *N′,N′,N′,N′-Tetramethylethylenediamine (TEMED)*: Store at 4 °C.
7. *Bromophenol blue*: Dissolve 0.05 g BPB in 100 mL ddH_2O.
8. *Sample buffer (1:2, v/v)*: 1.32 M Tris–HCl (pH 6.8), 20% (v/v) glycerol, 10% (w/v) SDS, and 0.05% (w/v) bromophenol blue. Mix well and finally add 2-MCE just before use.
9. *Electrode running buffer (10×)*: 25 mM Tris–HCl, pH 8.3, 192 mM glycine, 1% (w/v) SDS.
10. *Coomassie Brilliant Blue G-250 solution*: Dissolve 0.20 g Coomassie Brilliant Blue into a solution containing 40 mL methanol, 10 mL glacial acetic acid, and 50 mL ddH_2O to make up final volume to 100 mL.
11. *Destaining solution*: Add 30 mL methanol and 10 mL glacial acetic acid into 60 mL of ddH_2O. Final volume makes up to 100 mL.

2.3 DEAE Sephacel

Diethylaminoethyl-Sephacel microporous (40–160 μm) aqueous ethanol suspension (Sigma-Aldrich).

2.4 Anion-Exchange Glass Column

1.5 × 10 cm, glass chromatography column, maximum volume 18 mL (Econo column, Biorad).

2.5 Dialysis Buffer

40 mM Tris–HCl, pH 7.4, 5 mM EDTA, and 0.05% (v/v) 2-mercaptoethanol (2-MCE). Store at 4 °C until use and make fresh for every use.

2.6 Equilibration Buffer

40 mM Tris–HCl, pH 7.4, 0.5 mM EDTA, and 0.05% 2-MCE. Store at 4 °C until use and make fresh for every use.

2.7 Elution Buffer

Prepare solutions of 100 mM, 200 mM, 400 mM, and 2 M NaCl.

2.8 Equipment

1. Gel electrophoresis system along with accessories (spacer plate, short plate, casting frame, casting stand, buffer dam, electrophoresis module, tank, and lid).
2. Power supply.

3 Method

Carry out all of the procedures at 25 °C to avoid protein denaturation and preserve enzyme activity.

3.1 Preparation of Tissue Homogenate

Always collect fresh tissue samples immediately after slaughter to avoid autolytic degradation or denaturation of calpains and calpastatin. Prepare the sample extract under chilled conditions, or hold the tissues at −20 °C until use.

1. Weigh 0.3 g of finely cut tissue into a chilled mortar and pestle, and homogenize with 6 volumes of ice-cold tissue lysis buffer containing protease inhibitors (1.8 mL, pH 8.3) (*see* **Note 1**). This method was optimized for use with chicken muscle tissue but should be broadly applicable to other tissue types.
2. Centrifuge the tissue lysate at 12,000 rpm (20,000 × *g*) for 20 min at 4 °C.
3. Transfer the supernatant to a separate centrifuge tube and discard the pellet.
4. Centrifuge the supernatant once again as mentioned earlier to obtain purified supernatant. Then filter the purified supernatant through Whatman filter paper No. 1. This purified, filtered supernatant contains the calpain enzymes and is the sample that will be used for subsequent analysis (*see* **Note 2**).

3.2 Column Packing

1. An anion exchanger is used for packing the column.
2. Use a Pasteur pipette to draw up swollen DEAE Sephacel matrix, and then pour it carefully into the column without touching the sidewall.
3. Pack a column (1.77 cm inner dimension) with the matrix to about 5 cm in length, and wait for 40–50 min to settle down properly (*see* **Note 3**).

3.3 Dialysis

A dialysis step is required to remove unwanted protein molecules from the sample extract. Since the proteins of interest have MW of 18–80 kDa, a low molecular weight cutoff (MWC) filter should be used to remove those proteins. This will also help to get more available binding sites for the calpain and calpastatin domains in the anion exchanger.

1. Take 1 L freshly prepared working dialysis buffer solution (1× buffer solution, pH 7.4) in a graduated glass beaker.
2. Cut a dialysis tube (12 kDa MWC filter) to a length that can accommodate 20 mL of sample homogenate.
3. Place a knot on the one end of the dialysis tube with cotton thread while keeping the other end free.

4. Measure 20 mL sample homogenate, transfer to the dialysis tube, and place a knot at the free end to secure the solution.
5. Place the dialysis tube in the glass beaker containing the buffer solution, cover with aluminum foil, and keep at 4 °C for 24 h.
6. After 24 h of dialysis, take out the dialysis tube, open the thread carefully, and then transfer the sample into a 50 mL glass beaker.
7. Maintain the sample under chilled conditions either using ice or refrigeration.

3.4 Column Separation of Calpains and Calpastatin

1. Attach the prepacked column (as prepared earlier) with a stand.
2. Drain the excess ethanol without disturbing the column matrix. Handle the stopper carefully.
3. Wash the column two times with 20 mL 2 M NaCl solution (*see* **Note 4**).
4. Equilibrate the column three times using 15 mL working solution (equilibration buffer). For each time, after pouring off the solution, wait for 5–10 min, and then carefully drain out the excess solution. Please wait to dry up with gravity flow of the solution.
5. Load 15 mL meat homogenate into the column which is equivalent to 5 g of muscle. Wait for 40 min to bind the protein molecules with the matrix (*see* **Note 5**).
6. Pre-wash the column with 8 mL of equilibration buffer three times.
7. Elute with 12 mL of 100 mM NaCl eluting buffer. Collect the elutes in three different fractions of 4 mL each.
8. Now elute with 12 mL of 200 mM NaCl buffer, and collect it in three different fractions as mentioned above.
9. Finally elute with 12 mL of 400 mM NaCl buffer as mentioned above (*see* **Note 6**).
10. Store all the eluted fractions (3 + 3 + 3 = 12) at −20 °C until loading in the SDS-PAGE or using for biochemical assays.

3.5 SDS-PAGE

1. Prepare the resolving gel using 4.1 mL ddH_2O, 2.5 mL of resolving buffer, 3.3 mL polyacrylamide solution, 100 μL of SDS, 50 μL of APS, and 5 μL of TEMED. Mix the contents thoroughly and degas the solution for at least 15 min. Cast the gel in a gel cassette fitted with the casting stand. Allow space to add stacking gel by placing a comb completely into the assembled gel cassette and marking the glass plate 1 cm below the comb teeth. Remove the comb, pour the resolving gel solution gently up to the mark, and, finally, overlay with isobutanol (*see* **Note 7**).

2. Allow the gel to polymerize for at least 45 min. Rinse the gel surface completely with ddH_2O. Do not leave the alcohol overlay on the gel for more than 1 h since it will dehydrate the top of the gel (*see* **Note 8**).
3. Prepare the stacking gel monomer solution using 6.1 mL of ddH_2O, 2.5 mL of stacking gel, and 1.3 mL of polyacrylamide solution. Degas the solution for 15 min, and then add 100 μL of SDS, 50 μL of APS solution, and 10 μL of TEMED. Dry the top of the resolving gel with Whatman filter paper, smoothly pour the solution until the top of the short plate, and then place comb between the spacers. Wait for polymerization (*see* **Note 9**).
4. Set the clamping frame to the open position on a clean flat surface. Place the first gel cassette with the short plate facing inward on to the gel supports; gel supports are moulded into the bottom of the clamping frame assembly. The gel will now rest at a 30° angle, tilting away from the center of the clamping frame. Similarly place the second gel on the other side of the clamping frame (*see* **Note 10**).
5. Place and lock the gel cassettes gently, and if use an odd number, please use a buffer dam.
6. Fill the assembly upper chamber with electrode running buffer to just under the edge of the outer gel plate.
7. Mix the sample with the sample buffer (1:2, v/v), and heat all of the aliquots of 100 mM, 200 mM, and 400 mM NaCl fractions at 95 °C for 5 min.
8. Do not add lysis buffer to the prestained protein standard (marker), or subject it to heat. Spin the heated samples at 4000 × *g* for 30 s to bring down the condensate.
9. Load all of the samples on the SDS gel starting with 100 mM fraction, then 200 mM fraction, and finally 400 mM fraction. In the last well, load the protein standard (marker). Run the electrophoresis unit at 100 V in 1× electrode running buffer till the dye front reached to the bottom of the gel.
10. After the end of the electrophoresis, turn off the power supply, and disconnect the electrical lead. Remove the tank lid and carefully lift out the electrode assemblies. Pour off and discard the running buffer. Open the arms of the assembly, and remove the gel cassettes by gently separating the two plates (*see* **Note 11**).
11. Rinse the gel carefully with ddH_2O to remove traces of SDS-PAGE running buffer.
12. Stain the gel with CBB G-250 for 60 min, and then destain using destaining solution until bands are distinctly visible.

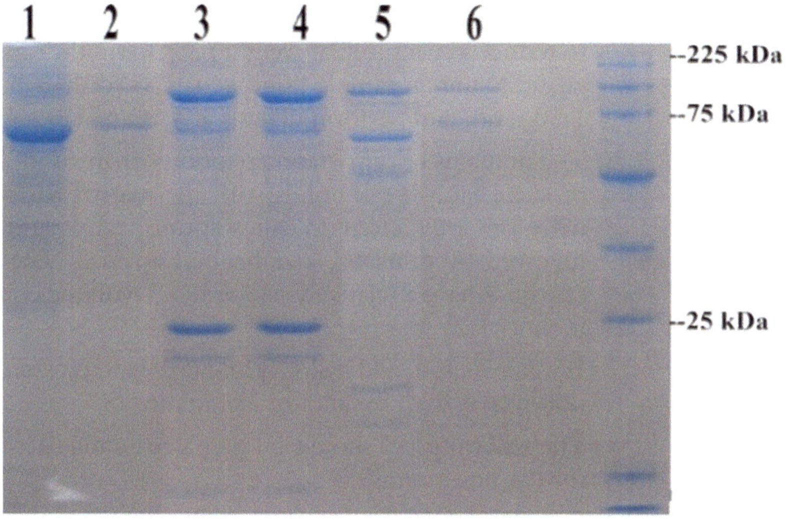

Fig. 1 SDS-PAGE showing different fractions of calpastatin, calpain-1, and calpain-2 derived from chicken muscle tissue lysates. Lane-1, calpastatin breast; Lane-2, calpain-1 breast; Lane-3, calpain-2 breast; Lane-4, calpain-2 thigh; Lane-5, calpastatin thigh; Lane-6, calpain-1 thigh. *Marker*: Novagen Perfect Protein Markers, 10–225 kDa, include protein sizes of 10, 15, 25, 35, 50, 75, 100, 150, and 225 kDa

13. Compare with the marker to identify calpastatin, calpain-1, and calpain-2 (Fig. 1).
14. Apply Western blotting for further confirmation of these proteins.

4 Notes

1. Never add the protease inhibitors directly to the un-lysed tissues. They are rapidly degraded in water, so they should be added to the tissue lysis buffer immediately before use.
2. Load the samples into the gel immediately after preparation for the best result.
3. A column with an inner dimension of 1.77 cm is sufficient to accommodate 15 mL of sample extract equivalent to extract from 5 g of tissue sample. Wash the column three times with 2 M NaCl solution before and after separation of calpains and calpastatin from the sample matrix, and while not under use, maintain the column matrix with 20% ethanol. Hold the column at 4 °C. Always keep a small quantity of ethanol in the column to avoid unwanted drying up of matrix during storage.
4. In column washing, gently pour the solution along the sidewall of the column without disturbing the column bed.
5. Dialyze the meat homogenate before centrifugation at 20,000 × *g*, filter through Whatman filter paper No. 42, and

then load into the column. Use 12 kDa MWC filter for dialysis to reduce low molecular weight proteins in the sample homogenate. Maintain sample homogenate vs. dialysis buffer at least 1:50 ratio, and perform dialysis at 4 °C.

6. The domains of calpastatin are poorly charged molecules, they produce fairly weak bonds with the matrix, and when eluted with low ionic strength salt solution, calpastatin elutes first, followed by calpain-1 and finally calpain-2. The domains of calpain-2 have a relatively strong bond with the column matrix; it elutes last.
7. If water is used, add it slowly and evenly to prevent mixing. Do not overlay with butanol or isobutanol.
8. The resolving gel should be prepared immediately before use for the best results.
9. Insert the desired comb between the spacers starting at the top of the spacer plate, making sure that the tabs at the ends of each comb are guided between the spacers. Seat the comb in the gel cassette by aligning the comb ridge with the top of the short plate.
10. Please use caution when placing the first gel, making sure that the clamping frame remains balanced and does not tip over. It is critical that gel cassettes are placed into the clamping frame with the short plate facing inward. Do not attempt to lock the green arms of the clamping frame, without first ensuring that the gel cassettes are perfectly aligned and stabilized against the notches on the green gaskets of the module.
11. Always pour off the electrode buffer before opening the arms of the assembly, to avoid spilling of the buffer. If the gel sticks to the plate, remove the gel by floating it off in transfer solution or even in gel staining solution.

Acknowledgment

This work is supported by DST (Govt. of India) grant SB/FT/LS-283/2012 and the Indian Council of Agricultural Research, New Delhi (Ministry of Agriculture).

References

1. Koohmaraie M, Geesink GH (2006) Contribution of post-mortem muscle biochemistry to the delivery of consistent meat quality with particular focus on the calpain system. Meat Sci 74:34–43
2. Voahanginirina R, Ingrid F (2010) The role of calpain in diabetes-associated platelet hyperactivation. Adv Pharmacol 59:235–257
3. Kripriyalini L (2015) Identification of calpains and calpastatin enzymes from turkey blood and their influence on post-mortem tenderization of skeletal muscle during refrigeration storage. M.V.Sc. Thesis, Deemed University, IVRI, India
4. DeMartino GN, Wachendorfer R, McGuire M, Croall DE (1988) Proteolysis of protein

inhibitor of calcium dependent proteases produces lower molecular weight fragments that retain inhibitory activity. Arch Biochem Biophys 262:189–198

5. Wilson K, Walker J (2000) Practical biochemistry-Principles and techniques. 5th edn. (Cambridge low-price edition). Replika Press Pvt. Ltd., New Delhi, India
6. Etherington DJ, Taylor MA, Dransfield E (1987) Conditioning of meat from different species. Relationship between tenderizing and the levels of cathepsin B, cathepsin L, calpain I, calpain II and β-glucuronidase. Meat Sci 20:1–18
7. Kretchmar DH, Hathaway MR, Epley RJ, Dayton WR (1989) In vivo effect of a b-adrenergic agonist on activity of calcium-dependent proteinases, their specific inhibitor, and cathepsins B and H in skeletal muscle. Arch Biochem Biophys 275:228–235
8. Koohmaraie M (1990) Quantification of Ca^{2+}-dependent protease activities by hydrophobic and ion-exchange chromatography. J Anim Sci 68:659–665
9. Geesink GH, Koohmaraie M (1999) Post-mortem proteolysis and calpain/calpastatin activity in callipyge and normal lamb *Biceps femoris* during extended post-mortem storage. J Anim Sci 77:1490–1501
10. Karlsson JO, Gustavsson S, Hall C, Nilsson E (1985) A simple one-step procedure for the separation of calpain I, calpain II and calpastatin. Biochem J 231:201–204
11. Biswas AK, Tandon S, Beura CK (2016) Simple extraction method for determination of different domains of calpain and calpastatin from chicken blood and their role in post-mortem ageing of breast and thigh muscles at 4±1 °C. Food Chem 200:315–321

Chapter 2

Bacterial Expression and Purification of Calpains

Christian-Scott E. McCartney and Peter L. Davies

Abstract

The production of recombinant proteins has been a cornerstone of the study of protein structure and function. As an example, the expression and purification of recombinant rat calpain-2 in bacteria was essential for solving the first crystal structures of the calpains in both calcium-free and calcium-bound forms. Here we describe the production and purification of recombinant rat calpain-2 from *Escherichia coli* using anion-exchange, affinity, and size-exclusion chromatographies. The heterodimeric enzyme is produced from a stable two-plasmid system. The order in which the protocol is carried out has been optimized to reduce unnecessary concentration and dialysis steps. The typical yield of this multi-domain enzyme from 4 L of *E. coli* culture is about 20 mg. The production of whole structures for the other calpain family members has been fraught with difficulty. To circumvent this roadblock, a certain amount of structure-function information can be gleaned about these other calpain isoforms by expressing just their protease core. These "mini-calpains" have been useful for X-ray co-crystallography with calpain inhibitors.

Here we also present a variation of the whole enzyme production and purification protocol optimized for the expression and purification of the calpain-1 and calpain-3 protease cores (mini-calpains).

Key words Calpain-2, Calpain-1, Calpain-3, Protease core, Protein purification, Recombinant protein, Protein expression, Affinity chromatography, Size-exclusion chromatography, Anion-exchange chromatography

1 Introduction

The large-scale expression and purification of recombinant rat calpain-2 has been tremendously impactful in resolving the structure-function relationships of the enzyme, for deducing its activation mechanism, and the mode of inhibition by its natural inhibitor, calpastatin [1]. Endogenous calpain-1 and calpain-2 have been purified from a number of animal tissues [2]. These preparations can be useful for some purposes, but for methods like X-ray crystallography, larger amounts of homogeneous protein are required. The advancement of non-native protein expression and cloning technologies has facilitated the production of foreign proteins in bacteria and made it possible to obtain a higher degree of protein purity and yield than can be achieved from natural sources.

Jeannette S. Messer (ed.), *Calpain: Methods and Protocols*, Methods in Molecular Biology, vol. 1915,
https://doi.org/10.1007/978-1-4939-8988-1_2,

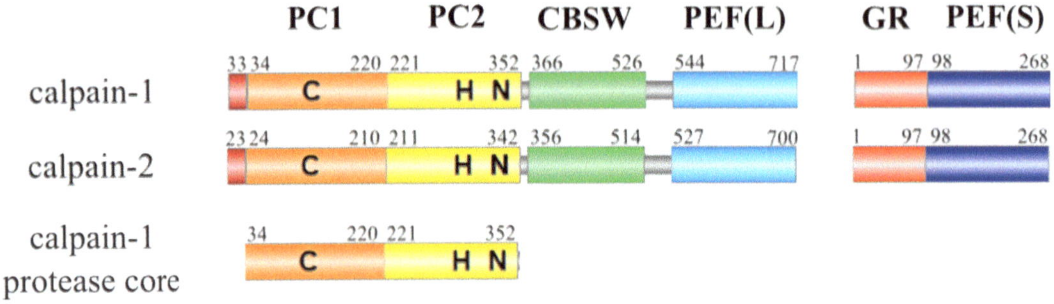

Fig. 1 Domain architecture of calpain-1, calpain-2, and calpain-1 protease core

Minor modifications to the DNA sequence of rat calpain-2, such as the truncation of the small 21 kDa subunit glycine-rich (GR) domain, the optional inactivation of the enzyme by replacement of the catalytic cysteine with serine or alanine, and the addition of a C-terminal His-tag to the large 80 kDa subunit, have helped to improve protein production and simplify its purification [3]. Although autoproteolysis and aggregation can complicate work on calcium-bound whole calpain-1 and calpain-2, the protease core (PC) and penta-EF-hand (PEF) (Fig. 1) domains of the major calpains seem to be relatively stable on their own. Within the whole enzyme, it is the calpain-type beta-sandwich (CBSW) domain that is very susceptible to proteolysis [4].

The discovery that the protease core domains PC1 and PC2 produce a functional "mini-calpain," free of the CBSW domain and resistant to proteolysis in the presence of Ca^{2+}, has been useful [5]. An advantage to the expression and purification of the protease core is that it alleviates the complication of working with the two-subunit system of the heterodimeric whole calpain-1 and calpain-2. Although considerably less active than the whole enzyme, the core does not autoproteolyze and so does not rapidly approach a plateau of activity. As a result, the core has been a useful reagent for co-crystallization with protease inhibitors. Another example of the usefulness of the "mini-calpains" has been the characterization of the skeletal muscle-specific calpain-3 (p94). Rapid autoproteolysis of its insertion sequences 1 and 2 has been linked to the chronic instability of the whole enzyme, making the production and purification of this calpain particularly difficult [6, 7]. However, the calpain-3 protease core is relatively stable as a recombinant protein in the absence of Ca^{2+} and is functional for some time after activation with Ca^{2+} [8].

We describe here a protocol for the expression and purification of recombinant rat calpain-2 and a modified protocol that can be used for the expression and purification of the calpain-1 and calpain-3 protease cores from *E. coli*.

2 Materials

2.1 Specialized Equipment

1. A sterile sparger or equivalent aerator.
2. A Sonic Dismembrator Model 500 (Fisher Scientific), or similar method of bacterial cell lysis.
3. An ÄKTA FPLC Purifier (GE Healthcare), or equivalent with a 150 mL FPLC Superloop (GE Healthcare) attachment.
4. Amicon Ultra 35 kDa centrifugal filters (Millipore) or equivalent protein concentration method.

2.2 Reagents and Solutions

1. cDNA clone for recombinant rat calpain-2 80 kDa large subunit: PET-24-80K-CHis$_6$ contains the cDNA encoding the rat calpain-2 80 kDa large subunit with a C-terminal His-tag and [3] has been cloned into the pET-24d expression vector (Novagen), which carries the gene for kanamycin resistance, the ColE1 origin of replication, and T7 promoter sequences (*see* **Note 1**).
2. cDNA clone for recombinant rat calpain-2 small subunit: pACpET-21k contains the cDNA encoding the rat calpain-2 small subunit with a C-terminal truncation of the GR domain [3]. This vector carries the ampicillin resistance gene, A15 origin of replication, and T7 promoter sequences. The small subunit is considerably overproduced in relation to the large subunit, which is why its DNA was inserted into a low copy number plasmid.
3. cDNA clones for recombinant calpain-1 protease core: The cDNA for the calpain-1 core domains, residues 29-356, was obtained by PCR amplification of rat calpain-1 DNA and inserted into pET-24d (Novagen) expression vector to include a C-terminal His-tag and kanamycin resistance [5].
4. cDNA clone for recombinant calpain-3 protease core: The cDNA encoding the protease core domains of human calpain-3, with an N-terminal truncation of residues 1-45 was achieved by RT-PCR of total RNA from human skeletal muscle (Stratagene). The cDNA was cloned into the *Nde*I and *Xho*I sites of the pET-24d expression vector (Novagen) [8, 9], which incorporates a C-terminal His-tag and kanamycin resistance.
5. *E. coli* strain BL21(DE3) without pLysS (Novagen) transformed with the above plasmids and maintained in media supplemented with 100 μg/mL kanamycin or 100 μg/mL ampicillin (when appropriate). Frozen stocks of the transformants kept at −80 °C in media supplemented with 0.1% (v/v) dimethyl sulfoxide (DMSO).
6. Growth medium (LB): Premixed powder for bacterial cultures dissolved at 25 g/L in water and autoclaved. Kanamycin

(BioShop Canada Inc.); ampicillin (BioShop Canada Inc.); antifoam (Sigma-Aldrich) or similar; isopropyl β-D-1-thiogalactopyranoside powder (IPTG) (Fisher Scientific).

7. Lysis buffer: 25 mM Tris–HCl, pH 7.5 (at room temperature), 5 mM EDTA, 5% (v/v) glycerol, 0.1 mM phenylmethanesulfonyl fluoride (PMSF) (added from a 0.1 M stock solution in isopropanol immediately before use), and 10 mM β-mercaptoethanol, also added just before use.
8. Ni-buffer (refrigerated): 50 mM Tris–HCl, pH 7.5, 150 mM NaCl, 2% (v/v) glycerol, 5 mM imidazole.
9. Ni elution buffer: Ni-buffer containing a final concentration of 250 mM imidazole.
10. Collection fluid (made fresh before use and kept on ice): 20 mM Tris–HCl, pH 7.5, 10 mM EDTA, 10 mM β-mercaptoethanol.
11. SEB buffer: 25 mM Tris–HCl, pH 7.5, 150 mM NaCl, 2% (v/v) glycerol, 10 mM β-mercaptoethanol (made fresh before use).
12. TE buffer: 25 mM Tris–HCl, pH 7.5, 5 mM EDTA, 10 mM β-mercaptoethanol (added just before use).
13. TE elution buffer (Q-Sepharose): TE buffer with 1 M NaCl.
14. TE elution buffer (DEAE-Sephacel): TE buffer with 0.75 M NaCl.
15. Chromatography resins (or equivalent): DEAE-Sephacel (GE Healthcare) Ni-NTA-agarose (Qiagen); HiLoad Q-Sepharose 16/10 high-performance FPLC column (GE Healthcare); size-exclusion column Sephacryl S-200 (GE Healthcare), or similar.
16. Materials for Tris-Tricine SDS polyacrylamide gel electrophoresis [10].

3 Methods

3.1 Bacterial Expression of Recombinant Calpain

We describe here the expression of a 4 L preparation of recombinant rat calpain-2 from *E. coli*. This method typically produces about 20 mg of highly pure protein. Its preparation takes approximately 1 week to complete and can be scaled to suit experimental needs. It is important to complete the purification without interruption, since calpain-2 has a tendency to autoproteolyze and aggregate.

The production of the protease core "mini-calpains" of rat calpain-1 and human calpain-3 is similar to that of whole rat calpain-2. The amount of protein produced by the expression of the core domains is greater than that of whole rat calpain-2, approximately 50 mg from 4 L of culture. DEAE-Sephacel anion-exchange chromatography can be performed when purifying the protease core; however, the more alkaline pH of the core domains causes the mini-calpain to elute along with higher concentrations of unwanted bulk proteins that have a tendency to form aggregates and precipitate. Therefore, when purifying the protease core, we omit this step in our purification protocol and proceed directly to Ni-NTA-agarose affinity chromatography.

1. Calpain-2 Day 1: Streak a fresh LB/agar plate containing kanamycin (100 μg/mL) and ampicillin (100 μg/mL) from a frozen stock of *E. coli* BL21 (DE3) carrying the PET-24-80K-$CHis_6$ plasmid encoding calpain-2 large subunit and pACpET-21 k plasmid encoding the small subunit genes, and grow overnight at 37 °C (*see* **Note 2**).
2. Protease Core Day 1: Streak a fresh LB/agar plate containing kanamycin (100 μg/mL) from a frozen stock of *E. coli* BL21 (DE3) carrying the pET24a plasmid encoding calpain-1 or calpain-3 protease core genes.
3. Calpain-2 Day 2: Autoclave 100 mL of LB in a 250 mL Erlenmeyer flask and 4 L of LB in a 6 L Erlenmeyer flask. The latter should be left to cool at room temperature overnight. To the 100 mL of LB, add 100 μg/mL kanamycin and 100 μg/mL ampicillin, and inoculate with a single colony from the plate. Grow the 100 mL culture at 37 °C overnight with vigorous shaking (~220 rpm).
4. Protease Core Day 2: Autoclave 100 mL of LB in a 250 mL Erlenmeyer flask and 4 × 1 L of LB in 2 L Erlenmeyer flasks. Cool the broth at room temperature overnight. To the 100 mL of LB, add 100 μg/mL kanamycin and a single colony. Grow the 100 mL culture at 37 °C overnight with vigorous shaking (~220 rpm).
5. Calpain-2 Day 3: Add 100 μg/mL kanamycin, 100 μg/mL ampicillin, and 0.5 mL of antifoam to 4 L of LB, and pre-warm in a 37 °C water bath. Slowly pour the 100 mL culture into pre-warmed 4 L LB. Grow the 4 L culture at 37 °C. Provide vigorous aeration by means of a sparger (*see* **Note 3**).
6. Protease Core Day 3: Add 100 μg/mL kanamycin and 25 mL of 100 mL culture to each pre-warmed 1 L of LB. Grow the 1 L cultures at 37 °C with vigorous shaking.
7. Monitor the growth of the 4 L culture by recording OD_{600} in regular intervals. This value should reach ~0.6–0.8 within 3–4 h.
8. Calpain-2: When OD_{600} reaches ~0.6, remove the flask from the water bath, and continue growth for 1 h at room temperature (*see* **Note 4**).

9. Protease Core: When OD_{600} reaches ~0.6, reduce the temperature to 18 °C, and continue to shake the flask for an additional hour before induction.
10. After 1 h, add 100 μg/mL IPTG and continue growth overnight (*see* **Note 5**).
11. The next day, recover the *E. coli* by centrifugation in 1 L bottles in a Beckman JS-42 series swing bucket rotor at 2500 × *g* for 25 min at 4 °C. Decant the supernatant (spent medium).
12. Suspend the cell pellets in 12.5 mL ice-cold lysis buffer, with freshly added PMSF and β-mercaptoethanol. Combine the cell suspensions in a 150–200 mL beaker on ice (*see* **Note 6**).

3.2 Purification of Bacterially Expressed Calpain

All procedures are carried out on ice or at 4 °C.

1. Place the beaker containing the cell suspension in an ice bath.
2. Sonicate the suspension with 4–5 bursts for 30–45 s, with intervals of cooling in the ice bath between bursts to prevent overheating. Make sure the suspension does not get warm. Adequate lysis can be determined by a change from an off-white dense suspension (contents viewed in a Pasteur pipette) to a brown, less opaque suspension (*see* **Note 7**).
3. Centrifuge the sonicated lysate at 25,000 × *g* in a Beckman JA-25.5 rotor or equivalent for 45 min.
4. Protease Core: Proceed directly to Subheading 3.3, Subsequent Column Steps (*see* **Note 8**).
5. Calpain-2: During the centrifugation, stir 150 mL (settled volume) of DEAE-Sephacel in 500 mL of chilled TE buffer with fresh 10 mM β-mercaptoethanol.
6. Pour the supernatant, about 50–100 mL, from the centrifuged lysate into the stirring DEAE-Sephacel suspension. Stir at 4 °C for 30–60 min, then stop the stirring, and allow the resin to settle (*see* **Note 9**).
7. Decant as much of the supernatant solution as possible, ensuring that you do not pour out any of the resin. Pour the remaining slurry into a 2.6/40 cm column, and allow the resin to settle.
8. Wash the column with TE buffer at a flow rate of 1 mL/min for approximately 60 min. Elute the bound proteins with a salt gradient in TE elution buffer from 0 to 0.75 M NaCl, in a total volume of 700 mL. The gradient flows at 1 mL/min overnight while collecting 10 mL fractions.
9. Calpain-2 typically elutes between fractions 35 and 45 in the conditions described, at a conductivity of 14–25 mmho (~0.25 M) (*see* **Note 10**) (Fig. 2). Next morning, screen the eluted fractions in this conductivity range for calpain-2 by SDS-PAGE and A280 or Bradford protein assay.

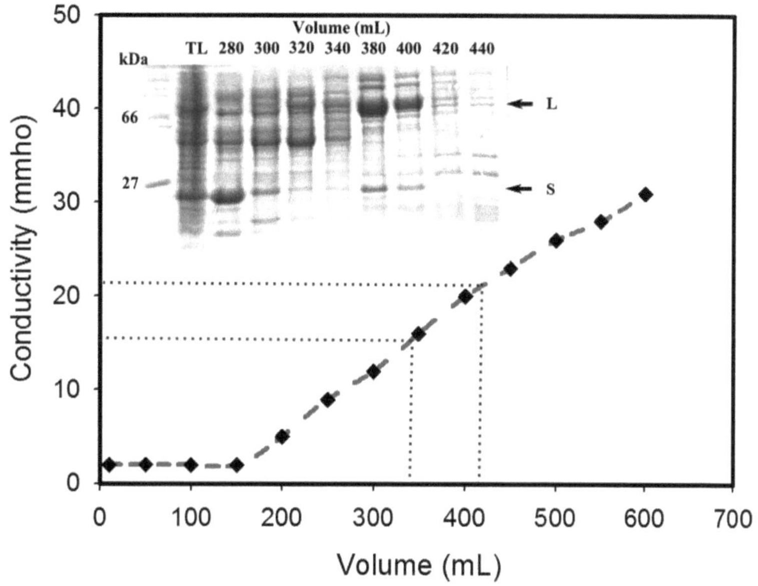

Fig. 2 Chromatography column 1: purification of rat calpain-2 on DEAE-Sephacel anion-exchange column. The elution range in conductivity (black squares and dashed gray line) and volume of recombinant rat calpain-2 are bounded by the dotted gray line intercepts. SDS-PAGE analysis (insert) stained with Coomassie Brilliant Blue was performed on the total cell lysate (TL) and elution fractions collected from the column. The presence of calpain-2 at the 380 and 400 mL elution volumes is indicated by distinct bands for both the large (80 kDa) and small (21 kDa) subunits, which are designated by arrows L and S, respectively. Excess 21 kDa small subunit (S) can be found eluting from the column at 280 mL

3.3 Ni-NTA-Agarose Affinity Chromatography (See Note 11)

1. Prepare a column (1 × 5 cm) of Ni-NTA-agarose resin in Ni-buffer.
2. Prepare the tubes for fraction collection by adding 200 μL of freshly prepared collection fluid to each tube.
3. Calpain-2: Pool DEAE-Sephacel fractions containing calpain-2 into a 250 mL beaker, and bring the volume up to 200 mL with Ni-buffer (*see* **Note 12**).
4. Protease Core: Pour the supernatant of the centrifuged cell lysate into a 250 mL beaker, and bring the volume up to 200 mL with Ni-buffer.
5. While stirring, slowly add 1 mL of 2 M $MgCl_2$ to saturate the EDTA (*see* **Note 13**) to the enzyme preparations in Ni-buffer.
6. Apply the enzyme solution to the column, and allow the solution to flow through by gravity at about 1 mL/min. Collect each fraction for SDS-PAGE or Bradford protein assay analysis.
7. Wash the column thoroughly 3–5 times with 20–30 mL ice-cold Ni-buffer. The resin should have a characteristic

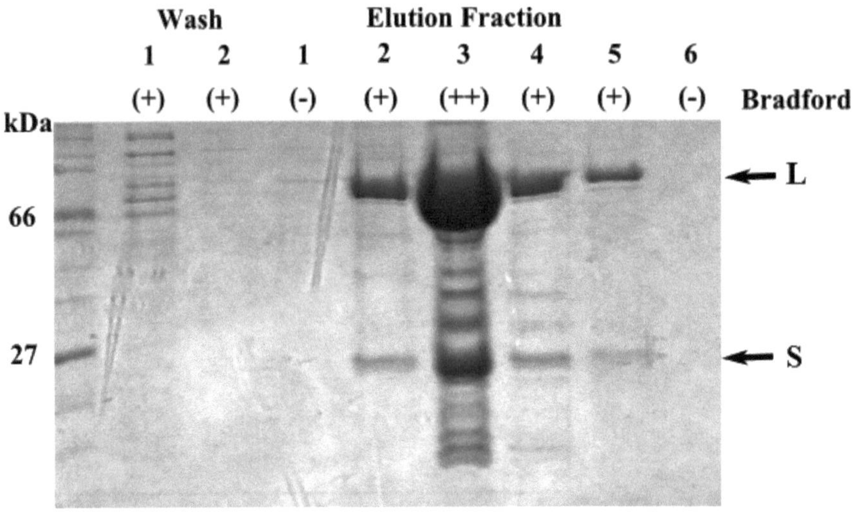

Fig. 3 Chromatography column 2: purification of rat calpain-2 by Ni-NTA-agarose affinity chromatography. SDS-PAGE and Bradford analysis of Ni-NTA-agarose affinity chromatography. Wash and elution fractions collected from the affinity purification of recombinant rat calpain-2 were subjected to a visual Bradford assay where (+) indicates a color shift and (−) indicates no color change. The gel was stained with Coomassie Brilliant Blue. The large (80 kDa) and small (21 kDa) subunits are designated by arrows L and S, respectively. Molecular weight markers are present in the first lane on the left

yellowish-brown color that gets fainter with each wash (*see* **Note 14**).

8. Elute the bound proteins with 35 mL of Ni-elution buffer by collecting 5 mL fractions of the eluate into 200 μL collection fluid. Immediately place the elution fractions on ice (*see* **Notes 15** and **16**).
9. The bound rat calpain-2 or protease cores of calpain-1/calpain-3 typically elute in fractions 2, 3, and 4 (*see* **Note 17**). The elution of bound proteins can be monitored by A_{280}, Bradford protein assay, and/or SDS-PAGE (Figs. 3 and 4).

3.4 Sephacryl S-200 Size-Exclusion Chromatography

1. Pool the 5 mL elution fractions obtained from Ni-NTA-agarose chromatography containing calpain as indicated by A_{280}, Bradford protein assay, or SDS-PAGE. If pooled fractions total more than 15 mL, concentrate to 15 mL total volume (*see* **Note 18**).
2. Equilibrate Sephacryl S-200 (500 mL) size-exclusion column with 500 mL of SEB buffer at a flow rate of 0.5 mL/min.
3. Load up to 15 mL of pooled Ni-NTA-agarose chromatography elution fractions.
4. Run the Sephacryl S-200 column overnight at a flow rate of 0.5 mL/min, collecting 7.5 mL fractions.
5. The next day, analyze the eluted fractions by A_{280}, Bradford protein assay, and/or SDS-PAGE (Figs. 5 and 6). The calpain

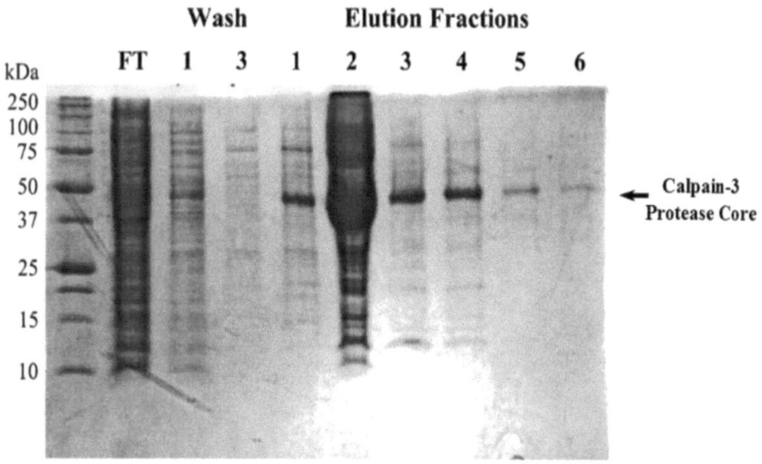

Fig. 4 Chromatography column 1: purification of human calpain-3 protease core by Ni-NTA-agarose affinity chromatography. SDS-PAGE analysis was performed on the flow-through (FT), wash and elution fractions from a Ni-NTA-agarose affinity column purification of the calpain-3 protease core (42 kDa). The gel was stained with Coomassie Brilliant Blue. Molecular weight markers are present in the first lane on the left

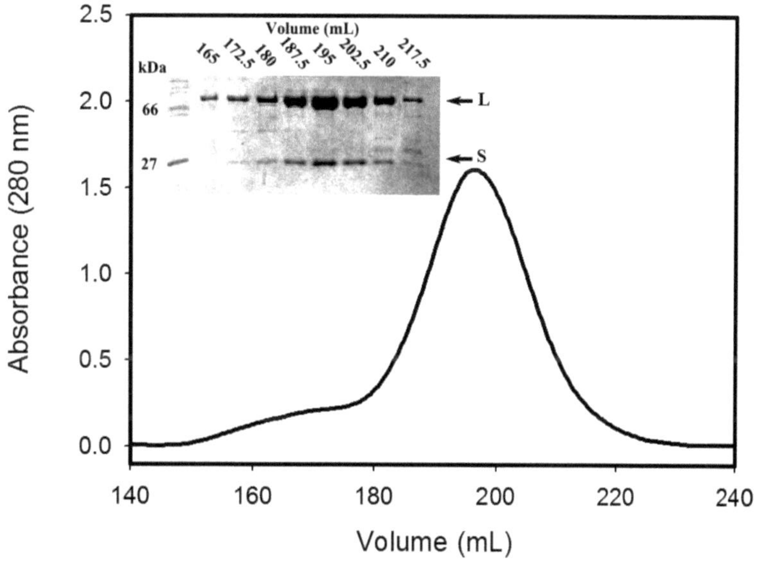

Fig. 5 Chromatography column 3: purification of rat calpain-2 by Sephacryl S-200 size-exclusion chromatography. The absorbance (280 nm) profile of rat calpain-2 eluted from the Sephacryl S-200 size-exclusion column is indicated by the black line. Elution fractions with high A_{280} values were collected from the purification of recombinant rat calpain-2 and assayed by SDS-PAGE and stained with Coomassie Brilliant Blue (insert). The large (80 kDa) and small (21 kDa) subunits are designated by arrows L and S, respectively. Molecular weight markers are present in the first lane on the left

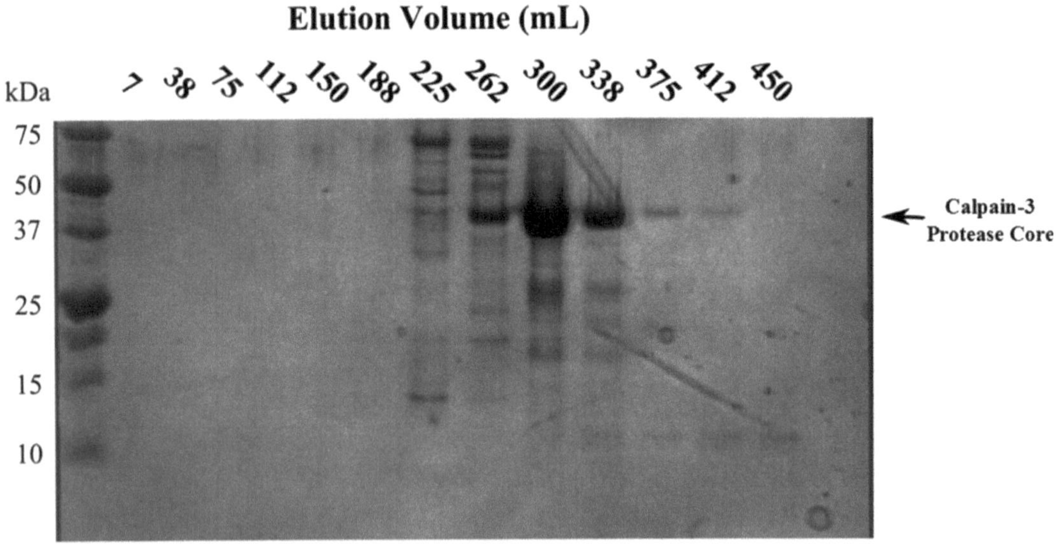

Fig. 6 Chromatography column 2: purification of human calpain-3 protease core by Sephacryl S-200 size-exclusion chromatography. SDS-PAGE analysis was performed on the elution fractions from an S-200 size-exclusion column purification of the calpain-3 protease core (42 kDa). The calpain-3 protease core elutes much later (262 mL) than whole rat calpain-2. The gel was stained with Coomassie Brilliant Blue. Molecular weight markers are present in the first lane on the left

protease core is typically observed at an elution volume of ~250 mL, while the rat calpain-2 heterodimer elutes at ~150 mL.

3.5 Q-Sepharose FPLC Anion-Exchange Chromatography

1. Equilibrate a Q16/10 Q-Sepharose anion-exchange column (Pharmacia), or equivalent, with TE buffer.
2. Pool up to 150 mL of Sephacryl S-200 size-exclusion column elution fractions into a 250 mL beaker on ice. Bring volume up to 150 mL with TE buffer if needed.
3. Load pooled fractions into a 150 mL FPLC Superloop (GE Healthcare), and apply to a Q16/10 Q-Sepharose anion-exchange column at 2 mL/min. Collect the flow-through and record its A_{280}.
4. Wash the column with TE buffer. Again, collect the flow-through and record A_{280}.
5. Elute the bound calpain with a gradient of 15%–50% (v/v) 1 M NaCl in TE elution buffer at a flow rate of 3 mL/min. The calpain protease core elutes early at ~20–25% (200 mM–250 mM NaCl) with a purity of >95%. Whole calpain-2 elutes at ~35–40% (350 mM–400 mM NaCl) (Figs. 7 and 8) (*see* **Note 19**).
6. This protocol produces whole rat calpain-2 and the protease cores of calpain-1 and calpain-3 that are stable for up to several

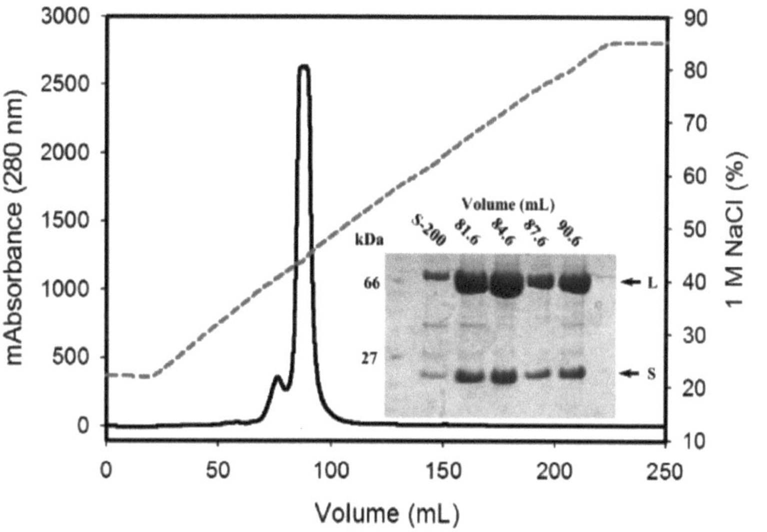

Fig. 7 Chromatography column 4: purification of rat calpain-2 by Q-Sepharose high-performance FPLC anion-exchange chromatography. The absorbance (280 nm) profile of rat calpain-2 eluted from the Q-Sepharose anion-exchange column (black line) using a 0.2–1.0 M NaCl gradient (dashed line) is plotted. Elution fractions collected from the purification of recombinant rat calpain-2 were analyzed by SDS-PAGE and stained with Coomassie Brilliant Blue (insert). The large (80 kDa) and small (21 kDa) subunits are designated by arrows L and S, respectively. Molecular weight markers are present in the first lane on the left

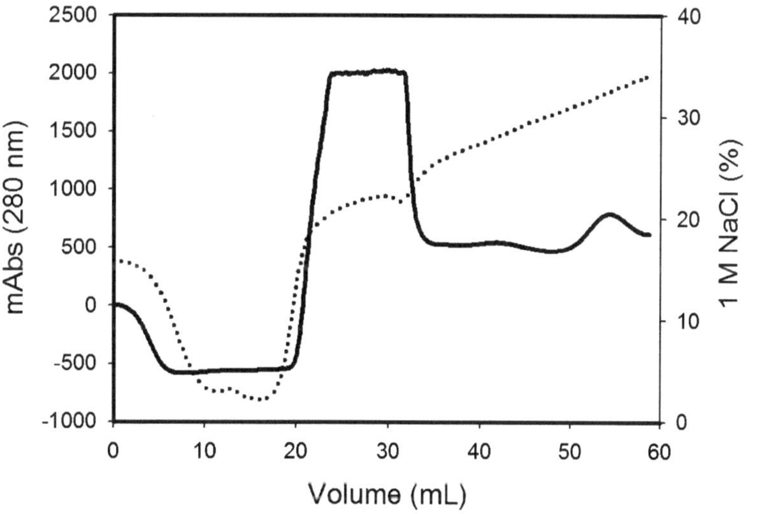

Fig. 8 Chromatography column 3: purification of human calpain-3 protease core by Q-Sepharose anion-exchange chromatography. The absorbance (280 nm) profile of the calpain-3 protease core eluted from the Q-Sepharose anion-exchange column (black line) using a 0.2–1.0 M NaCl gradient (dashed line). The calpain-3 core elutes much earlier (20% NaCl) from the column as compared to whole rat calpain-2

days at 4 °C with added β-mercaptoethanol. It is important to note that whole rat calpain-2 and calpain-3 protease core autoproteolyze in the presence of Ca^{2+}. For long-term storage, buffer exchange to 50 mM HEPES, pH 7.6, and 10 mM 1,4-dithiothreitol (DTT), and snap freeze. If need be, concentrate the enzyme during the buffer exchange step.

4 Notes

1. Other expression systems, such as baculovirus-infected insect cells, have also been used to produce calpains [11]. Improved protein production may be achieved by codon optimization and RNA secondary structure reduction to better suit your desired expression system.
2. The use of newly transformed cells and/or freshly streaked plates maximizes protein expression and yield.
3. If an air delivery system is unavailable, it will be necessary to shake the cultures vigorously in 1 L volumes (as was done with the protease core cultures), since thorough aeration is imperative.
4. If induction occurs at temperatures greater than 30 °C, inclusion bodies often result. Growth in the range 16–25 °C works well. Cooler temperatures may be helpful with isoforms or mutants that are harder to fold and keep soluble.
5. A concentration of 0.4 M IPTG is commonly recommended. Similar to temperature, the concentration of IPTG can be manipulated to suit your experimental needs and manage folding or solubility issues.
6. The cell suspension can be sonicated immediately or can be frozen in lysis buffer for temporary storage for a few days at −20 °C. If frozen, thaw the cell suspension while swirling it in a warm water bath just until all the ice is melted. Add fresh PMSF and β-mercaptoethanol prior to sonication.
7. To maximize the yield of the protein product, it is important to ensure that the cell suspension is fully sonicated. A Bradford protein assay can be a useful way to check for sufficient sonication of your cell suspensions. After each burst, add 1–2 μL of the cell suspension to 200 μL of Bradford reagent. As the sonication bursts liberate the protein from the cell, the intensity of the Bradford reagent color shift increases until the point at which the cells have been fully lysed. Insufficient cell lysis can also be observed by SDS-PAGE, where inadequate breakage of the cells produces a similar complexity of the bands in both the supernatant and pellet.

8. The initial DEAE-Sephacel anion-exchange chromatography can be performed on crude lysate supernatant when purifying the protease core; however, due to its more alkaline pH, the core is often found to elute with unwanted bulk protein that has a tendency to aggregate and precipitate.
9. Crude lysate supernatant can be flowed over a prepacked DEAE-Sephacel column. However, we prefer batch absorption, which removes much of the unwanted bulk proteins and produces a cleaner elution pattern. The DEAE-Sephacel resin can be cleaned following the elution, with 1 M NaOH followed by 1 M HCl and then regenerated in fresh TE buffer at pH 7.6.
10. Rat calpain-2 elutes from the DEAE-Sephacel resin after most of the bulk protein. Tris-Tricine SDS-PAGE stained with Coomassie Brilliant Blue is recommended to confirm the elution profile of the intact whole calpain, because excess small subunit often elutes before the whole enzyme, and is not apparent by A_{280} or Bradford protein assay (Fig. 2).
11. There are many variations of chromatographic methods that have been used to successfully purify calpain. We have found that Ni-nitrilotriacetic acid (NTA)-agarose affinity chromatography followed by size-exclusion and high-resolution Q-Sepharose anion-exchange works well. This sequence of chromatographies conveniently avoids concentration and/or dialysis steps between columns. For a shortcut, size-exclusion can be omitted from the rat calpain-2 procedure; however this may result in some high molecular weight contaminants and aggregated material in the final product.
12. Literature entries sometimes suggest a wash of 40 mM imidazole prior to eluting the Ni-NTA-agarose column, but this is too high for calpain. The column can tolerate up to 10 mM β-mercaptoethanol. Dilution of the total soluble protein with Ni-buffer prior to applying it to the column is advised.
13. The presence of EDTA in the lysis buffer and pooled DEAE elution fractions is sufficient to fully remove the Ni^{2+} chelated to the NTA-agarose resin. This problem can be avoided by precipitating the protein in 60% (w/v) ammonium sulfate and redissolving the pellet in Ni-buffer or by adding sufficient Mg^{2+} to fully saturate the EDTA. The chelation of Mg^{2+} by EDTA releases H^+, which drops the pH and must be adequately buffered. A pH below 6.0 tends to cause protein precipitation and impede column binding.
14. A Bradford protein assay can be used to ensure sufficient washing of the Ni-NTA-agarose column. After each column volume, add 10–20 μL of the eluent to 200 μL of Bradford reagent. Continue washing the column until no Bradford reagent color shift occurs.

15. A gradient of imidazole from 4 mM to 250 mM can be used to elute bound proteins. This may improve purity at this step but adds to the elution volume of the recovered fractions and may require additional concentration steps.
16. The Ni-NTA-agarose resin may be washed and regenerated according to the manufacturer's instructions, by adding 20 mM EDTA and 8 M urea, followed by recharging with excess $NiSO_4$ solution and washing with Ni-buffer.
17. Calpain has a tendency to precipitate when stored for long periods in Ni-elution buffer supplemented with collection fluid. For storage, buffer exchange into a more suitable buffer is recommended; otherwise continue immediately with size-exclusion chromatography.
18. To produce a sharp elution profile from the size-exclusion column, it is important not to load a large volume of protein solution (<15 mL). Protein precipitation in 60% (w/v) ammonium sulfate or ultrafiltration in an Amicon Ultra 35 kDa centrifugal filter (Millipore) can be used to concentrate protein samples.
19. The main peak of eluted rat calpain-2 is preceded in some cases by a small shoulder (Fig. 7) consisting of some minor contaminants or partial oxidation products. This is not usually observed for the protease core, since it elutes earlier from the column.

Acknowledgments

We thank Dr. Laurie Graham for her suggestion for monitoring the success of sonication using the Bradford assay. This work was funded by the Canadian Institute of Health Research. PLD holds a Canada Research Chair in Protein Engineering.

References

1. Campbell RL, Davies PL (2012) Structure-function relationships in calpains. Biochem J 447(3):335–351. https://doi.org/10.1042/BJ20120921
2. Goll DE, Thompson VF, Li H, Wei W, Cong J (2003) The calpain system. Physiol Rev 83 (3):731–801. https://doi.org/10.1152/physrev.00029.2002
3. Elce JS, Hegadorn C, Gauthier S, Vince JW, Davies PL (1995) Recombinant calpain II: improved expression systems and production of a C105A active-site mutant for crystallography. Protein Eng 8(8):843–848
4. Chou JS, Impens F, Gevaert K, Davies PL (2011) m-Calpain activation in vitro does not require autolysis or subunit dissociation. Biochim Biophys Acta 1814(7):864–872. https://doi.org/10.1016/j.bbapap.2011.04.007
5. Moldoveanu T, Hosfield CM, Lim D, Elce JS, Jia Z, Davies PL (2002) A Ca(2+) switch aligns the active site of calpain. Cell 108(5):649–660
6. Sorimachi H, Toyama-Sorimachi N, Saido TC, Kawasaki H, Sugita H, Miyasaka M, Arahata K, Ishiura S, Suzuki K (1993) Muscle-specific calpain, p94, is degraded by autolysis immediately after translation, resulting in disappearance

from muscle. J Biol Chem 268 (14):10593–10605

7. Federici C, Eshdat Y, Richard I, Bertin B, Guillaume JL, Hattab M, Beckmann JS, Strosberg AD, Camoin L (1999) Purification and identification of two putative autolytic sites in human calpain 3 (p94) expressed in heterologous systems. Arch Biochem Biophys 363(2):237–245. https://doi.org/10.1006/abbi.1998.1091
8. Rey MA, Davies PL (2002) The protease core of the muscle-specific calpain, p94, undergoes Ca2+−dependent intramolecular autolysis. FEBS Lett 532(3):401–406
9. Diaz BG, Moldoveanu T, Kuiper MJ, Campbell RL, Davies PL (2004) Insertion sequence 1 of muscle-specific calpain, p94, acts as an internal propeptide. J Biol Chem 279 (26):27656–27666. https://doi.org/10.1074/jbc.M313290200
10. Schagger H, von Jagow G (1987) Tricine-sodium dodecyl sulfate-polyacrylamide gel electrophoresis for the separation of proteins in the range from 1 to 100 kDa. Anal Biochem 166(2):368–379
11. Strobl S, Fernandez-Catalan C, Braun M, Huber R, Masumoto H, Nakagawa K, Irie A, Sorimachi H, Bourenkow G, Bartunik H, Suzuki K, Bode W (2000) The crystal structure of calcium-free human m-calpain suggests an electrostatic switch mechanism for activation by calcium. Proc Natl Acad Sci U S A 97 (2):588–592

Part II

Determination of Calpain Localization, Expression, and Activity

Chapter 3

Casein Zymography for Analysis of Calpain-1 and Calpain-2 Activity

Ashim Kumar Biswas and Samarth Tandon

Abstract

Casein zymography is used to detect calpain activity in cell or tissue lysates. In this technique, lysates are loaded into a polyacrylamide gel containing casein, and the enzymes within the lysates are electrophoretically separated. The casein embedded in the gel acts as a substrate for calpains in the lysates, and its degradation reflects the activity of these enzymes. Calpain degradation of the casein is visualized as distinct bands that are devoid of dye when stained with Coomassie Brilliant Blue G-250. We describe here how calpains can be extracted from muscle tissue and assayed for activity using this technique. This technique is also generally applicable to lysates from other types of cells or tissues.

Key words Casein zymography, Chicken, Muscle tissues, Substrate zymography, Electrophoresis

1 Introduction

Zymography is a protein chemistry technique used for measuring non-denatured enzyme activity. It was first introduced in 1978 by Granelli-Piperno and Reich and has been subsequently adapted for use with a number of different enzymes [1–3]. Different types of zymography include (1) substrate based (gelatine, casein, collagen, heparin, etc.), (2) reserved zymography, or (3) in situ zymography. Casein zymography is a type of substrate zymography. In this technique, enzymes within a sample are separated by polyacrylamide gel electrophoresis under denaturing, nonreducing conditions. After electrophoresis, the enzymes are allowed to renature and then consume casein within the polyacrylamide gel. The gel is then stained with Coomassie Blue R-250 to identify bands corresponding to enzymatic degradation of the substrate. The location and size of clear bands within the gel provide information about which enzyme degraded the substrate and how much activity was present in the sample.

Casein is enzymatically degraded by calpains. Calpains are calcium-dependent cysteine proteases that are widely expressed

Jeannette S. Messer (ed.), *Calpain: Methods and Protocols*, Methods in Molecular Biology, vol. 1915,
https://doi.org/10.1007/978-1-4939-8988-1_3, © Springer Science+Business Media, LLC, part of Springer Nature 2019

throughout the body. There are ubiquitous and tissue-specific isoforms of calpains. Many of these calpains have been implicated in important cellular processes including cell motility, signal transduction pathways, apoptosis, cell differentiation, and regulation of the cytoskeleton [4]. In muscle tissue, calpain activity is an important component of meat quality [5]. Lastly, calpains have been implicated in a number of human and animal diseases [6]. These factors make characterization of calpain activity in tissues or cells important for understanding its roles in cellular homeostasis and disease. In this chapter we describe a method to perform casein zymography. Casein zymography is useful for studying calpain-1 and calpain-2 activity and inhibitory agents of calpains. This technique can detect calpain activity in samples from tissues or cultured cells, even in the presence of endogenous calpastatin [7].

2 Materials

All the solvents and reagents required for this experiment should be of analytical quality. Protease inhibitors and substrate (casein) should be highly purified forms. Prepare all solutions using double-distilled water (ddH_2O), and store at 4 °C until further use, unless otherwise indicated.

2.1 Tissue Lysis Buffer

50 mM Tris–HCl, (pH 8.3), 10 mM ethylenediaminetetraacetic acid (EDTA), and 0.05% (v/v) 2-mercaptoethanol (2-MCE). Add a cocktail mixture of protease inhibitors from the following stock solutions (phenylmethane sulfonyl fluoride (PMSF), leupeptin, and ovomucoid) to the tissue lysis buffer immediately before use (*see* **Note 1**).

1. *PMSF* (100 mM stock solution): Weigh 174 mg of PMSF (*see* **Notes 2** and **3**), and transfer to a 15 mL conical plastic tube. Add 10 mL of isopropanol, dissolve well, and store at −20 °C. Add 195 μL of stock solution to 10 mL of tissue lysis buffer immediately before use.
2. *Leupeptin* (5 mg/mL stock solution): Weigh 10 mg of leupeptin dry powder (*see* **Notes 4** and **5**), transfer to a micro-centrifuge tube, and dissolve with 2 mL of ddH_2O. Store the solution at −20 °C. Add 12 μL of stock solution to 10 mL of tissue lysis buffer immediately before use.
3. *Ovomucoid* (300 mg/mL stock solution): Weigh 150 mg of ovomucoid dry powder, transfer to a micro-centrifuge tube, and dissolve with 500 μL of ddH_2O. The solution is stable for up to 2 months at −20 °C. Add 66 μL of stock solution to 10 mL of tissue lysis buffer immediately before use.

2.2 Casein Zymography Gel

1. *Resolving gel buffer*: 1.5 M Tris–HCl, pH 8.8. Store at 4 °C.
2. *Stacking gel buffer*: 1.32 M Tris–HCl, pH 6.8. Store at 4 °C.
3. *Polyacrylamide stock solution*: 4:0.16 Acrylamide to bis-acrylamide. Place 4 g of acrylamide and 0.16 g bis-acrylamide (*see* **Note 6**) into a 50 mL Falcon tube, and add 5 mL of ddH_2O to dissolve the chemicals. After the chemicals are dissolved, increase the volume to 10 mL using ddH_2O. Filter the solution through Whatman filter paper 1 (*see* **Note 7**), and then use it immediately, or store it at 4 °C, protected from light. The stock solution can be used for up to 1 month (*see* **Notes 8** and **9**).
4. *Ammonium persulfate solution (APS)*: 10% ammonium persulfate (w/v) in ddH_2O. Prepare fresh whenever required.
5. *N′,N′,N′,N′-Tetramethylethylenediamine (TEMED)*: Store at 4 °C.
6. *Casein (purified)*: Store at 4 °C.
7. *Sample buffer (1:3, v/v)*: 150 mM Tris–HCl (pH 6.8), 20% (v/v) glycerol, 0.04% (w/v) bromophenol blue, and 0.75% (v/v) 2-MCE. Add the 2-MCE immediately before use.
8. *Electrode running buffer* (10×): 25 mM Tris–HCl, pH 8.3, 192 mM glycine, 1 mM EGTA [ethylene glycol-bis (β-aminoethyl ether)-N′,N′,N′,N′-tetraacetic acid], EDTA, and 0.05% 2-MCE (*see* **Note 10**).
9. *Proteolysis buffer (10×)*: 20 mM Tris–HCl (pH 7.4), 10 mM DTT [dithiothreitol], and 4 mM $CaCl_2$. Store at 4 °C (*see* **Note 11**).
10. *Coomassie Brilliant Blue G-250 solution*: Dissolve 0.20 g Coomassie Brilliant Blue into a solution containing 40 mL methanol, 10 mL glacial acetic acid, and 50 mL ddH_2O to make final volume of 100 mL.
11. *Destaining solution*: Add 30 mL methanol and 10 mL glacial acetic acid into 60 mL of ddH_2O. Final volume makes up to 100 mL.

2.3 Equipment

1. Gel electrophoresis system along with accessories (spacer plate, short plate, casting frame, casting stand, buffer dam, electrophoresis module, tank, and lid).
2. Power supply.

3 Methods

Carry out all of the procedures at 25 °C to avoid protein denaturation, and preserve enzyme activity.

3.1 Sample Preparation

The calpain enzymes are very heat sensitive and quickly undergo autolytic degradation. To avoid unwanted loss of activity due to environmental exposures, tissue samples should be collected immediately after slaughtering operations are over. The samples should be labeled properly and chilled quickly.

1. Weigh 0.3 g of finely cut tissue into a chilled mortar and pestle, and homogenize with 6 volumes of ice-cold tissue lysis buffer containing protease inhibitors (1.8 mL, pH 8.3) (*see* **Note 12**). This method was optimized for use with chicken muscle tissue but should be broadly applicable to other tissue types.
2. Centrifuge the tissue lysate at 20,000 × *g* for 20 min at 4 °C.
3. Transfer the supernatant to a separate centrifuge tube and discard the pellet.
4. Centrifuge the supernatant once again as mentioned earlier to obtain purified supernatant. Then filter the purified supernatant through Whatman filter paper No 1. This purified, filtered supernatant contains the calpain enzymes and is the sample that will be used for subsequent analysis (*see* **Note 13**).

3.2 Gel Preparation and Sample Electrophoresis

1. Prepare the resolving gel using 4.9 mL ddH_2O, 2.5 mL resolving gel buffer, 2.5 mL polyacrylamide solution, 0.21 mg casein, 40 μL of APS, and 28 μL of TEMED. Mix the contents thoroughly and degas the solution for at least 15 min. Cast the gel in a gel cassette fitted with the casting stand. Allow space to add the staking gel by placing a comb completely into the assembled gel cassette and marking the glass plate 1 cm below the comb teeth. Remove the comb, and pour the resolving gel solution gently into the cassette using a disposable plastic pipette, stopping at the mark. Overlay the gel with t-amyl alcohol or water (*see* **Note 14**).
2. Allow the gel to polymerize for at least 45 min. Rinse the gel surface completely with ddH_2O. Do not leave the alcohol overlay on the gel for more than 1 h since it will dehydrate the top of the gel (*see* **Note 15**).
3. Prepare the stacking gel solution using 6.5 mL ddH_2O, 2.5 mL stacking gel buffer, and 1 mL polyacrylamide solution. Degas the solution for 15 min and then add 50 μL of APS and 10 μL of TEMED. Dry the top of the resolving gel with Whatman filter paper, smoothly pour the stacking gel into the reserved space in the gel cassette, and then place a 10-well comb between the spacers. Allow polymerization as above (*see* **Note 16**).
4. Set the clamping frame to the open position on a clean flat surface. Place the first gel cassette with the short plate facing inward on to the gel supports; gel supports are molded into the bottom of the clamping frame assembly. The gel will now rest

at a 30° angle, tilting away from the center of the clamping frame. Similarly place the second gel on the other side of the clamping frame (*see* **Note 17**).

5. Place and lock the gel cassettes gently, and if using an odd number, you should use a buffer dam.
6. Fill the assembly upper chamber with electrode running buffer to just under the edge of the outer gel plate.
7. Mix the sample with the sample buffer (3:1, v/v), and load into the casein polyacrylamide gel. Load the samples into the assemblies, while it is inside the tank.
8. Run the electrophoresis unit at 100 V for 160 min at 4 °C in 1× electrode running buffer.
9. After the end of the electrophoresis, turn off the power supply, and disconnect the electrical lead. Remove the tank lid and carefully lift out the electrode assemblies. Pour off and discard the running buffer. Open the arms of the assembly, and remove the gel cassettes by gently separating the two plates (*see* **Note 18**).
10. Incubate the gel in 1× proteolysis buffer for 18–24 h at 20 °C.
11. Stain the gel with Coomassie Brilliant Blue G-250 for 30 min, and then destain using the destaining solution until bands are clearly visible.
12. The presence of clear bands indicates activity of calpain-1 and calpain-2 (Fig. 1). These are confirmed using certified reference material (CRM). In the region of presence of activated calpain, the casein is digested into small fragments that diffuse out of the gel.

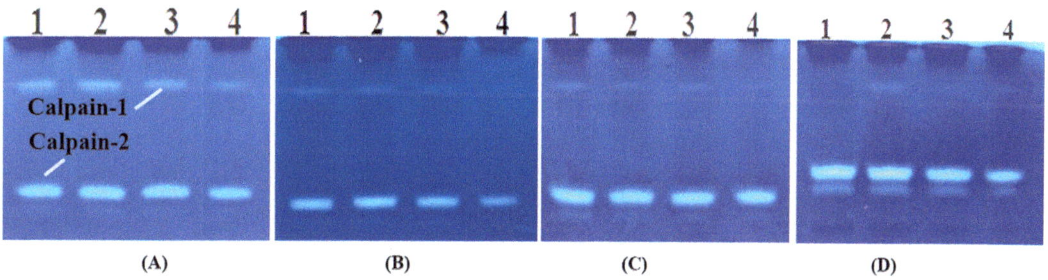

Fig. 1 Casein zymogram showing clear bands of calpain-1 and calpain-2 extracted from chicken tissue samples. Postmortem aging times (h) indicating rate of degradation of calpain-1 and calpain-2. Male = (♂); female = (♀). (**a**) *0 h*: Lane-1, breast muscle (♂); Lane-2, thigh muscle (♂); Lane-3, breast muscle (♀); Lane-4, thigh muscle (♀). (**b**) *6 h*: Lane-1, breast muscle (♂); Lane-2, thigh muscle (♂); Lane-3, breast muscle (♀); Lane-4, thigh muscle (♀). (**c**) *18 h*: Lane-1, breast muscle (♂); Lane-2, thigh muscle (♂); Lane-3, breast muscle (♀); Lane-4, thigh muscle (♀). (**d**) *24 h*: Lane-1, breast muscle (♂); Lane-2, thigh muscle (♂); Lane-3, breast muscle (♀); Lane-4, thigh muscle (♀)

4 Notes

1. Use freshly made 2-mercaptoethanol for all solutions.
2. PMSF is a serine protease inhibitor, but it does not inhibit all serine proteases. It is rapidly degraded in water (half-life 10 min at pH 7.0 and 35 min at pH 8.0), so it should be added to the tissue lysis buffer immediately before use [8]. The working concentration of PMSF lies in the range of 0.1–1.0 mM.
3. PMSF is a toxic chemical and should be handled carefully, in a fume hood. The LD_{50} for this compound is less than 500 mg/kg.
4. Leupeptin is an organic protease inhibitor produced by actinomycetes. It can inhibit trypsin-like proteases and some cysteine proteases including endoproteinase, Lyp-C, kallikrein, papain, thrombin, cathepsin B, and trypsin [9]. The working concentration of leupeptin is in the range of 10–100 μM.
5. Leupeptin is stable for several hours in water at 30 °C. The reconstituted solution can be stored at −20 °C for up to 6 months. Avoid repeated freezing and thawing.
6. Wear protective equipment including gloves and a mask while handling acrylamide and bis-acrylamide. These chemicals should be weighed in a covered balance and handled inside a fume hood.
7. It is better to use a Corning filter, but if one is not available, Whatman filter paper No 1 will suffice.
8. Since acrylamide is highly light sensitive and rapidly oxidized, always wrap the bottle containing the solution with aluminum foil, and minimize its exposure to light.
9. It is best to prepare the polyacrylamide solution fresh. However, it can be stored at 4 °C for about 1 month.
10. Always dilute the native electrode running buffer (10×) to 1× before using it. Do not use SDS, but instead use 1 mM EGTA and 1 mM EDTA. 2-MCE should be added to the 1x buffer immediately before use.
11. It is important to use the precise concentration of calcium chloride indicated since it acts as a source of Ca^{2+}, which is required for activation of calpain-1 and calpain-2 [10]. For the best results, always prepare fresh proteolysis buffer.
12. Never add the protease inhibitors directly to the un-lysed tissues. They should be added after the tissue is minced and at the time that it is homogenized in the mortar and pestle. Otherwise, the efficacy of the protease inhibitors will be reduced due to binding with other organic matrices.

13. Load the samples into the casein gel immediately after preparation for the best measure of calpain activity. Freezing the samples is likely to reduce the calpain enzymatic activity.
14. If water is used, add it slowly and evenly to prevent mixing. Do not overlay with butanol or isobutanol.
15. The resolving gel should be prepared immediately before use for the best results.
16. Insert the desired comb between the spacers starting at the top of the spacer plate, making sure that the tabs at the ends of each comb are guided between the spacers. Seat the comb in the gel cassette by aligning the comb ridge with the top of the short plate.
17. Please use caution when placing the first gel, making sure that the clamping frame remains balanced and does not tip over. It is critical that the gel cassettes are placed into the clamping frame with the short plate facing inward. Do not attempt to lock the green arms of the clamping frame, without first ensuring that the gel cassettes are perfectly aligned and stabilized against the notches on the green gaskets of the module.
18. Always pour off the electrode buffer before opening the arms of the assembly, to avoid spilling the buffer. If the gel sticks to the plate, remove the gel by floating it off in transfer solution or even in the proteolysis buffer.

Acknowledgment

This work is supported by DST (Govt. of India) grant SB/FT/LS-283/2012 and the Indian Council of Agricultural Research, New Delhi (Ministry of Agriculture).

References

1. Granelli-Piperno A, Reich E (1978) A study of proteases and protease-inhibitor complexes in biological fluids. J Exp Med 148:223–234
2. Huessen C, Dowdle EB (1980) Electrophoretic analysis of plasminogen activator in polyacrylamide gels containing sodium dodecylsulfate and co-polymerized substances. Anal Biochem 102:196–202
3. Ono Y, Sorimachi H (2012) Calpains—An elaborate proteolytic system. Biochim Biophys Acta Proteins Proteomics 1824(1):224–236
4. Rajesh RK (2013) Zymography: enzymes in action. Sci Int 1:70–75
5. Biswas AK, Tandon S, Sharma D (2016) Identification and characterization of different domains of calpain and their influence on post-mortem ageing of goat meat during holding at 4±1 °C. LWT-Food Sci Technol 71:60–68
6. Potz BA, Abid MR, Sellke FW (2016) Role of calpain in pathogenesis of human disease processes. J Nat Sci 2:e218
7. Raser KJ, Posner A, Wang KK (1995) Casein zymography: A method to study μ-Calpain, m-calpain and their inhibitory agents. Arch Biochem Biophys 319:211–216

8. James GT (1978) Inactivation of the protease inhibitor phenylmethylsulfonyl fluoride in buffers. Anal Biochem 86(2):574–579
9. Beynon RAYJ, Bond JS (1989) Proteolytic enzymes: a practical approach. IRL Press, Oxford
10. Biswas AK, Tandon S, Beura CK (2016) Simple extraction method for determination of different domains of calpain and calpastatin from chicken blood and their role in post-mortem ageing of breast and thigh muscles at 4±1 °C. Food Chem 200:315–321

Chapter 4

FRET-Based Assays to Determine Calpain Activity

Christian-Scott E. McCartney and Peter L. Davies

Abstract

Calpains are signaling proteases that have relatively little sequence specificity but some preferences for certain residues on either side of the scissile bond. As with most proteases, they mainly cut unstructured or extended regions of their target proteins. The tendency for concentrated calpain to rapidly autoproteolyze when activated by calcium complicates the kinetic assessment of calpain activity. As calpain autoproteolyzes, the amount of fully active enzyme continuously decreases until all of the calpain molecules have been cut and their activity reduced to a tiny fraction of the starting rate. To accurately measure calpain kinetics, only the initial rate of substrate hydrolysis, where autoproteolysis is minimal, can be used. To accomplish this, a method for rapid, quantifiable determination of substrate cleavage is required. Many of the existing assays are lacking in their sensitivity to accurately quantify calpain activity within this timeframe. However, the FRET peptide substrates developed by Cuerrier et al. have been shown to have sufficiently high affinity between substrate and enzyme to accurately measure the initial enzyme reaction velocity at substrate concentrations above the K_m value. With a suitably sensitive fluorimeter, sufficient data can be obtained to evaluate calpain kinetics and inhibition. Here we describe a facile, reliable calpain assay based on the continuous monitoring of FRET fluorescence from the highly sensitive calpain-specific substrate, (EDANS)-EPLFAERK-(DABCYL). We illustrate some difficulties associated with determining kinetic constants of whole calpains that are simultaneously undergoing autoproteolysis and how the assay can be used to help characterize calpain-specific inhibitors. We also present a variation of this fluorescence-based assay for high-throughput screening using the calpain protease core and a fluorescence plate reader.

Key words Calpain activity, Cysteine protease, Peptide substrate, Calpain inhibitor, Enzyme kinetics, FRET, Fluorescence, PLFAER

1 Introduction

Calpains play a variety of roles in intracellular calcium signaling by catalyzing the limited cleavage of a wide range of protein substrates [1]. Limited, rather than degradative, proteolysis by calpain effects changes in the regulation and function of its substrates. These changes can contribute to the control of a diverse array of cellular functions including cell fusion, cell motility, signal transduction, cell cycle progression, gene expression, apoptosis, and necrosis [2]. Since substrate processing by calpain is an irreversible process, it follows that calpain activation must be tightly controlled. Indeed,

Jeannette S. Messer (ed.), *Calpain: Methods and Protocols*, Methods in Molecular Biology, vol. 1915,
https://doi.org/10.1007/978-1-4939-8988-1_4,

unregulated or prolonged calcium influx leads to overactivation of calpain, which is associated with many pathologies including Alzheimer's disease [3–5], Parkinson's disease [6], asthma [7], muscular dystrophy [8], myocardial infarction [9], multiple sclerosis [10, 11], colitis [12], amyotrophic lateral sclerosis [13], neuronal ischemia [3, 14], cancer [15–17], type 2 diabetes [18], pulmonary fibrosis [19], pulmonary arterial hypertension [20], and traumatic spinal cord/brain injury [21–25]. Therefore, calpains are of great biomedical and therapeutic interest. The scarcity of calpain-specific inhibitors and substrates results in part from its mechanistic and structural similarities with the other cysteine proteases, like the cathepsins. The enzyme's tendency to aggregate and autoproteolyze when it is isolated and concentrated in vitro has also made this a challenging area to study.

A number of assays have been developed to investigate and quantify calpain activity. Calpain zymography with casein, an in-gel substrate, has been a popular method for detecting active calpain. One advantage of this method is that it can distinguish the activity of calpain-1 from calpain-2 and the calpains from the cathepsins or the proteasome [26, 27]. A disadvantage of the method is that it is not quantitative. The release of trichloroacetic acid (TCA)-soluble peptides from substrates in solution has also been used to measure calpain activity. However, endpoint assays like these suffer from imprecise quantification. The tendency for concentrated calpain to rapidly autoproteolyze when activated by calcium further complicates the kinetic assessment of calpain activity. As calpain autoproteolyzes, the amount of fully active enzyme continuously decreases until all of the enzymes have been cut and their activity reduced to a fraction of the starting rate that is likely coming from the residual protease core.

To accurately measure calpain kinetics, only the initial rate of substrate hydrolysis, where autoproteolysis is minimal, can be used. To accomplish this, a method for rapid, quantifiable determination of substrate cleavage is required.

Increasingly, fluorescent compounds have been used to continuously monitor calpain proteolysis [28–33]. Many of these constructs use small molecule fluorophores linked by a short calpain-sensitive peptide and are measured using Förster resonance energy transfer (FRET). Concerted efforts have been made to determine calpain cleavage site specificity by targeting the unprimed side of the scissile bond [34, 35]. However, little work had been done on primed side preferences until these were explored by Cuerrier et al. (following the protocol of Turk et al.). Using a peptide library-based approach, an optimal peptide cleavage sequence, PLFAER, was discovered [36, 37]. The incorporation of PLFAER into an internally quenched fluorescent substrate with an N-terminal donor (EDANS) and a C-terminal quencher (DABCYL) (Fig. 1) was used to make real-time measurements of substrate hydrolysis.

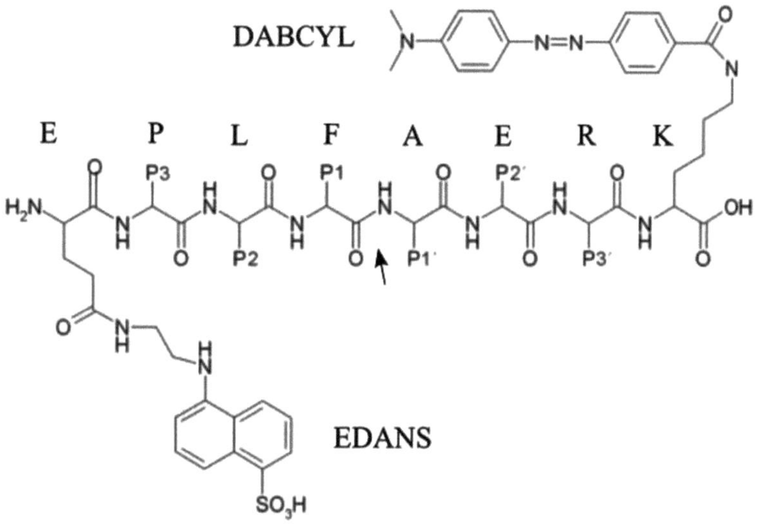

Fig. 1 (EDANS)-EPLFAERK-(DABCYL) structure. The PLFAER cleavage motif is centered between the fluorophore EDANS and quencher DABCYL (*modified from* [36]). The scissile bond is indicated by an arrow

Experiments showed that PLFAER is preferentially cleaved 18 times more rapidly than the cleavage site of a naturally occurring substrate hexapeptide from alpha-spectrin (EVYGMM) and 300-fold faster than the unprimed side-targeting amino methylcoumarin-based substrate (SLY-MCA). Furthermore, PLFAER binds with sufficiently high affinity ($K_m = 41.4 \pm 2.3$ μM) to mitigate the effects of autoproteolysis in the early reaction time period, while making it possible to monitor rapid substrate cleavage at a variety of concentrations exceeding the K_m with a high signal to noise, a condition needed for accurate Michaelis-Menten kinetic analyses.

Because of the therapeutic potential of downregulating calpains to treat a wide range of pathophysiological conditions, many inhibitor candidates have been described for this enzyme. These consist of peptidyl calpain inhibitors from mammals, and plants, as well as synthetic compounds [38–40], the majority of which are active site-directed. They include reversible and irreversible agents with varying selectivity for calpain over other cysteine proteases. Many of the established inhibitors and their derivatives occupy the unprimed side of the catalytic cleft in the S1, S2, and S3 subsites, where it is difficult to differentiate calpains from the other cysteine proteases, leading to poor calpain selectivity. Inhibitors that extend beyond S3 into the S4 subsite area would likely make contacts with the calpain-type beta-sandwich domain (CSBW), which is unique to the calpain family. Contacts made with the CBSW domain could prove to be more discriminatory; for this reason whole calpain-1 or calpain-2 is of greater physiological relevance than assays that use the protease core. However, in the past, problems with calpain autoproteolysis and aggregation have complicated assays that employ the full-length calpains; therefore, the protease core (domains PC1 and PC2) of calpain-1 can be used to mimic the function of the whole enzyme because it is not susceptible to autoproteolytic degradation but otherwise has a the same specificity, which is useful for high-throughput calpain inhibitor screening.

Here, we describe a protocol for the accurate measurement of recombinant whole rat calpain-2 enzyme kinetics in vitro using the highly sensitive FRET substrate, (EDANS)-EPLFAERK-(DABCYL). We also describe a useful high-throughput inhibitor screening method using the autoproteolysis-resistant calpain-1 protease core and a fluorescent plate reader.

2 Materials

2.1 Equipment

1. A LS 55 PerkinElmer fluorescence spectrometer, or equivalent, with a stir plate accessory, set to an excitation wavelength at 335 nm and an emission wavelength at 500 nm (10 nm slit width). Data were accumulated and processed using the program FL WinLAB in the development of this protocol.
2. A 500 μL Quartz SUPRASIL Macro/Semi-micro Cell (PerkinElmer) with a stir bar and stirring motor chamber, or equivalent.
3. A fluorescent microplate reader (Molecular Devices SpectraMax M2), or similar, with an excitation wavelength set to 335 nm and an emission wavelength at 500 nm. Data were accumulated using SoftMax Pro 5.3.
4. A 120 μL half-area 96-well Corning plate (black), or equivalent.

2.2 Reagents and Solutions

1. FRET substrate: 3 mM and 30 μM substrate stock solutions of (EDANS)-EPLFAERK-(DABCYL) (Biomer Technology) made in DMSO and stored at −20 °C.
2. EDANS: 50 μM stock solution of free EDANS (Biomer Technology) dissolved in dimethyl sulfoxide (DMSO) and stored at −20 °C.
3. Reaction buffer: 50 mM HEPES-NaOH, pH 7.5, and 10 mM 1,4-dithiolthreitol (DTT) added just before use.
4. Purified calpain/calpain protease core, or equivalent:
   (a) Recombinant rat calpain-2: expressed and purified from *E. coli* using column chromatography. Concentrated to 50 μM aliquots, snap frozen and stored at −80 °C in 50 mM HEPES-NaOH, pH 7.5, and 10 mM DTT.

(b) Recombinant calpain-3 protease core: expressed and purified from *E. coli* using column chromatography. Concentrated to 50 μM aliquots, snap frozen, and stored at −80 °C in 50 mM HEPES-NaOH, pH 7.5, and 10 mM DTT.

5. Enzyme solution: 200 μM of purified calpain-3 protease core, 50 mM HEPES-NaOH, pH 7.5, and 2 mM DTT, prepared immediately from frozen stocks before use.
6. Substrate solution: 80 μM PLFAER substrate in 5 mM $CaCl_2$, prepared immediately before use.

3 Methods

3.1 Correcting the Inner Filter Effect

The FRET substrate, (EDANS)-EPLFAERK-(DABCYL), consists of a protease-sensitive peptide linked at one end to a fluorescence donor and at the other end to a quenching acceptor. The intrinsic fluorescence of the donor, EDANS (5-[2′-aminoethyl)amino] naphthalenesulfonic acid), is dramatically reduced because of the intramolecular resonance energy transfer to the quenching group, DABCYL (4-[[4′-(dimethylamino)phenyl]azo]benzoic acid). Since FRET transfer becomes insignificant beyond distances of 10 nm, the fluorescence of the donor group is restored upon enzymatic cleavage of the linking peptide. However, under some conditions the fluorescent light is absorbed by neighboring quenching groups, either on neighboring substrates or as cleaved product, such that a fraction of the emitted light is never measured by the detector [41]. This phenomenon is known as the inner filter effect, the magnitude of which varies according to wavelength, path length, and concentration of quencher. The inner filter effect must be determined for each instrumental configuration in the assay in order to obtain accurate fluorescence measurements (Fig. 2).

The method, which we describe here, has been adapted from Liu et al. [41] and involves the measurement of free EDANS fluorescence at various intact substrate concentrations containing the quencher DABCYL. The reduction in the fluoresced light between free EDANS and free EDANS in the presence of DABCYL reflects the magnitude of the inner filter effect at each of those specific substrate concentrations.

1. Using a LS 55 PerkinElmer fluorescence spectrometer running FL WinLAB, or a similar program, input the settings outlined in Table 1. For optimal results, initiate the stir plate accessory for continuous mixing of the sample. Select the "Read" application, or equivalent, for recording single time-point fluorescence measurements.

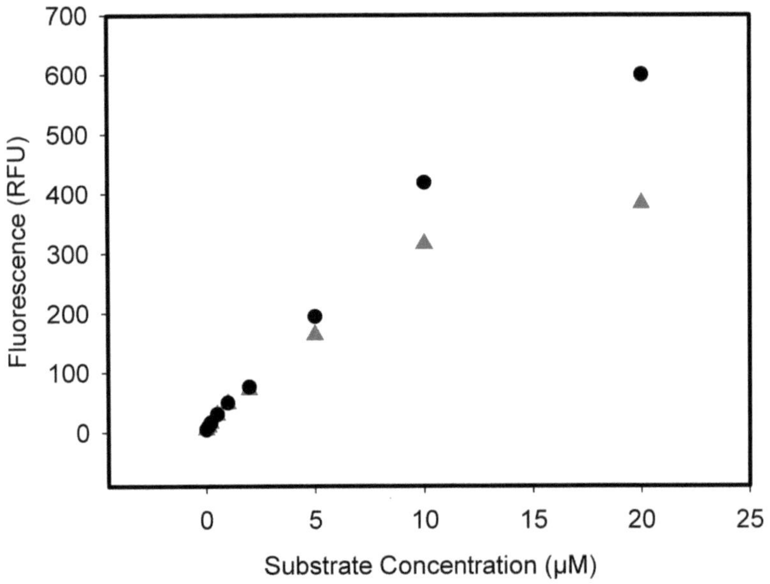

Fig. 2 Inner filter effect: DABCYL quenching of free EDANS. Free EDANS fluorescence (Ex, 335 nm; Em, 500 nm) was measured in the presence of varying concentrations of intact (EDANS)-EPLFAERK-(DABCYL) substrate (gray triangles) and compared to the inner filter effect corrected substrate fluorescence (black circles)

Table 1
The fluorescence parameters for EDANS inner filter effect corrections

	Wavelength	Slit width
Excitation	335 nm	10 nm
Emission	500 nm	10 nm

2. Using 3 mM and 30 μM stock solutions of (EDANS)-EPLFAERK-(DABCYL) substrate, generate 2 mL of 0, 0.1, 0.2, 0.5, 1, 2, 5, 10, 20, 50, and 100 μM substrate sample solutions in reaction buffer (*see* **Note 1**).
3. Pipette 0.5 mL of substrate sample solution into a clean 500 μL Quartz cuvette with stir bar and place into the fluorimeter (*see* **Note 2**). Thoroughly mix the solution for approximately 30 s.
4. Record the fluorescence emission of the substrate sample, and designate *f*(S).
5. Next, add 2.5 μL of 50 μM free EDANS stock solution to the substrate sample while stirring. After about 30 s, measure and record the fluorescence of the free EDANS plus substrate, and designate *f*(S + EDANS). As greater substrate concentrations are measured, a noticeable decrease in free EDANS

fluorescence will occur, resulting from the quenching of free EDANS by the DABCYL group of the intact substrate.

6. Vigorously wash the 500 μL Quartz cuvette, and stir bar, with 0.1% acetic acid, deionized water, and 70% ethanol (*see* **Note 3**).
7. Repeat **steps 3–6**, in triplicate, for each substrate concentration.

Only a fraction of the free EDANS fluorescence will be recorded by the detector when mixed with quencher. To correct for the inner filter effect, subtract the fluorescence of the intact substrate *f*(S) from the total fluorescence of substrate mixed with free EDANS, *f*(S + EDANS), to determine the true fluorescence of free EDANS *f*(EDANS) at each substrate concentration.

$$f(\mathrm{S} + \mathrm{EDANS}) - f(\mathrm{S}) = f(\mathrm{EDANS})$$

To generate a correction factor, divide *f*(EDANS) at each substrate concentration by *f*(EDANS) ([S] = R μM) in the absence of substrate ([S] = 0 μM).

$$\mathrm{Corr\%} = \frac{f(\mathrm{EDANS})\mathrm{at}\,[\mathrm{S}] = \mathrm{R}\,\mu M}{f(\mathrm{EDANS})\mathrm{at}\,[\mathrm{S}] = 0\ \ \mu M}$$

The inner filter effect becomes important when conducting kinetic experiments using multiple concentrations of FRET substrate (Fig. 2). Using the inner filter effect correction factor Corr%, the true fluorescence of free EDANS generated by the hydrolysis of (EDANS)-EPLFAERK-(DABCYL) can be calculated and the true initial reaction velocities determined.

3.2 Kinetic Analysis of Recombinant Rat Calpain-2 Activity

As mentioned in Subheading 1, calpain autolysis causes a sharp decline in enzyme activity (Fig. 3). To circumvent this, the calpain-sensitive substrate (EDANS)-EPLFAERK-(DABCYL) is used for rapid, quantifiable determination of substrate cleavage. Here we describe the kinetic assessment of recombinant whole rat calpain-2 purified from *E. coli*. However, whole calpain from other sources, such as porcine erythrocyte calpain-1, may also be used. As stated previously, the protease core might also prove useful. However, it should be noted that the activity of the protease core is around a tenth to one 20th that of the full-length calpain-2.

To accurately measure calpain kinetics, a range of 1–100 μM (EDANS)-EPLFAERK-(DABCYL) is typically used. The initial reaction velocity, within the first 10–20 s, is determined by fitting a straight line to the linear portion of the progress curve (Fig. 3), where autoproteolysis is less problematic. The resulting velocities are adjusted using the inner filter effect correction factor calculated in Subheading 3.1, and the data are then examined by fitting them to a Michaelis-Menten plot. IC_{50} values are only useful for

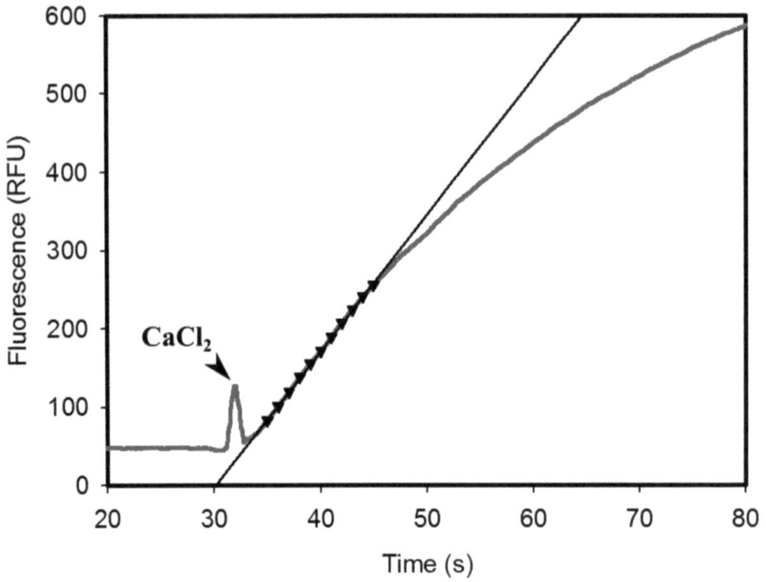

Fig. 3 Progress curve for the hydrolysis of FRET substrate by recombinant rat calpain-2. Cleavage of (EDANS)-EPLFAERK-(DABCYL) (gray line) was initiated with 5 mM $CaCl_2$ after 30 s, as indicated. The initial rate of reaction follows a linear trend for 10 s between 35 and 45 s (black triangles and regression line), where the effects of calpain autoproteolysis are minimal. After about 50 s, the reaction begins to deviate from linearity and will eventually form a plateau, an indication of rapid autoproteolytic cleavage (Ex, 335 nm; Em, 500 nm)

Table 2
The fluorescence parameters for calpain hydrolysis of EDANS-EPLFAERK-DABCYL

Excitation	335 nm	Slit width	10 nm
Emission	500 nm	Slit width	10 nm
Time	300 s		
Interval	0.1 s		

comparing similar inhibitors under similar conditions, such as substrate concentration and affinity. Generally, an IC_{50} value is easier to obtain than a K_i value; however, K_i values are preferred because they are not restricted to specific reaction conditions, which allows for more robust comparisons to be made between compounds. Additionally, determination of IC_{50} values does not provide information on the mode of inhibition of a particular compound (competitive, noncompetitive, or mixed). Following the methods outlined here will not only provide trustworthy K_i values for new and existing calpain inhibitors but also provide an accurate determination of substrate K_m for newly discovered cleavage sequences.

1. Using a LS 55 PerkinElmer fluorescence spectrometer running FL WinLAB, or a similar program, input the settings outlined in Table 2. For optimal results, initiate the stir plate accessory for continuous mixing of the sample. Select the "Time-Drive" application, or equivalent, for recording continuous fluorescence measurements (*see* **Note 4**).
2. Using 3 mM and 30 μM stock solutions of (EDANS)-EPLFAERK-(DABCYL) substrate, generate 2 mL of 0, 0.1, 0.2, 0.5, 1, 2, 5, 10, 20, 50, and 100 μM substrate sample solutions in reaction buffer. Ensure that the range of substrate concentrations matches those used for the inner filter effect corrections (*see* **Note 5**).
3. Vigorously wash the 500 μL Quartz cuvette, and stir bar, with 0.1% acetic acid, deionized water, and 70% ethanol as outlined in **Note 3**.
4. Pipette 0.5 mL of substrate sample solution into a clean 500 μL Quartz cuvette with the stir bar, and place into the fluorimeter (*see* **Note 2**). Thoroughly mix the solution for approximately 30 s. Add 1 μL of 50 μM fresh recombinant rat calpain-2, or equivalent (*see* **Note 6**).
5. Record the initial substrate fluorescence for 30 s before initiating the reaction to ensure thorough mixing and a steady background. If an unstable background is produced, this may indicate cross contamination has occurred, so clean the cuvette, and repeat **steps 2–4** (*see* **Note** 7).
6. At precisely 30 s, quickly add 5 mM $CaCl_2$ to initiate the calpain hydrolysis reaction (*see* **Note 8**). When calcium is added, whole calpain rapidly cleaves the PLFAER peptide and, due to rapid autoproteolysis, approaches a plateau of activity within several minutes (Fig. 4). The protease core may be substituted in these studies, which produces a linear response because of its resistance to autoproteolysis. However, up to 1–2 μM protease core may be required, since the protease core has only a fraction of the activity of the whole enzyme. Continue to measure the reaction for 10 min.
7. Repeat **steps 2–5** for each substrate concentration in triplicate to acquire enough data for calpain kinetic analysis. Inhibitors dissolved in DMSO may be added to the solutions prepared in **step 1**, using a DMSO equivalent as a control. The entire protocol should be repeated for each compound, at several different inhibitor concentrations (*see* **Note 5**).
8. Calculate the rate of reaction in the first 10–20 s following initiation with $CaCl_2$ for each sample. Use the inner filter effect

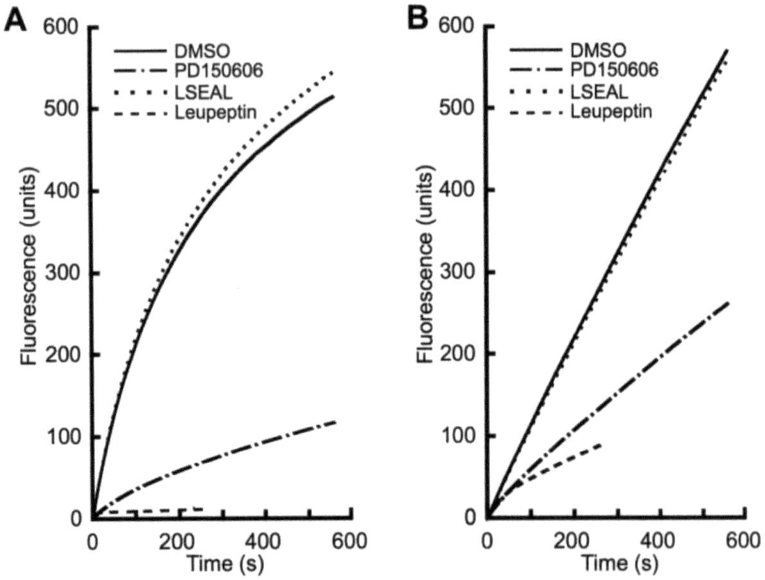

Fig. 4 The progress curves of inhibition of calpain-2 and calpain-1 protease core by the mercaptoacrylate calpain inhibitor, PD150606. The lines represent the fluorescence-based substrate cleavage assays of (EDANS)-EPLFAERK-(DABCYL), by either calpain-2 (**a**), which shows slowing of the reaction rate due to autoproteolysis, or calpain-1 protease core (**b**), which produces a linear response. Assays were done in the absence of inhibitors (solid line) and in the presence of 50 μM PD150606 (dash-dot line), 100 μM LSEAL peptide (dotted line), or 10 μM leupeptin (dashed line) (*Modified from* [42])

correction factor Corr% from Subheading 3.1 to correct for the inner filter effect in the reaction velocities. Plot the corrected initial reaction velocities on a Michaelis-Menten curve, and calculate K_m, K_i, and V_{max}, by fitting the data to a nonlinear regression (Fig. 5) (*see* **Note 9**).

3.3 High-Throughput Screening of Inhibitors Using the Calpain-1 Core

When working on whole calpain-1 and calpain-2, where only single reactions in a cuvette can be made at one time, the process of screening many inhibitor candidates can be tedious and impractical. As described previously, a useful property of the "mini-calpains" is that they mimic the function of, and have a similar substrate specificity to, the whole enzyme without quickly approaching a plateau of activity (Fig. 4). Because the rate of substrate turnover by the protease core is linear for longer, the reaction can be monitored using a fluorescent plate reader, where the interval between measurements is greater. This method can be particularly valuable for inhibitors that target the subsites immediately flanking the active site cysteine but may not be as useful for compounds that extend into the S4 region and beyond. Therefore, follow-up assays with the whole enzyme are highly recommended.

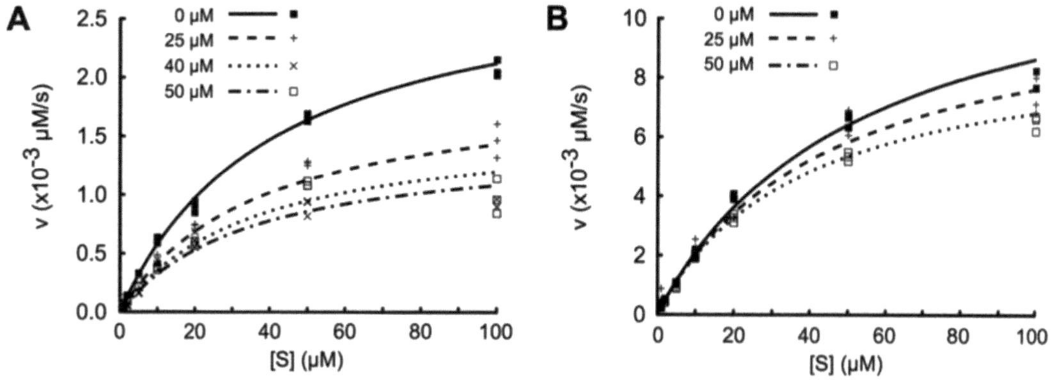

Fig. 5 Enzyme kinetic curves for calpain-2 and calpain-1 protease core inhibited by PD150606. Substrate cleavage was performed with increasing concentrations of (EDANS)-EPLFAERK-(DABCYL) and PD150606 (0, 25, 40, and 50 μM as indicated). Progress curves for calpain-2 (**a**) were analyzed using the linear segment of the curve to determine an initial reaction velocity. These data were fitted to the Michaelis-Menten equation. Progress curves for calpain-1 protease core (**b**) were analyzed, and the data were fitted in a similar manner (*Modified from* [42])

Table 3
Fluorescent plate reader setup parameters

Excitation	335 nm
Emission	500 nm
Time	30 min
Temperature	37 °C
Interval	20 s

Here we describe a protocol for the high-throughput screening of inhibitors using the protease core on a fluorescent plate reader as a means for detecting lead compounds for inhibitor development.

1. Turn on the fluorescent microplate reader (Molecular Devices SpectraMax M2) according to manufacturer specific instructions, and allow the lamp to warm up. Select the continuous read application, or equivalent, for continuous measurement of fluorescence. Set the temperature to 37 °C and wait until the instrument warms up. Input the fluorescence parameters outlined in Table 3.
2. Prepare enzyme (200 μM of purified calpain-3 protease core, 50 mM HEPES-NaOH, pH 7.5, and 2 mM DTT) and substrate (80 μM PLFAER substrate and 5 mM $CaCl_2$) solutions, and keep on ice.

3. To a 120 μL half-area 96-well Corning plate, in triplicates, add 30 μL of deionized water, DMSO, or desired inhibitor solution to each well.
4. To each well add 60 μL of enzyme solution and mix well. With the fluorescent microplate reader ready to read, using a multi-channel pipette, quickly add 30 μL of substrate solution to activate the reaction, and quickly insert the plate into the fluorescent microplate reader, mix by shaking for 5 s and begin reading the plate. Note that for these assays, the reaction is started by the addition of substrate and $CaCl_2$, rather than just the addition of $CaCl_2$.
5. Read the plate continuously for 30 min–1 h at 37 °C.
6. Repeat **steps 3–6** until data for all desired inhibitors at several different concentrations, in triplicate, has been acquired (*see* **Note 10**).
7. Calculate the rate of reaction in the first few minutes following initiation with substrate solution for each well, where the segment of the reaction produces a linear rate of hydrolysis (*see* **Note 11**). For a quick comparison, contrast the rate of the inhibited reactions at different inhibitor concentrations with the rate of the uninhibited control to determine the percent activity/inhibition of each compound (Fig. 6).

High-throughput assays using the protease core are practical for screening compounds for inhibition, but the use of a plate reader severely limits the amount of data that can be acquired in the initial stages of the reaction. For this reason, it is not recommended that whole calpain be used for screening assays using a plate reader nor do we recommend that enzyme kinetics be assessed using this method.

4 Notes

1. Consult your manufacturer's literature to determine the optimal range of substrate concentrations suitable for your equipment. Please note that substrate concentrations in excess of 100 μM may become insoluble and distort fluorescence measurements. Similarly, a range of substrate concentrations that do not meet or exceed the substrate K_m may produce higher associated error when calculating the kinetic constants.
2. When loading samples into the 0.5 mL fluorescent/luminescent cuvette, ensure that air bubbles do not accumulate in the stir bar chamber. Air bubbles in this chamber will prevent thorough mixing of the solution and can have an impact on fluorescence measurements. Gently tapping the side of the cuvette can help to release the air from the chamber.

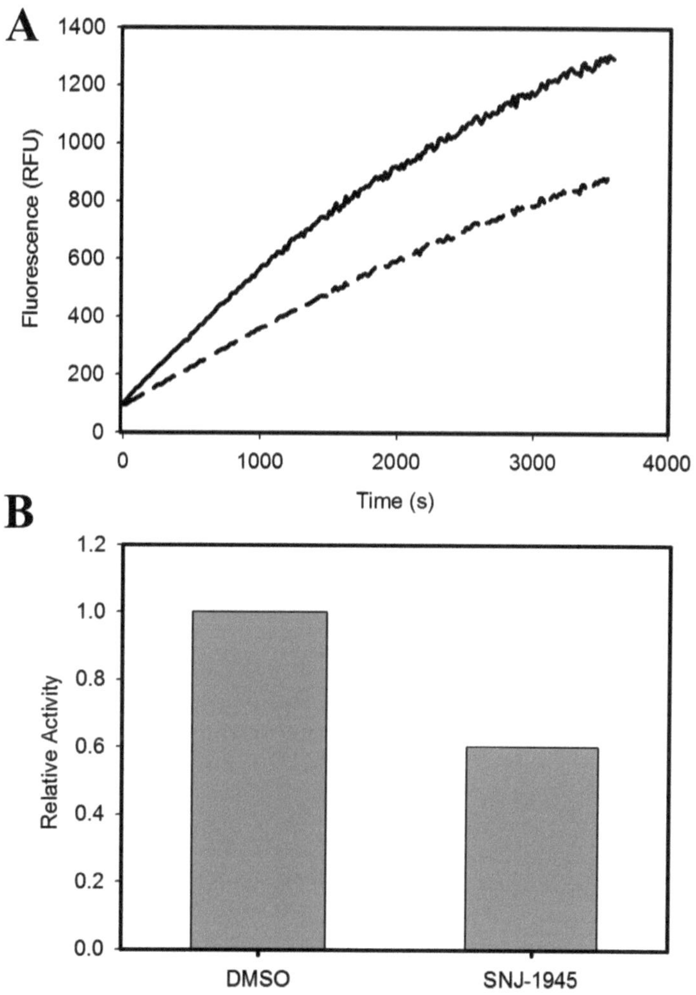

Fig. 6 (**a**) Progress curves for inhibitor screening using calpain-1 protease core and inhibitor, SNJ-1945. The lines represent the fluorescence-based substrate cleavage assays of (EDANS)-EPLFAERK-(DABCYL) with calpain-1 protease core, which produces an initial linear response, in the absence of inhibitors (solid line) and in the presence of 10 μM SNJ-1945 (dashed line), in a 120 μL half-area flat bottom 96-well Corning plate at 37 °C for 1 h. Note that, after about 15 min, substrate depletion begins to cause a curvature in the progress curve. (**b**) The initial rates of reaction, when the progress curves are linear, were compared to determine the relative activity of the calpain-1 protease core in the presence of 10 μM SNJ-1945

3. To thoroughly wash the cuvette, inject several cuvette volumes of 0.1% acetic acid with a squirt bottle. Use a magnet to retain the small stir bar in the chamber, and flick the cuvette to remove the wash solution from the stir bar chamber. Following the same method, wash the cuvette with deionized water and subsequently with 95% ethanol. Pressurized air can be used to

rapidly dry the cuvette. Vigorous washing of the cuvette is necessary to remove all traces of substrate, inhibitors, protein, and $CaCl_2$.

4. Only the initial 10–20 s of the reaction provides reliable data free from the effects of autoproteolysis. In order to maximize the amount of data collected in the initial stages of the reaction, consult your manufacturer's literature, and adjust the settings to reduce the interval between readings as much as possible. Insufficient data gathered during this portion of the progress curve can lead to unreliable reaction velocities and kinetics data that are difficult to reproduce.
5. Inhibitors are typically dissolved in DMSO and can be added to the solutions prepared in Subheading 3.2, **step 1**. To accurately assess inhibition, several different concentrations of inhibitor should be tested. It must be noted that treatment with an inhibitor can reduce both the substrate turnover rate and the rate of calpain autoproteolysis. Therefore, only data accumulated in the first 10–20 s of the reaction, when the slope is linear, should be used for enzyme analysis. In some instances, treatment with an inhibitor may seem to produce an "activation" of calpain (Fig. 4); however, caution should be used when interpreting these data, a more likely cause may be the inhibition of autoproteolysis resulting in a longer enzymatic lifetime.
6. Kinetic assays on one batch of enzyme should be performed without interruption to minimize potential oxidation, aggregation, and autoproteolysis. Fresh reducing agent should be applied to each enzyme stock solution prior to use. If recombinant protein is used, it has been noted that aliquots of pure enzyme prepared from the same source (i.e., the same protein purification) minimize variance in the measured initial reaction velocities.
7. The fluorescence background is measured for 30 s before initializing with $CaCl_2$. This step allows the user to accurately time the addition of $CaCl_2$ as well as notifying the user of any potential cross contamination from insufficient cuvette washings. If an unstable background is produced in this interval, clean the cuvette and repeat **steps 2–4**.
8. A complication of working with purified whole calpain is its tendency to aggregate, particularly when divalent ion concentrations are high and monovalent cation concentrations are low [43].
9. The initial reaction velocities are typically determined in RFU (relative fluorescence units). To convert RFU(s) to μM (EDANS)(s), the user will need to determine the amount of EDANS fluorescence (corrected for the inner filter effect) from

a predetermined amount of fully cleaved substrate. This can be accomplished by allowing the reaction to progress to completion, typically 3 h at room temperature with the protease core, and measuring the endpoint fluorescence.

10. As with cuvette measurements, inner filter effect corrections must also be calibrated for each substrate concentration measured in the multi-well plate. This can be done according to Liu et al. [41]. Similarly, a conversion factor from RFU(s) to μM (EDANS)(s) must also be determined.
11. The protease core produces a constant rate of activity, approximately one tenth of the whole calpain-2 enzyme. However, toward the end of the reaction, the activity may begin to plateau as a result of the inner filter effect and/or substrate depletion. For this reason, use the data within the linear portion of the curve (1–5 min) to calculate turnover rate.

Acknowledgments

CM would like to acknowledge the instruction and guidance provided by Dr. Kristin Low. This work was supported by the Canadian Institutes of Health Research. PLD holds a Canada Research Chair in Protein Engineering.

References

1. Liu Z, Cao J, Gao X, Ma Q, Ren J, Xue Y (2011) GPS-CCD: a novel computational program for the prediction of calpain cleavage sites. PLoS One 6(4):e19001. https://doi.org/10.1371/journal.pone.0019001
2. Goll DE, Thompson VF, Li H, Wei W, Cong J (2003) The calpain system. Physiol Rev 83 (3):731–801. https://doi.org/10.1152/physrev.00029.2002
3. Saido TC, Yokota M, Nagao S, Yamaura I, Tani E, Tsuchiya T, Suzuki K, Kawashima S (1993) Spatial resolution of fodrin proteolysis in postischemic brain. J Biol Chem 268 (33):25239–25243
4. Trinchese F, Fa M, Liu S, Zhang H, Hidalgo A, Schmidt SD, Yamaguchi H, Yoshii N, Mathews PM, Nixon RA, Arancio O (2008) Inhibition of calpains improves memory and synaptic transmission in a mouse model of Alzheimer disease. J Clin Invest 118(8):2796–2807. https://doi.org/10.1172/JCI34254
5. Higuchi M, Iwata N, Matsuba Y, Takano J, Suemoto T, Maeda J, Ji B, Ono M, Staufenbiel M, Suhara T, Saido TC (2012) Mechanistic involvement of the calpain-calpastatin system in Alzheimer neuropathology. FASEB J 26(3):1204–1217. https://doi.org/10.1096/fj.11-187740
6. Amini M, Ma CL, Farazifard R, Zhu G, Zhang Y, Vanderluit J, Zoltewicz JS, Hage F, Savitt JM, Lagace DC, Slack RS, Beique JC, Baudry M, Greer PA, Bergeron R, Park DS (2013) Conditional disruption of calpain in the CNS alters dendrite morphology, impairs LTP, and promotes neuronal survival following injury. J Neurosci 33(13):5773–5784. https://doi.org/10.1523/JNEUROSCI.4247-12.2013
7. Rao SS, Mu Q, Zeng Y, Cai PC, Liu F, Yang J, Xia Y, Zhang Q, Song LJ, Zhou LL, Li FZ, Lin YX, Fang J, Greer PA, Shi HZ, Ma WL, Su Y, Ye H (2017) Calpain-activated mTORC2/Akt pathway mediates airway smooth muscle remodelling in asthma. Clin Exp Allergy 47 (2):176–189. https://doi.org/10.1111/cea.12805
8. Richard I, Broux O, Allamand V, Fougerousse F, Chiannilkulchai N, Bourg N, Brenguier L, Devaud C, Pasturaud P, Roudaut C et al (1995) Mutations in the proteolytic

enzyme calpain 3 cause limb-girdle muscular dystrophy type 2A. Cell 81(1):27–40

9. Sandmann S, Yu M, Unger T (2001) Transcriptional and translational regulation of calpain in the rat heart after myocardial infarction–effects of AT(1) and AT(2) receptor antagonists and ACE inhibitor. Br J Pharmacol 132(3):767–777. https://doi.org/10.1038/sj.bjp.0703860
10. Trager N, Smith A, Wallace Iv G, Azuma M, Inoue J, Beeson C, Haque A, Banik NL (2014) Effects of a novel orally administered calpain inhibitor SNJ-1945 on immunomodulation and neurodegeneration in a murine model of multiple sclerosis. J Neurochem 130 (2):268–279. https://doi.org/10.1111/jnc.12659
11. Shields DC, Schaecher KE, Saido TC, Banik NL (1999) A putative mechanism of demyelination in multiple sclerosis by a proteolytic enzyme, calpain. Proc Natl Acad Sci U S A 96 (20):11486–11491
12. Huang Z, Rose AH, Hoffmann FW, Hashimoto AS, Bertino P, Denk T, Takano J, Iwata N, Saido TC, Hoffmann PR (2013) Calpastatin prevents NF-kappaB-mediated hyperactivation of macrophages and attenuates colitis. J Immunol 191(7):3778–3788. https://doi.org/10.4049/jimmunol.1300972
13. Yamashita T, Hideyama T, Hachiga K, Teramoto S, Takano J, Iwata N, Saido TC, Kwak S (2012) A role for calpain-dependent cleavage of TDP-43 in amyotrophic lateral sclerosis pathology. Nat Commun 3:1307. https://doi.org/10.1038/ncomms2303
14. Wang Y, Briz V, Chishti A, Bi X, Baudry M (2013) Distinct roles for mu-calpain and m-calpain in synaptic NMDAR-mediated neuroprotection and extrasynaptic NMDAR-mediated neurodegeneration. J Neurosci 33 (48):18880–18892. https://doi.org/10.1523/JNEUROSCI.3293-13.2013
15. Storr SJ, Carragher NO, Frame MC, Parr T, Martin SG (2011) The calpain system and cancer. Nat Rev Cancer 11(5):364–374. https://doi.org/10.1038/nrc3050
16. Storr SJ, Thompson N, Pu X, Zhang Y, Martin SG (2015) Calpain in breast cancer: role in disease progression and treatment response. Pathobiology 82(3-4):133–141. https://doi.org/10.1159/000430464
17. Flevaris P, Stojanovic A, Gong H, Chishti A, Welch E, Du X (2007) A molecular switch that controls cell spreading and retraction. J Cell Biol 179(3):553–565. https://doi.org/10.1083/jcb.200703185
18. Horikawa Y, Oda N, Cox NJ, Li X, Orho-Melander M, Hara M, Hinokio Y, Lindner TH, Mashima H, Schwarz PE, del Bosque-Plata L, Horikawa Y, Oda Y, Yoshiuchi I, Colilla S, Polonsky KS, Wei S, Concannon P, Iwasaki N, Schulze J, Baier LJ, Bogardus C, Groop L, Boerwinkle E, Hanis CL, Bell GI (2000) Genetic variation in the gene encoding calpain-10 is associated with type 2 diabetes mellitus. Nat Genet 26(2):163–175. https://doi.org/10.1038/79876
19. Li FZ, Cai PC, Song LJ, Zhou LL, Zhang Q, Rao SS, Xia Y, Xiang F, Xin JB, Greer PA, Shi HZ, Su Y, Ma WL, Ye H (2015) Crosstalk between calpain activation and TGF-beta1 augments collagen-I synthesis in pulmonary fibrosis. Biochim Biophys Acta 1852 (9):1796–1804. https://doi.org/10.1016/j.bbadis.2015.06.008
20. Kovacs L, Han W, Rafikov R, Bagi Z, Offermanns S, Saido TC, Black SM, Su Y (2016) Activation of calpain-2 by mediators in pulmonary vascular remodeling of pulmonary arterial hypertension. Am J Respir Cell Mol Biol 54(3):384–393. https://doi.org/10.1165/rcmb.2015-0151OC
21. Saatman KE, Creed J, Raghupathi R (2010) Calpain as a therapeutic target in traumatic brain injury. Neurotherapeutics 7(1):31–42. https://doi.org/10.1016/j.nurt.2009.11.002
22. Schoch KM, Evans HN, Brelsfoard JM, Madathil SK, Takano J, Saido TC, Saatman KE (2012) Calpastatin overexpression limits calpain-mediated proteolysis and behavioral deficits following traumatic brain injury. Exp Neurol 236(2):371–382. https://doi.org/10.1016/j.expneurol.2012.04.022
23. Yang J, Wu Z, Renier N, Simon DJ, Uryu K, Park DS, Greer PA, Tournier C, Davis RJ, Tessier-Lavigne M (2015) Pathological axonal death through a MAPK cascade that triggers a local energy deficit. Cell 160(1-2):161–176. https://doi.org/10.1016/j.cell.2014.11.053
24. Rubenstein R, Wang KK, Chiu A, Grinkina N, Sharma DR, Agarwal S, Lin F, Yang Z (2016) PrPC expression and calpain activity independently mediate the effects of closed head injury in mice. Behav Brain Res 340:29–40. https://doi.org/10.1016/j.bbr.2016.04.041
25. Kobeissy FH, Liu MC, Yang Z, Zhang Z, Zheng W, Glushakova O, Mondello S, Anagli J, Hayes RL, Wang KK (2015) Degradation of betaII-spectrin protein by calpain-2 and caspase-3 under neurotoxic and traumatic brain injury conditions. Mol Neurobiol 52(1):696–709. https://doi.org/10.1007/s12035-014-8898-z

26. Arthur JS, Mykles DL (2000) Calpain zymography with casein or fluorescein isothiocyanate casein. Methods Mol Biol 144:109–116. https://doi.org/10.1385/1-59259-050-0:109
27. Jiang ST, Wang JH, Chang T, Chen CS (1997) A continuous method for measuring calpain activity. Anal Biochem 244(2):233–238. https://doi.org/10.1006/abio.1996.9872
28. Tompa P, Schad E, Friedrich P (2000) A sensitive and continuous fluorometric activity assay using a natural substrate. Microtubule-associated protein 2. Methods Mol Biol 144:137–141. https://doi.org/10.1385/1-59259-050-0:137
29. Guttmann RP, Johnson GV (2000) Measurement of calpain activity in vitro and in situ using a fluorescent compound and tau as substrates. Methods Mol Biol 144:143–150. https://doi.org/10.1385/1-59259-050-0:143
30. Vanderklish PW, Krushel LA, Holst BH, Gally JA, Crossin KL, Edelman GM (2000) Marking synaptic activity in dendritic spines with a calpain substrate exhibiting fluorescence resonance energy transfer. Proc Natl Acad Sci U S A 97(5):2253–2258. https://doi.org/10.1073/pnas.040565597
31. Gerencser AA, Mark KA, Hubbard AE, Divakaruni AS, Mehrabian Z, Nicholls DG, Polster BM (2009) Real-time visualization of cytoplasmic calpain activation and calcium deregulation in acute glutamate excitotoxicity. J Neurochem 110(3):990–1004. https://doi.org/10.1111/j.1471-4159.2009.06194.x
32. Takatsuka K, Ishii TM, Ohmori H (2005) A novel Ca2+ indicator protein using FRET and calpain-sensitive linker. Biochem Biophys Res Commun 336(1):316–323. https://doi.org/10.1016/j.bbrc.2005.08.089
33. Stockholm D, Bartoli M, Sillon G, Bourg N, Davoust J, Richard I (2005) Imaging calpain protease activity by multiphoton FRET in living mice. J Mol Biol 346(1):215–222. https://doi.org/10.1016/j.jmb.2004.11.039
34. Tompa P, Buzder-Lantos P, Tantos A, Farkas A, Szilagyi A, Banoczi Z, Hudecz F, Friedrich P (2004) On the sequential determinants of calpain cleavage. J Biol Chem 279(20):20775–20785. https://doi.org/10.1074/jbc.M313873200
35. Sasaki T, Kikuchi T, Yumoto N, Yoshimura N, Murachi T (1984) Comparative specificity and kinetic studies on porcine calpain I and calpain II with naturally occurring peptides and synthetic fluorogenic substrates. J Biol Chem 259 (20):12489–12494
36. Cuerrier D, Moldoveanu T, Davies PL (2005) Determination of peptide substrate specificity for mu-calpain by a peptide library-based approach: the importance of primed side interactions. J Biol Chem 280(49):40632–40641. https://doi.org/10.1074/jbc.M506870200
37. Turk BE, Huang LL, Piro ET, Cantley LC (2001) Determination of protease cleavage site motifs using mixture-based oriented peptide libraries. Nat Biotechnol 19(7):661–667. https://doi.org/10.1038/90273
38. Donkor IO (2000) A survey of calpain inhibitors. Curr Med Chem 7(12):1171–1188
39. Donkor IO (2011) Calpain inhibitors: a survey of compounds reported in the patent and scientific literature. Expert Opin Ther Pat 21 (5):601–636. https://doi.org/10.1517/13543776.2011.568480
40. Donkor IO (2015) An updated patent review of calpain inhibitors (2012–2014). Expert Opin Ther Pat 25(1):17–31. https://doi.org/10.1517/13543776.2014.982534
41. Liu Y, Kati W, Chen CM, Tripathi R, Molla A, Kohlbrenner W (1999) Use of a fluorescence plate reader for measuring kinetic parameters with inner filter effect correction. Anal Biochem 267(2):331–335. https://doi.org/10.1006/abio.1998.3014
42. Low KE, Karunan Partha S, Davies PL, Campbell RL (2014) Allosteric inhibitors of calpains: reevaluating inhibition by PD150606 and LSEAL. Biochim Biophys Acta 1840 (12):3367–3373. https://doi.org/10.1016/j.bbagen.2014.08.014
43. Pal GP, Elce JS, Jia Z (2001) Dissociation and aggregation of calpain in the presence of calcium. J Biol Chem 276(50):47233–47238. https://doi.org/10.1074/jbc.M105149200

Chapter 5

In Vivo Monitoring of Calpain Activity by Forster Resonance Energy Transfer

William Lostal, Daniel Stockholm, and Isabelle Richard

Abstract

Calpains are a 15-member class of calcium-activated nonlysosomal neutral proteases. They are involved in many cellular processes and are highly upregulated in pathological conditions. Some are ubiquitously expressed (CAPN1, CAPN2, CAPN4, CAPN5, CAPN7, and CAPN10), but others are thought to be localized in specific tissues. The monitoring of in vivo calpain activity is required for physiological, pathological, and therapeutic evaluations. This past decade, a tool for monitoring calpain activity in such conditions was developed using Forster resonance energy transfer (FRET). Studies showed that the level of calpain activity correlates with a decrease in FRET between the two fluorescent proteins. This chapter describes the methodologies from the design of the construct to the imaging procedure and analysis to evaluate ubiquitous calpain activity in vivo.

Key words Calpain activity, Muscle fibers, In vivo, FRET, Imaging

1 Introduction

Calpains are a 15-member class of calcium-activated nonlysosomal neutral proteases which are localized to the cytosol and mitochondria [1]. Some calpains are ubiquitously expressed (CAPN1, CAPN2, CAPN4, CAPN5, CAPN7, and CAPN10), but others are specifically expressed in dedicated tissues (3 in skeletal muscle, 6 in placenta, 8 in smooth muscle, 9 in stomach, 11 in testes, 12 in skin, and 13 in testes and lung) [2]. They are mainly involved in regulation of key cellular mechanisms such as cytoskeleton reorganization through proteolysis of their substrates [2, 3]. Ubiquitous calpains are highly upregulated in many tissues during disease including in the brain during Alzheimer's disease [4, 5], in ischemic retina of the eye [6], and in heart [7], lung [8], pancreas [9], kidney [10], vascular system [11], and skeletal muscle [12] dysfunction. In a more important manner, diabetes leads to a systemic upregulation of calpains, with overexpression associated with reduced vision [13], cardiomyopathy [7], beta cell dysfunction in the pancreas

Jeannette S. Messer (ed.), *Calpain: Methods and Protocols*, Methods in Molecular Biology, vol. 1915,
https://doi.org/10.1007/978-1-4939-8988-1_5,

[14], and vasculopathy [11]. For the past decade, we have focused on calpain activity regulation in skeletal muscle, where calpains, and especially calpain 3, play a crucial role in muscle functions and in cytoskeleton remodeling [15]. The ability to follow the in vivo calpain activity is a requirement to evaluate physiological, pathological, and therapeutic aspects of these enzymes. Due to ubiquitous expression of calpains, inhibition of their activity has been suggested to be a potent therapeutic strategy for many diseases affecting cardiac muscle, skeletal muscle, and brain tissues [16–18].

To study calpain-related pathophysiological mechanisms and evaluate therapeutic approaches, monitoring of calpain activity is required. Most of the methods currently used are based on purification of calpain proteins through chromatography and quantification of the catalytic activity using casein substrate [19] or on isolation of living cells by flow cytometry based on peptide substrate activity [20, 21]. Another method is fluorescence resonance energy transfer (FRET), which allows monitoring of calpain activity in living animals. Fluorescence resonance energy transfer between genetically encoded fluorescent proteins (FP) [22] is a powerful and highly versatile tool for the spatiotemporal monitoring of cellular functions [23]. This technique has been used in living cells for calcium sensing [24], detection of protein-protein interactions [25], detection of enzymatic activities [26], and monitoring of conformational transitions in single molecules [27]. Ubiquitous calpain activity was detected for the first time in neuronal cells in vitro using a FRET sensor in 2000 [28].

In our studies, we generated a calpain sensor based on an identified cleavage site present in alpha-fodrin [29] as described in Fig. 1. In the absence of calcium, calpains are inactive, and FRET is generated through excitation of eCFP, transferring energy to eYFP and emitting light at a specific wavelength. When activated by calcium, calpains cleave the fodrin cleavage site, separating the fluorescent proteins and leading to the disappearance of FRET.

Previously, we showed that such strategy can be used in vivo by delivering the construct by electrotransfer to monitor calpain proteolytic activity in the living muscle of mice [30]. Then, a mouse model, CAFI mouse (calpain activity monitored by FRET imaging), was created to ubiquitously express a fluorescent FRET-sensor reporter [31]. Using this mouse model, changes in level of activity were observed in pathological conditions compared to the wild-type condition showing that CAFI mice and the FRET technique are valuable tools to follow calpain activity at physiological levels and in disease conditions.

In this paper, we present the detailed procedure to perform and apply the FRET technique in vivo in skeletal muscle tissue from the reporter to be used to mode of injection and methods of imaging and analysis. However, this method can also be applied to any other physiological condition.

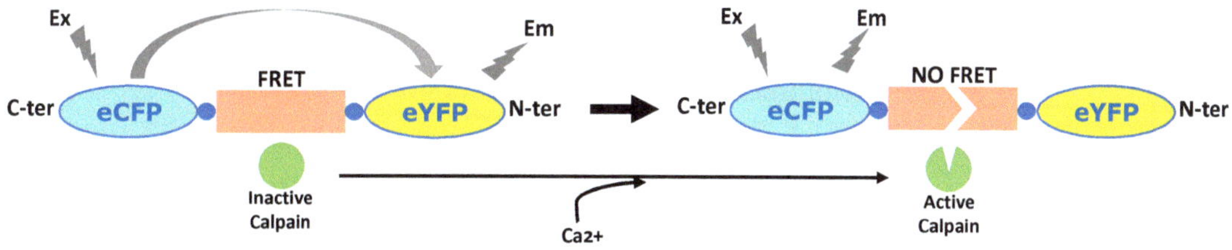

Fig. 1 Scheme of the calpain FRET sensor

2 Material

2.1 Generating the Cis-Plasmid for the FRET-Sensor Construct

1. The coding sequence of eCFP (*see* **Note 1**).
2. The coding sequence of eYFP (*see* **Note 1**).
3. The coding sequences of ubiquitous calpains sensor (*see* **Note 2**).

2.2 In Vivo DNA Plasmid Delivery in Skeletal Muscle Fibers

2.2.1 For Plasmid Delivery in Living Animals (See Note 3)

1. A wild-type mouse model; C57BL/6 mouse was used in our studies.
2. FRET-sensor plasmid.
3. Anesthesia solution: 1 mL Rompun 2% (xylazine, Centravet, France) and 2 mL ketamine 1000FL (Centravet) in 7 mL of 150 mM NaCl to obtain a final concentration of 2 mg/mL of xylazine and 20 mg/mL of ketamine. Store at +4 °C.
4. Injection syringes and needles:
   (a) Terumo 1 mL sterile syringe (VWR International, West Chester, PA, USA) with 27G*3/4 needle (20 mm length) for intraperitoneal injection of anesthesia solution.
   (b) Hamilton 50 μL syringe (NH Bio, Massy, France) with 30G needle (27 mm length, point style pst 4) (NH Bio) for intramuscular injection of rAAV preparations.
5. Electropulsator ECM-830 (BTX, Holliston, MA, USA).
6. In vivo electrodes: Tweezertrodes electrodes (BTX).

2.2.2 For Transgenic Animals

The plasmid FRET construct can be used to generate transgenic animals, leading to a mouse model expressing ubiquitous FRET sensor (*see* **Note 4**).

2.3 In Vivo Imaging (See Note 5)

1. Anesthesia solution (as described above).
2. A Terumo 1 mL sterile syringe (VWR International, West Chester, PA, USA) with 27G*3/4 needle (20 mm length) for intraperitoneal injection of anesthesia solution.
3. Scalpel (*see* **Note 6**).

4. Calpain-activation solution: 100 mM CaCl2, 15 mM ionomycin, 7% fetal calf serum, and 5 mM Hepes (pH 7.4) in a final volume of 50 mL (*see* **Note 7**).
5. A glass coverslide.
6. A Radiance 2100 MP scan head (Bio-Rad) confocal microscope equipped with a mode-locked titanium-sapphire laser system (Coherent Verdi-Mira) pumped at 5 W and tuned to a 800 nm excitation with 100 fs pulses at 76 MHz (*see* **Note 8**).
7. An inverted microscope Nikon TE300 (Nikon Instech Co.) with Nikon objectives, dry CFI Plan APO, 20! NA0.75 (*see* **Note 9**).

2.4 Measurements and Analysis

1. Software capable of acquiring two simultaneous confocal microscopic images (donor image « D » and the FRET image « F »).
2. ImageJ (National Health Institute—Wayne Rasband).
3. Custom-made plug-in « FRET Compute Environment » (freely available upon request from stockho@genethon.fr).

3 Methods

3.1 Generating the Cis-Plasmid for the FRET-Sensor Construct (See Note 10)

The final construct can be easily generated by gene synthesis as soon as all of the required sequences are known (*see* **Note 11**) and inserted in an expression vector. Extract plasmid DNA using a maxi-prep DNA extraction NucleoBond Xtra EF column kit.

3.2 In Vivo DNA Plasmid Delivery in Skeletal Muscle Fibers

3.2.1 For Plasmid Delivery in Living Animals

Electroporation is currently used to evaluate the expression and consequences of expression in local delivery. All procedures on animals must be carried out in accordance with the animal use and care regulations for the country in which they are performed. In Europe, the procedures follow the directive of 24 November 1986 (86/609/EEC) of the Council of the European Communities and are to be approved by local ethics committee.

For Intramuscular (IM) Injection

1. Anesthetize 4- to 6-week-old mice by intraperitoneal injection of ketamine (100 mg/kg)/xylazine (10 mg/kg).
2. Dilute 100 μg of endotoxin-free plasmid into a final volume of 25 μL of saline.
3. Shave the hind limb of the mouse using a lab animal clipper. Slowly inject the plasmid preparation into the middle of the *tibialis anterior* (TA) muscle using a 30G needle on a Hamilton 50 μL syringe.

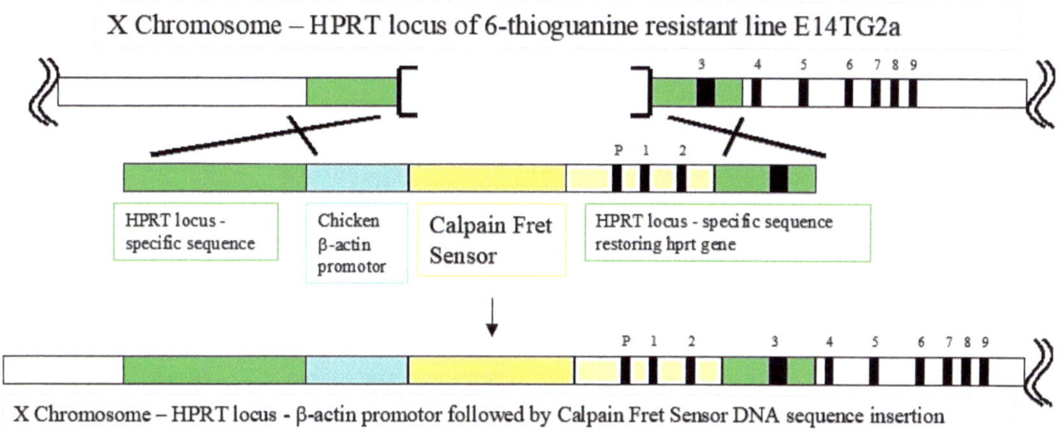

Fig. 2 Homologous recombination scheme to generate the transgenic « Calpain sensor » mouse

4. Immediately after injection in TA, place the electrodes on either side of the hind limb.
5. Apply transcutaneous electric pulses by the electropulsator with 200 V/cm, a pulse length of 20 ms, and a frequency of pulse delivery of 2 Hz.
6. After anesthesia, place the animal on a heating pad and monitor until it recovers.

3.2.2 For Transgenic Animals

CAFI transgenic mice were obtained using our construct by inserting the cassette in the HPRT (hypoxanthine phosphoribosyltransferase) [32] locus as depicted in Fig. 2.

3.3 In Vivo Imaging

1. Anesthetize mice by intraperitoneal injection of anesthesia solution with 100 ‘ per 20 mg, long enough for image acquisition, ~4–5 h (*see* **Note 12**).
2. With a scalpel blade, perform a skin incision over the *tibialis anterior* muscle and expose the muscle.
3. Position the glass coverslide on the tissue, and attach using common tape.
4. Place the animal and coverslide on the microscope for direct observation.

3.4 Measurements and Analysis

1. Acquire preliminary control images Dd (for the donor filter) and Fd (for the FRET filter) from cells expressing only the donor fluorophore (eCFP) in order to calibrate the FRET efficacy computation. The level of expression of the control cells must be similar to the level of expression of the FRET sensor (*see* **Note 13**).
2. Acquire images D (for the donor filter) and F (for the FRET filter) of the sample of interest expressing the FRET sensor with

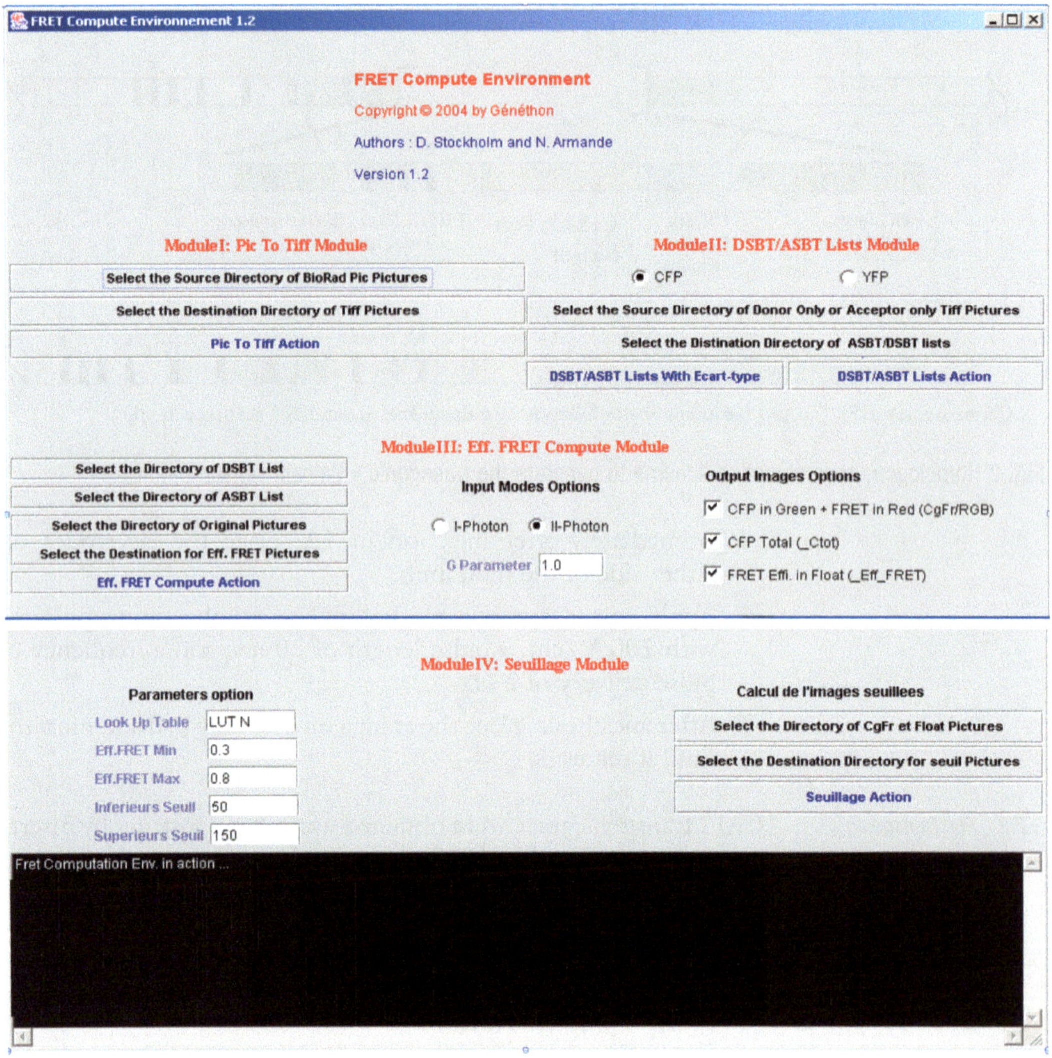

Fig. 3 User interface of the plug-in FRET Compute Environmen

the same configuration (laser intensity, photo-multiplicator tube gains, etc.).

3. Use the plug-in "FRET Compute Environment" to compute first the donor bleed-through (DSBT) coefficients list by loading the directory containing the Dd and Fd images (Fig. 3).
4. Place the images of the sample of interest in a separate folder.
5. Indicate to the module III of the plug-in by clicking on the « Select the Directory of Original Pictures ».
6. Click on "Eff. FRET Compute Action" to initiate the computation.
7. Measure the FRET efficiency of specific region of interest (ROI) by opening the 32 bits (float) images by ImageJ.

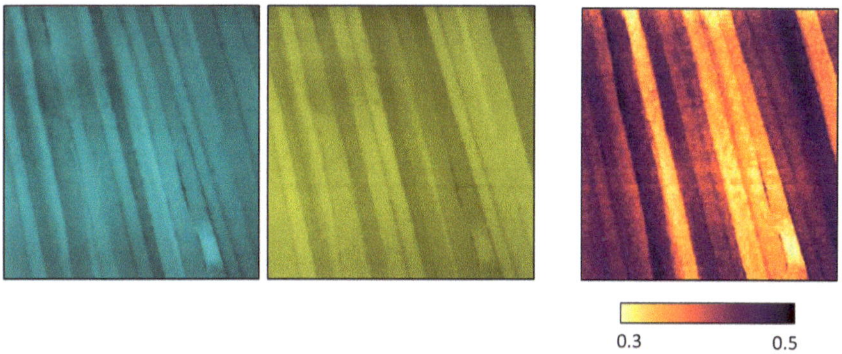

Fig. 4 Left panels: RG images of the donor filter image (CFP) and the FRET filter image (YFP) (after 800 nm length wave excitation) of a CAFI muscle after a running exercise. Right panel: pseudo color images of FRETeff (Inferno lookup table). The pseudo color scale is indicated below. The brighter the color, the stronger the calpain activity

8. Draw ROI with the polygon selection tool.
9. Display the mean of the ROI by applying "Measure" or "Histogram" from the "Analyze" tool.
10. Generate false color images of the FRET efficiency by using the module IV of the plug-in (Fig. 4).

4 Notes

1. FRET is a mechanism of energy transfer between molecules in proximity to each other. We used in our setup the classical pair eCFP and eYFP, but derivates or other combinations are possible ([33]).
2. The calpain sensor site corresponds to the cleavage site of the murine alpha-fodrin protein described as QQEVYGAMPRD [30]. Here we describe a global strategy to evaluate ubiquitous calpain activity. Nevertheless, the same strategy can be used for any other protease as soon as the cleavage site is identified. In this method calpain 3 activity in skeletal muscle was evaluated.
3. The calpain sensor construct is ubiquitous. In this protocol, we choose to focus on skeletal muscle. Here is presented a method to deliver a FRET construct to muscle fibers.
4. Plasmid delivery using an electrotransfer approach can be used as the proof-of-concept strategy due to the ease and the quickness of this process. However, a more robust and reliable approach is needed for stable and long-term expression. For this purpose, the construct can be delivered by adeno-associated virus (AAV), or a transgenic mouse model can be generated. In our studies, the FRET-sensor construct was used to generate a transgenic strain mouse. The details of the

generation and characterization of our mouse model, CAFI (calpain activity monitored by FRET imaging), were presented in [31].

5. To monitor calpain activity is a widespread issue in in vivo approaches. This methodology can be used for many applications. However, small changes in calpain activity may be difficult to detect.
6. Follow-up of the animals is feasible by sewing up the wound.
7. Depending on the experimental parameters, calpain activation can occur in many different ways. As a positive control, the calpain-activation solution can be used at the described concentration. A high concentration of calcium is necessary considering that the volume injected is diluted in the muscle volume and that the muscle fibers are imaged at a distance from the injection site.
8. Checking of the excitation parameters is done through the use of BEAM conditioning and Pockels cell units (Bio-Rad) that enable pulse length to be controlled and output power levels to be attenuated below a range so that no damage can be observed on the tissue (typically 10–46 mW depending on the objective lens). In our studies, power measurements exiting the microscope objective were made with a Coherent LaserMate model 33-0191 power meter. Two-photon fluorescence was detected by a non-descanned detector (Bio-Rad) after passing optical filters. Dual emission of eCFP and FRET used a dichroic mirror 500DCLPXR (Bio-Rad) and two band-pass emission filters: HQ450/80 for eCFP (donor filter; Bio-Rad) and 545AF35 for eYFP (FRET filter; Omega Optical). The detector gain settings were fixed to 50% for each detector, and the laser power with a 20× objective was set to 10 mW. The term FRET filter corresponds to a combination of an acceptor filter with a donor excitation.
9. We present, in this study, a methodology using a multiphoton confocal microscope. Nevertheless, the data acquisition and analysis can be done using a conventional confocal microscope as well.
10. FRET-sensor (#36182) and FRET-control (#36262) plasmids are available on Addgene.
11. The synthetized fragment includes the calpain sensor site flanking by eCFP in 5′ part and eYFP in 3′ part. The sensor site is encoded by GGAAGCGGCCAGCAGGAGGTG-TATGGTGCGATGCCCAGGGATGGCTCAGGG, with underlined sequences corresponding to GSG linker.

At the end of the experiment, the wound can be sutured closed for follow-up study; or mice may be sacrificed at the desired time point (e.g., 8 days for electroporation; as soon as 15 days for short-term study with rAAV; time of interest for Tg animals). Briefly, dissect the muscles of interest, and mount them into tragacanth gum on cork pieces, and snap freeze them in liquid nitrogen-cooled isopentane. Muscles are stored at −80 °C until use.

12. Anesthetized animals should be kept in a temperature-controlled microscope chamber (Solent Scientific) at 30 °C during all MPM experiments.

13. For the FRET calculation, we use the equation described by Gordon et al. [34] FRETeff = Ffa/(Ffa + G.Df). Details can be found in our previously published paper [30].

References

1. Smith MA, Schnellmann RG (2012) Calpains, mitochondria, and apoptosis. Cardiovasc Res 96:32–37
2. Potz BA, Abid MR, Sellke FW (2016) Role of calpain in pathogenesis of human disease processes. J Nat Sci 2:e218
3. Goll DE, Thompson VF, Li H, Wei W, Cong J (2003) The calpain system. Physiol Rev 83:731–801
4. Baudry M, Chou MM, Bi X (2013) Targeting calpain in synaptic plasticity. Expert Opin Ther Targets 17:579–592
5. Kurbatskaya K, Phillips EC, Croft CL, Dentoni G, Hughes MM, Wade MA, Al-Sarraj S, Troakes C, O'Neill MJ, Perez-Nievas BG, Hanger DP, Noble W (2016) Upregulation of calpain activity precedes tau phosphorylation and loss of synaptic proteins in Alzheimer's disease brain. Acta Neuropathol Commun 4:34
6. Nakajima E, Hammond KB, Rosales JL, Shearer TR, Azuma M (2011) Calpain, not caspase, is the causative protease for hypoxic damage in cultured monkey retinal cells. Invest Ophthalmol Vis Sci 52:7059–7067
7. Kumar S, Kain V, Sitasawad SL (2012) High glucose-induced Ca2+ overload and oxidative stress contribute to apoptosis of cardiac cells through mitochondrial dependent and independent pathways. Biochim Biophys Acta 1820:907–920
8. Li FZ, Cai PC, Song LJ, Zhou LL, Zhang Q, Rao SS, Xia Y, Xiang F, Xin JB, Greer PA, Shi HZ, Su Y, Ma WL, Ye H (2015) Crosstalk between calpain activation and TGF-beta1 augments collagen-I synthesis in pulmonary fibrosis. Biochim Biophys Acta 1852:1796–1804
9. Panico P, Salazar AM, Burns AL, Ostrosky-Wegman P (2014) Role of calpain-10 in the development of diabetes mellitus and its complications. Arch Med Res 45:103–115
10. Covington MD, Arrington DD, Schnellmann RG (2009) Calpain 10 is required for cell viability and is decreased in the aging kidney. Am J Physiol Renal Physiol 296:F478–F486
11. Stalker TJ, Gong Y, Scalia R (2005) The calcium-dependent protease calpain causes endothelial dysfunction in type 2 diabetes. Diabetes 54:1132–1140
12. Gallardo E, de Andres I, Illa I (2001) Cathepsins are upregulated by IFN-gamma/STAT1 in human muscle culture: a possible active factor in dermatomyositis. J Neuropathol Exp Neurol 60:847–855
13. Ahn YJ, Kim MS, Chung SK (2016) Calpain and caspase-12 Expression in lens epithelial cells of diabetic cataracts. Am J Ophthalmol 167:31–37
14. Huang CJ, Gurlo T, Haataja L, Costes S, Daval M, Ryazantsev S, Wu X, Butler AE, Butler PC (2010) Calcium-activated calpain-2 is a mediator of beta cell dysfunction and apoptosis in type 2 diabetes. J Biol Chem 285:339–348
15. Sorimachi H, Ono Y (2012) Regulation and physiological roles of the calpain system in muscular disorders. Cardiovasc Res 96:11–22
16. Cagmat EB, Guingab-Cagmat JD, Vakulenko AV, Hayes RL, Anagli J (2015) Potential use of calpain inhibitors as brain injury therapy. In: Kobeissy FH (ed) Brain neurotrauma: molecular, neuropsychological, and rehabilitation aspects. Frontiers in Neuroengineering, Boca Raton (FL)
17. Hollinger K, Selsby JT (2013) The physiological response of protease inhibition in dystrophic muscle. Acta Physiol (Oxf) 208:234–244

18. Potz BA, Sabe AA, Abid MR, Sellke FW (2016) Calpains and coronary vascular disease. Circ J 80:4–10
19. Stroev EA, Riazanova EA, Tavintsev VD (1993) A method of serial determination of calpain activity in biological material. Vopr Med Khim 39:56–58
20. Farr C, Berger S (2010) Measuring calpain activity in fixed and living cells by flow cytometry. J Vis Exp 41:2050
21. Niapour M, Berger S (2007) Flow cytometric measurement of calpain activity in living cells. Cytometry A 71:475–485
22. Miyawaki A, Llopis J, Heim R, McCaffery JM, Adams JA, Ikura M, Tsien RY (1997) Fluorescent indicators for Ca2+ based on green fluorescent proteins and calmodulin. Nature 388:882–887
23. Jares-Erijman EA, Jovin TM (2003) FRET imaging. Nat Biotechnol 21:1387–1395
24. Truong K, Sawano A, Mizuno H, Hama H, Tong KI, Mal TK, Miyawaki A, Ikura M (2001) FRET-based in vivo Ca2+ imaging by a new calmodulin-GFP fusion molecule. Nat Struct Biol 8:1069–1073
25. Mahajan NP, Linder K, Berry G, Gordon GW, Heim R, Herman B (1998) Bcl-2 and Bax interactions in mitochondria probed with green fluorescent protein and fluorescence resonance energy transfer. Nat Biotechnol 16:547–552
26. Nagai Y, Miyazaki M, Aoki R, Zama T, Inouye S, Hirose K, Iino M, Hagiwara M (2000) A fluorescent indicator for visualizing cAMP-induced phosphorylation in vivo. Nat Biotechnol 18:313–316
27. Neininger A, Thielemann H, Gaestel M (2001) FRET-based detection of different conformations of MK2. EMBO Rep 2:703–708
28. Vanderklish PW, Krushel LA, Holst BH, Gally JA, Crossin KL, Edelman GM (2000) Marking synaptic activity in dendritic spines with a calpain substrate exhibiting fluorescence resonance energy transfer. Proc Natl Acad Sci U S A 97:2253–2258
29. Harris AS, Croall DE, Morrow JS (1988) The calmodulin-binding site in alpha-fodrin is near the calcium-dependent protease-I cleavage site. J Biol Chem 263:15754–15761
30. Stockholm D, Bartoli M, Sillon G, Bourg N, Davoust J, Richard I (2005) Imaging calpain protease activity by multiphoton FRET in living mice. J Mol Biol 346:215–222
31. Bartoli M, Bourg N, Stockholm D, Raynaud F, Delevacque A, Han Y, Borel P, Seddik K, Armande N, Richard I (2006) A mouse model for monitoring calpain activity under physiological and pathological conditions. J Biol Chem 281:39672–39680
32. Bronson SK, Plaehn EG, Kluckman KD, Hagaman JR, Maeda N, Smithies O (1996) Single-copy transgenic mice with chosen-site integration. Proc Natl Acad Sci U S A 93:9067–9072
33. Bajar BT, Wang ES, Zhang S, Lin MZ, Chu J (2016) A guide to fluorescent protein FRET pairs. Sensors (Basel) 16:E1488
34. Gordon GW, Berry G, Liang XH, Levine B, Herman B (1998) Quantitative fluorescence resonance energy transfer measurements using fluorescence microscopy. Biophys J 74:2702–2713

Chapter 6

Determining Temporal and Spatial Expression of Calpains in Amphibians

Anna Charalambous, Ioanna Antoniades, Neophytos Christodoulou, Sara Zanardelli, and Paris A. Skourides

Abstract

Calpains are a family of calcium-dependent intracellular cysteine proteases that regulate important physiological processes by substrate cleavage. Despite the fact that Calpains have been identified in the *Xenopus* genome, their expression patterns and role have not been characterized. Therefore, herein, we describe two methods to determine temporal and spatial expression of Calpain 2 during *Xenopus* development, namely, RT-PCR and whole-mount in situ hybridization (WISH). In addition, indirect immunofluorescence (IF) is described to determine translocation to the plasma membrane, which correlates with activity levels of Calpain 2.

Key words *Xenopus*, Calpain, RT-PCR, Whole-mount in situ hybridization, Immunofluorescence

1 Introduction

Calpains are a family of calcium-dependent intracellular cysteine proteases that regulate important physiological processes (i.e., signal transduction, cell proliferation, apoptosis, and cell motility) by substrate cleavage [1]. The most extensively studied members of this family are Calpains 1 and 2; these proteases are heterodimers, composed of an 80 kDa *catalytic* subunit and a common 30 kDa regulatory subunit. They are very similar with respect to substrate specificity and subcellular localization but require different Ca^{2+} concentrations for their activity, ~10–50 μM and >500 μM, respectively [2, 3]. These concentrations are considerably higher than physiological intracellular concentrations of Ca^{2+} (around <1 μM), suggesting the presence of regulatory mechanisms that may lower this requirement. Indeed Calpain 2 binds to phospholipids (particularly phosphoinositide biphosphate, PIP2) and thus

Anna Charalambous and Ioanna Antoniades contributed equally to this work.

Jeannette S. Messer (ed.), *Calpain: Methods and Protocols*, Methods in Molecular Biology, vol. 1915, https://doi.org/10.1007/978-1-4939-8988-1_6, © Springer Science+Business Media, LLC, part of Springer Nature 2019 localizes at the membrane close to its substrates/cofactors and calcium channels [4].

Herein, we describe the use of RT-PCR and whole-mount in situ hybridization (WISH) to determine the temporal and spatial expression of Calpain 2 during *Xenopus* development. In addition, indirect immunofluorescence (IF) experiments are described to determine the protein's intracellular distribution and specifically its translocation to the plasma membrane, which is known to correlate well with activity [5, 6].

2 Materials

Prepare all solutions using analytical/molecular biology grade reagents.

Diligently follow all waste disposal regulations when disposing waste materials.

2.1 *Xenopus laevis* Embryos: Collection, Microinjections, and Fixation

1. *Xenopus laevis* eggs collected from induced spawning of adult females, in vitro fertilized using testes from adult males [7, 8].
2. Human chorionic gonadotropin (hCG), 800 units.
3. 10% benzocaine.
4. Testis medium: 90% Leibovitz's L-15 medium, 10% fetal calf serum (FCS), and 50 μg/ml gentamicin.
5. Petri dishes.
6. Microinjector, micromanipulator, forceps, and glass capillary needles.
7. 10× Marc's modified Ringer's, pH 7.8 (MMR): 0.1 M NaCl, 1.8 mM KCl, 2 mM $CaCl_2$, 1 mM $MgCl_2$, 5 mM HEPES (4-(2-hydroxyethyl)-1-piperazineethanesulfonic acid), 0.1 mM EDTA (ethylenediaminetetraacetic acid), and 0.05 mg/ml gentamicin gel.
8. 1.8% L-cysteine in MMR (*see* **Note 1**).
9. 4% Ficoll in MMR.
10. 1× MEMFA (MOPS (3-(*N*-morpholino)propanesulfonic acid), EGTA (ethylene glycol tetraacetic acid), $MgSO_4$, formaldehyde): 0.1 M MOPS pH 7.4, 2 mM EGTA, 1 mM $MgSO_4$, and 1% formaldehyde.

2.2 RNA Isolation and RT-PCR

1. PCR machine.
2. 10× phosphate buffered saline (PBS), pH 7.4.
3. TRIzol reagent.
4. Chloroform.
5. Isopropanol.
6. 70% ethanol (in nuclease-free H_2O).

7. Reverse transcriptase kit.
8. PCR kit.
9. Nuclease-free H_2O.

2.3 Agarose Gel Electrophoresis and Imaging

1. UltraPure 10× Tris-acetate-EDTA (TAE) running buffer: A pH-adjusted blend of ultrapure grade TAE buffer which, when diluted to a 1× working concentration, contains 40 mM Tris-acetate and 1 mM Na_2EDTA.
2. Agarose: Certified free of RNases, DNases, and inhibitors.
3. Ethidium bromide.
4. 100 bp or 1 kb DNA ladder.
5. Nuclease-free H_2O or diethyl pyrocarbonate (DEPC)-H_2O.
6. Agarose gel electrophoresis machine.
7. Gel documentation imaging system.

2.4 Generation of Constructs

1. PCR machine.
2. Primers.

Primer	Sequence
Capn2RT F	GGACCATTGTTGACCTGCTG
Capn2RT R	AGTGTCCATCTTGAGCGGC
Capn2probe F	ACCTTGAATTCGGACCATTGTTGACCTGC TG
Capn2probe R	AGATGCGGCCGCAGTGTCCATC TTGAGCGGC
Capn2ResF	AAGCGGCCGCATGAGCGGAG TCGCAGACAGACTTGCTAAG
Capn2HAR	AACTCGAGTCA*AGCGTAATCTGGAACATCG TATGGGT*AGAT CACAGTCAGGGAAAGCCAAG

3. Plasmids: pCS108, pCMV-Sport6 plasmid containing *Xenopus* Calpain2.
4. Restriction enzymes: NotI, XhoI, EcoRI.
5. T4 DNA ligase.
6. Chemically competent bacteria.
7. Mini-prep kit for DNA extraction.

2.5 Whole-Mount In Situ Hybridization (WISH)

1. RNA probe synthesis: SP7 promoter kit.
2. Phenol/chloroform/isoamyl alcohol (25:24:1).
3. 100% and 70% ethanol (in nuclease-free H_2O).
4. 3M CH_3COONa.

5. 70% isopropanol (in nuclease-free H_2O).
6. RNA purification kit.
7. Nuclease-free H_2O.
8. 1× MEMFA.
9. Methanol.
10. Proteinase K (10 mg/ml).
11. PBTw: 1× PBS, 0.1% Tween-20.
12. 0.1 M triethanolamine (TEO), pH = 7–8.
13. 4% formaldehyde, in DEPC-H_2O.
14. 2× saline-sodium citrate (SSC) buffer: 0.3 M sodium chloride and 30 mM trisodium citrate in DEPC-H_2O, pH 7.
15. Hybridization solution (Hyb): 50% formamide, 5× SSC, 1 mg/ml torula RNA, 100 μg/ml heparin, 1× Denhardt's, 0.1% Tween-20, 0.1% 3-[(3-cholamidopropyl)dimethylammonio]-1-propanesulfonate (CHAPS), and 10 mM EDTA.
16. RNase A, 20 mg/ml.
17. Maleic acid buffer (MAB): 100 mM maleic acid and 150 mM NaCl.
18. Boehringer Mannheim blocking solution (BMB): 2% BMB reagent in MAB.
19. Anti-digoxigenin (DIG)-alkaline phosphatase (AP) antibody.
20. AP buffer: 100 mM Tris pH 9.5, 50 mM $MgCl_2$, 100 mM NaCl, 0.1% Tween-20, and 2 mM levamisole.
21. BM purple AP substrate precipitating.
22. Bleaching solution: 0.5× SSC, 5% formamide, 1.2% peroxide.

2.6 *In Vitro* Transcription of mRNA

1. Sp6 promoter kit.

2.7 Whole-Mount Immunofluorescence

1. 1× MEMFA.
2. Permeabilization buffer: 1× PBS, 1% dimethyl sulfoxide (DMSO), and 0.5% Triton-X100.
3. Ethanol.
4. Serum blocking solution: 10% normal donkey serum in permeabilization buffer.
5. Benzyl benzoate/benzyl alcohol (BB:BA): 2 volumes BB/1 volume BA.
6. Antibodies.

Antibody	Dilution
Anti-DIG-AP	1:3000 in BMB blocking solution
Anti-HA (rabbit)	1:500 in serum blocking solution
Anti-C-cadherin (mouse)	1:300 in serum blocking solution
Anti-rabbit Alexa 488	1:500 in serum blocking solution
Anti-mouse Cy3	1:500 in serum blocking solution

2.8 Imaging

1. Forceps, slides, and cover slips.
2. Bright-field imaging: Fluorescent stereomicroscope.
3. Fluorescence imaging: Bright-field fluorescent microscope or laser scanning confocal microscope (LSM) equipped with the appropriate filter sets, camera, and the appropriate imaging analysis software.

3 Methods

3.1 Obtaining Embryos

1. Induce ovulation of female frogs (*Xenopus laevis*) by injecting hCG into the dorsal lymph sac. Keep female frogs at room temperature for 9–10 h until egg collection.
2. Sacrifice a male frog by immersion in 0.05% benzocaine for 1 h. Isolate testes and keep in testis medium at 4 °C.
3. The eggs collect in the sac near the cloaca so simultaneous lateral and vertical pressure should release them. Collect the eggs manually by massaging the belly with one thumb over a clean petri filled with 0.33× MMR.
4. Fertilize eggs immediately after collection (*see* **Note 2**). Cut a small piece of testis with scissors and macerate it with forceps so it is evenly distributed throughout the media. Leave the embryos for about 20 min. The eggs should rotate within the vitelline membrane so the animal hemisphere faces upward.
5. About 2 h after fertilization, the first cleavage is visible. De-jelly the embryo for microinjection by adding 1.8% cysteine for 3–5 min and then wash with 0.33× MMR. Keep the embryos in 0.33× MMR [9].

3.2 Examination of thc Tcmporal Expression Pattern of Calpain2 in Xenopus by Reverse Transcriptase PCR (RT-PCR)

This involves two major steps: (1) the isolation of RNA and cDNA synthesis and (2) the amplification and detection of Calpain2 cDNA.

1. Collect embryos of different developmental stages in sterile 1.5 ml centrifuge tubes (*see* **Note 3**).
2. Wash the embryos three times with sterile 1× PBS and add 500 μl cold TRIzol reagent. Homogenize embryos by pipetting up and down, and then allow samples incubate at room temperature for 10 min (at this step the cells are being lysed).

3. Spin the samples at 21,000 × *g* for 10 min at 4 °C. At this point the RNA is found in the supernatant. Remove the supernatant, place it in a clean sterile 1.5 ml tube, and add equal volume (as supernatant) of chloroform.
4. Vortex the samples for 15 s and then let them sit at room temperature for 3 min.
5. Spin the samples at 16,000 × *g* for 15 min at 4 °C. Two layers appear after spin. The top clear layer contains the RNA.
6. Remove the top layer and place it into a clean sterile 1.5 ml tube and add 500 μl isopropanol. Mix the samples by inverting the tubes and let them at room temperature for 10 min.
7. Spin the samples at 16,000 × *g* for 10 min at 44 °C. At this point the pellet contains the RNA. Discard the supernatant and wash the pellet with 500 μl RNase-free 70% ethanol (*see* **Note 4**).
8. Recover the pellet by centrifugation, at 9000 × *g* for 5 min at 4 °C. Remove the supernatant and let the pellet dry at room temperature for ~10 min. Dissolve the pellet in 30 μl sterile water (*see* **Note 5**).
9. For cDNA synthesis use a reverse transcriptase kit, and set up the reaction using the extracted RNA, according to manufacturer's instructions.
10. Use the cDNA immediately or store at −20 °C.
11. For each developmental stage, set up a PCR to amplify a Calpain2 cDNA-specific fragment, and detect the expression of Calpain2 at the mRNA level. Set up a reaction according to the PCR-kit manufacturer's instructions using primers Capn2RT F and Capn2RT R (*see* **Note 6**), at a final concentration of 0.1 μM.
12. Use the following PCR program: 95 °C for 3 min, 25 cycles each of 95 °C for 30 s, 52 °C for 30 s, 72 °C for 30 s followed by 72 °C for 3 min and 4 °C until reaction termination (*see* **Note** 7).
13. Run and analyze the PCR products (~350 bp) on a 1% agarose gel (Fig. 1a).

3.3 Examination of the Spatial Expression Pattern of Calpain2 in Xenopus by Whole-Mount In Situ Hybridization (WISH) (See Note 6)

This involves two major steps: (1) the preparation of a DIG-labeled Calpain2 RNA probe and (2) then whole-mount in situ hybridization [10].

1. Set up a PCR to amplify the cDNA fragment between nucleotides 1713 and 2081 of Calpain2 (*see* **Note 8**). Set up the PCR using 10 ng of the *Xenopus* Calpain2 pCMV-Sport6 plasmid as the template and the primers Capn2probe F and Capn2probe R (*see* **Note 9**), at a final concentration of 0.1 μM.

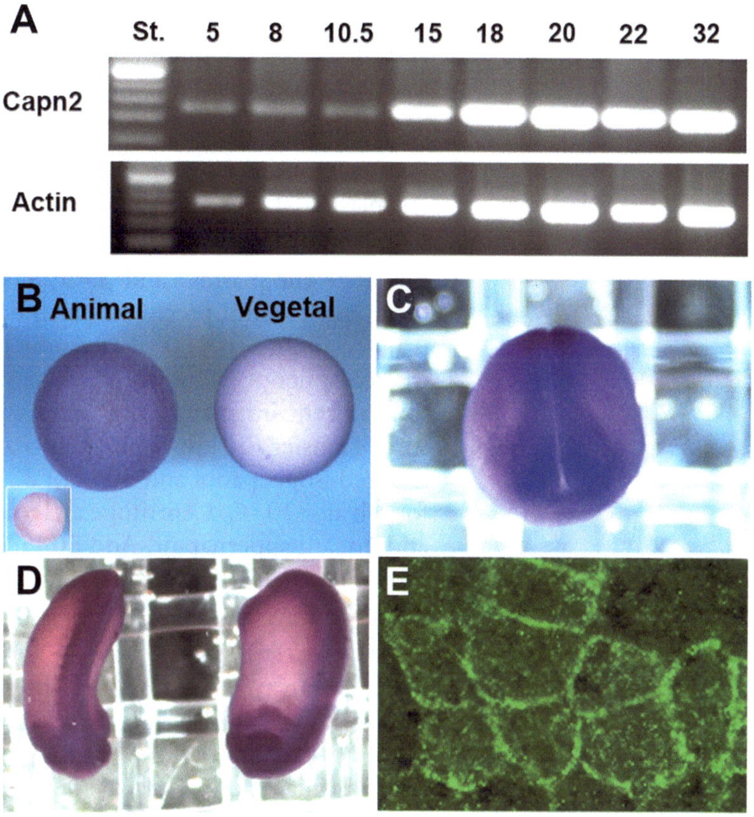

Fig. 1 Calpain2 temporal and spatial expression patterns during *Xenopus* development. (**a**) RT-PCR was performed using specific primers for Calpain2 at different stages of development. Actin was used as a loading control. (**b–d**) Expression pattern of Calpain2 in *Xenopus* by WISH. (**b**) Animal (left) and vegetal (right) view of a stage 9 embryo. Control hybridization with the sense probe is shown in the inset. (**c**) Dorsal view of a stage 17 embryo, anterior to the bottom. (**d**) Dorsal (left) and lateral (right) view of a stage 22 embryo, anterior to the bottom. (**e**) HA-Calpain2 localization in neural ectodermal cells of a stage 13 embryo. Scale bar = 20 μM. Adapted from Zanardelli et al. (2013) Dev Biol 384 (1):83–100 (https:/creativecommons.org/licenses/by/3.0)

2. Use the following PCR program: 95 °C for 5 min, 35 cycles each of 95 °C for 15 s, 60 °C for 30 s, 68 °C for 30 s followed by 68 °C for 10 min and 4 °C until reaction termination (*see* **Note** 7).
3. Run the PCR product (~370 bp band corresponding to the amplified Calpain2 fragment) on a 1% agarose gel. Use any PCR purification kit and follow the manufacturer's recommendations to purify the DNA.
4. Set up a digestion reaction for the PCR product and for 5 μg of pCS108 plasmid using EcoRI and NotI restriction enzymes. Run the digested PCR product and plasmid on a 1% agarose gel and extract as before.

5. Set up a ligation reaction between the digested PCR product and the pCS108 plasmid, using T4 DNA ligase.
6. Transform chemically competent bacteria using 2 μl of the ligation reaction.
7. Pick colonies and set up mini-preps for DNA extraction. Use any plasmid-DNA purification kit to purify DNA following the manufacturer's recommendations (*see* **Note 10**).
8. Linearize the plasmid and purify DNA using phenol/chloroform and ethanol precipitation under RNase-free conditions. Adjust volume to 100 μl and add the equivalent amount of ice-cold phenol/chloroform/isoamyl alcohol (25:24:1). Vortex and centrifuge at 12,000 × *g* for 10 min at room temperature. Transfer the upper phase into a new tube and add 500 μl ice-cold isopropanol and 50 μl of 3 M CH_3COONa. Incubate for 1 h at −20 °C. Centrifuge at 15,000 × *g* for 30 min at 4 °C and discard supernatant. Add 500 μl 70% ice-cold ethanol and centrifuge at 15,000 × *g* for 10 min at 4 °C. Discard supernatant and allow pellet to dry. Resuspend DNA in 20 μl of RNase-free H_2O (*see* **Note 11**).
9. Set up an in vitro transcription reaction using an appropriate transcription kit. Use 1 μg of linearized DNA, 2 μl 10× reaction buffer, 2 μl 10× NTP labeling mix containing digoxigenin labeled (Dig)-UTP, 1 μl RNase inhibitor, 2 μl T7 RNA polymerase, and RNase-free H_2O up to 20 μl. Incubate at 37 °C for 2 h, add 1 μl DNase, and incubate at 37 °C for 15 more min.
10. Use any RNA purification kit and follow the manufacturer's recommendations to purify the RNA probe. Store probe at −80 °C until use (*see* **Note 12**).
11. Fix embryos of different developmental stages (*see* **Note 13**) in 1× MEMFA, and then dehydrate them in 100% methanol for 1 h (changing methanol every 15 min), in 4 ml glass vials.
12. Rehydrate embryos through a series of methanol solutions (100%, 75%, 50%, and 25% methanol in H_2O), 5 min each at room temperature, and then wash three times in PBTw, 5 min each (*see* **Note 14**).
13. Permeabilize embryos using 10 μg/ml proteinase K in PBTw for 5 min at room temperature.
14. Wash embryos in 0.1 M TEO twice (5 min each), then in TEO with acetic anhydride (2 × 5 min), and then in PBTw (2 × 5 min) at room temperature.
15. Postfix embryos in 4% formaldehyde in PBTw for 20 min at room temperature, and then wash five times (5 min each) in PBTw.

16. Incubate embryos in hybridization solution for 6 h at 65 °C, and then incubate overnight at 65 °C with the probe diluted in hybridization solution at a final concentration 1 mg/ml.
17. Incubate the embryos with the hybridization solution (without probe) for 3 min at 60 °C, and then wash with 2× SSC at 60 °C (2 × 3 min and 3 × 20 min).
18. Treat embryos with 20 μg/ml RNase A in 2× SSC for 30 min at 37 °C, wash with 2× SSC at 60 °C for 10 min, and then wash twice with 0.2× SSC for 30 min at 60 °C.
19. Incubate embryos in MAB solution for 30 min at room temperature, block in BMB blocking solution for 1 h at room temperature, and then incubate with anti-DIG-AP antibody overnight at 4 °C.
20. Wash embryos in MAB solution (5 × 1 h) at room temperature followed by two washes, 5 min each, in AP buffer.
21. Incubate embryos in BM purple and check over time for color development. The visualization of color at a specific region determines the presence of mRNA. Stop the chromogenic reaction by washing the embryos in MAB solution at room temperature, and postfix the embryos overnight in 1× MEMFA at room temperature.
22. Bleach embryos in bleaching solution. Embryos can be used for imaging using a fluorescent stereomicroscope for bright-field images (Fig. 1b, c, and d) or a fluorescent microscope for the acquisition of fluorescent images, using customized filter sets for excitation (365–465 nm) and detection of BM purple precipitates (emission at 795 nm).

3.4 Examination of Calpain2 Subcellular Localization by Whole-Mount Immunofluorescence (See Note 15)

This procedure consists of five major parts: (1) generation of an HA-tagged *Xenopus* Calpain2 DNA (Capn2R construct) by PCR, (2) introduction of the construct into the pCS108 vector, (3) in vitro transcription of mRNA encoding *Xenopus* Calpain2, (4) injection of the *Xenopus* HA-Capn2 mRNA into *Xenopus* embryos, and (5) whole-mount immunofluorescence.

1. Set up a PCR to amplify the DNA fragment encoding *Xenopus* Calpain 2. Set up the reaction using 10 ng of the *Xenopus* Calpain2 pCMV-Sport6 plasmid as the template and the primers Capn2ResF and Capn2HAR (*see* **Note 16**), at a final concentration of 0.1 μM.
2. Use the following PCR program: 95 °C for 5 min, 35 cycles each of 95 °C for 15 s, 65 °C for 30 s, 68 °C for 2.5 min followed by 68 °C for 10 min and 4 °C until reaction termination (*see* **Note** 7).
3. Run the PCR product (~2100 bp band corresponding to HA-tagged Calpain2) on a 1% agarose gel, and use any PCR

purification kit to purify the DNA, following the manufacturer's recommendations.

4. Set up digests for the PCR product and for 5 μg of pCS108 plasmid using NotI and XhoI restriction enzymes (*see* **Note 17**). Run the digested PCR product and plasmid on a 1% agarose gel and extract as before.
5. Set up a ligation reaction between the digested PCR product corresponding to the HA-tagged Calpain2 and the pCS108 plasmid, using T4 DNA ligase.
6. Transform chemically competent bacteria using 2 μl of the ligation reaction.
7. Pick colonies and set up a mini-prep for DNA extraction (*see* **Note 18**).
8. Perform in vitro transcription of capped RNA from linearized pCS108 plasmid DNA, containing the HA-Capn2 construct. Use any commercially available kit and follow the manufacturer's instructions for in vitro transcription.
9. Use any RNA purification kit and follow the manufacturer's recommendations to purify the synthesized mRNA. Aliquot and store at −80 °C until injection (*see* **Note 19**).
10. Place *Xenopus* embryos in 4% Ficoll, in 0.33× MMR solution, and inject using a glass capillary pulled needle, forceps, a micromanipulator, and microinjector. (This is a standard procedure in any lab working with *Xenopus* embryos.)
11. Inject embryos, at the 2-cell stage with 200 pg of HA-Capn2R mRNA per blastomere. Allow injected *Xenopus* embryos until stage 8 in 4% Ficoll in 0.33× MMR. Then wash and maintain in 0.1× MMR until the desired stage, and proceed to whole-mount immunofluorescence staining for HA-Capn2R (*see* **Note 20**) [11].
12. Transfer embryos into glass vials, and fix with 1× MEMFA for 2 h at room temperature, under continuous rotation, and dehydrate in absolute ethanol.
13. Rehydrate embryos using a series of ethanol solutions (75%, 50%, and 25% ethanol) diluted in permeabilization buffer, 5 min each, and permeabilize in 100% permeabilization buffer for 4–5 h at room temperature.
14. Dissect embryos (sagittal sections) and block for 30 min in serum blocking buffer.
15. Incubate with the primary antibodies anti-HA (rabbit) and anti-C-cadherin (mouse), overnight at 4 °C under continuous rotation (*see* **Note 21**).

16. Wash embryos in permeabilization buffer several times (5 × 15 min), and incubate with the secondary antibodies anti-rabbit Alexa 488 and anti-mouse Cy3 for 2 h.
17. Wash embryos in permeabilization buffer several times (5 × 15 min), postfix in 1× MEMFA for 20 min, dehydrate in methanol for 30 min (3 × 10 min), and clear in BB:BA.
18. Image embryos on a confocal microscope (Fig. 1e).

4 Notes

1. All solutions are prepared using a final concentration of 0.33× MMR (from a 10× MMR stock solution).
2. The excess 0.33× MMR has to be removed from the petri containing the eggs when they are ready to fertilize.
3. We used embryos of developmental stages 5, 8, 10.5, 15, 18, 20, 22, and 32.
4. Make all nuclease (DNase/RNase) free solutions using nuclease-free H_2O.
5. Subsequent to isolation the amount and quality of RNA are examined using a spectrophotometer and gel electrophoresis, respectively.
6. The Capn2RT F and Capn2RT R primers are designed to amplify a Calpain2 specific region, between nucleotides 1730 and 2076, that is absent from other members of the Calpain-protein family. The size of the PCR product is ~350 bp.
7. Cycle temperatures and times may vary depending on PCR kit used.
8. Use an RNA probe complementary to a region of Calpain2 mRNA that is not conserved in other members of the Calpain-protein family.
9. Capn2probe F primer was designed starting from nucleotide 1713 and includes the EcoRI restriction enzyme site (underlined). Capn2probe R primer was designed to end at nucleotide 2081 and includes the NotI restriction enzyme site (underlined). EcoRI and NotI were used to insert the amplified DNA fragment into pCS108 which contains a T7 promoter for in vitro transcription.
10. Verify sequence by DNA sequencing.
11. Confirm complete linearization by running 1–2 μl of DNA on agarose gel, and then measure DNA concentration with a spectrophotometer.
12. Examine RNA quality and concentrations using gel electrophoresis and a spectrophotometer, respectively.

13. We used cleavage, blastula, gastrula, neurula, and tail bud stage embryos.
14. DEPC-H_2O is used to make all the solutions used in WISH.
15. Because of the unavailability of antibodies able to recognize the different isoforms of Calpain, the localization of Calpain2 was examined using an HA-tagged Calpain2 construct.
16. Capn2ResF primer was designed starting from the ATG and including silent mutations in the coding sequence, shown in bold. It also includes the NotI restriction enzyme site (underlined). Capn2HAR primer includes the sequence that encodes a C-terminal HA tag. It also includes the XhoI restriction enzyme site (underlined). The sequence in italics corresponds to the sequence of HA tag.
17. The pCS108 plasmid is used since it carries an Sp6 promoter that allows in vitro mRNA transcription.
18. Verify sequence by DNA sequencing.
19. Subsequent to in vitro transcription, the amount and quality of RNA are examined using a spectrophotometer and gel electrophoresis, respectively.
20. We examined the localization in ectodermal cells of stage 9 embryos and in ectodermal and mesodermal cells of stage 12 and 13 embryos.
21. The anti-C-cadherin antibody is used to stain the cell boundaries.

References

1. Wells A, Huttenlocher A, Lauffenburger DA (2005) Calpain proteases in cell adhesion and motility. Int Rev Cytol 245:1–16. https://doi.org/10.1016/S0074-7696(05)45001-9
2. Croall DE, DeMartino GN (1991) Calcium-activated neutral protease (calpain) system: structure, function, and regulation. Physiol Rev 71(3):813–847
3. Sorimachi H, Tsukahara T, Okada-Ban M, Sugita H, Ishiura S, Suzuki K (1995) Identification of a third ubiquitous calpain species--chicken muscle expresses four distinct calpains. Biochim Biophys Acta 1261(3):381–393
4. Leloup L, Shao H, Bae YH, Deasy B, Stolz D, Roy P, Wells A (2010) m-Calpain activation is regulated by its membrane localization and by its binding to phosphatidylinositol 4,5-bisphosphate. J Biol Chem 285 (43):33549–33566. https://doi.org/10.1074/jbc.M110.123604
5. Suzuki K, Hata S, Kawabata Y, Sorimachi H (2004) Structure, activation, and biology of calpain. Diabetes 53(Suppl 1):S12–S18
6. Zanardelli S, Christodoulou N, Skourides PA (2013) Calpain2 protease: A new member of the Wnt/Ca(2+) pathway modulating convergent extension movements in Xenopus. Dev Biol 384(1):83–100. https://doi.org/10.1016/j.ydbio.2013.09.017
7. Antoniades I, Stylianou P, Christodoulou N, Skourides PA (2017) Addressing the functional determinants of FAK during ciliogenesis in multiciliated cells. J Biol Chem 292 (2):488–504. https://doi.org/10.1074/jbc.M116.767111
8. Antoniades I, Stylianou P, Skourides PA (2014) Making the connection: ciliary adhesion complexes anchor basal bodies to the actin cytoskeleton. Dev Cell 28(1):70–80. https://doi.org/10.1016/j.devcel.2013.12.003

9. Charalambous A, Andreou M, Antoniades I, Christodoulou N, Skourides PA (2012) In vivo, site-specific, covalent conjugation of quantum dots to proteins via split-intein splicing. Methods Mol Biol 906:157–169. https://doi.org/10.1007/978-1-61779-953-2_11
10. Charalambous A, Koyioni M, Antoniades I, Pegeioti D, Eleftheriou I, Michaelidou SS, Amelichev SA, Konstantinov LS, Rakitin OA, Koutentis PA, Skourides PA (2015) 1,2,3-Dithiazoles - new reversible melanin synthesis inhibitors: a chemical genomics study. MedChemComm 6(5):935–946. https://doi.org/10.1039/c5md00052a
11. Charalambous A, Antoniades I, Christodoulou N, Skourides PA (2011) Split-inteins for simultaneous, site-specific conjugation of quantum dots to multiple protein targets in vivo. J Nanobiotechnol 9:37. https://doi.org/10.1186/1477-3155-9-37

Chapter 7

Immunohistochemical Localization of Calpains in the Amphibian *Xenopus laevis*

Jean-Marie Exbrayat, Elara N. Moudilou, and Claire Brun

Abstract

Though histochemical techniques have been used for decades, they are still very important in basic research. They make it possible to work on fixed tissues and provide a large amount of information in a relatively short time and at a low cost. Here we describe methods for indirect immunohistochemistry and immunofluorescence on sections of tadpoles and tissues of adult amphibians belonging to the species *Xenopus laevis*. The objective is to localize calpains within tissues in order to understand their involvement in cellular processes.

Key words Immunofluorescence, Immunohistochemistry, Calpains, Amphibians, Embryonic development

1 Introduction

Xenopus laevis is the animal of choice for the study of molecular mechanisms involved in embryogenesis. Calpains have been implicated in numerous cellular processes (cell differentiation, cell proliferation, apoptosis, and cell migration) and are essential to the development process of invertebrates and vertebrates [1–9]. Characterization of the classical calpain family member shows that there are many isoforms of these proteases in all vertebrates. The amphibians possess a single gene encoding for CAPN 1, 2, 8, and 11 and several specific genes encoding for CAPN 3, 9, 12, 13, and 17 [10, 11]. Like in other vertebrates, calpains are involved in several physiological mechanisms, and more especially, an autolytic activity of amphibian m-calpain has been studied [12]. In *Xenopus laevis*, several studies using different molecular approaches have shown the involvement of the calpain XCl-2 in the morphogenetic movements of gastrulation and neurulation [13], the role of CAPN1 (mμ/I) and CAPN8b (XCl-2) in organogenesis and at the onset of metamorphosis [14], and the involvement of the calpain/calpastatin system in controlling elongation of notochord and somite differentiation [15]. Immunohistochemical methods

Jeannette S. Messer (ed.), *Calpain: Methods and Protocols*, Methods in Molecular Biology, vol. 1915,
https://doi.org/10.1007/978-1-4939-8988-1_7,

on transverse or sagittal sections can also be used to follow the expression profiles of calpains during embryonic development [16].

Calpains play an important role in the cellular process of apoptosis that occurs during embryonic development [17]. Among the factors that make it possible to modulate physiological apoptosis are those responsible for the changes in the cytosolic calcium concentration. These calcium changes can be induced by an overstimulation of the ionotropic receptors of glutamate [18, 19]. So, these receptors play an important role in triggering apoptosis via calpain expression, contributing to the sculpture of the embryo.

The study of the localization of the main calpains (CP1, CP2/X-CL2, and CP3) can provide information on the role of these cysteine proteases in *Xenopus laevis* embryonic development and their possible involvement in cell death following the exposure of tadpoles to drugs.

The expression of calpains was investigated throughout the development of *Xenopus laevis*. Some analysis of calpain distribution was performed by use of immunohistological methods on transverse or sagittal sections [16, 20]. In order to know the relationships which could exist between the calpains and apoptosis throughout the development [17, 21], the expression of these proteases was researched on eggs, tadpoles, and adult tissues of individuals in contact with several drugs acting on glutamate receptors [22]. Apoptotic cells and several kinds of glutamate receptors were visualized previously on the same material [18, 19, 23–25]. The presence of calpains was first researched on the skin and gills, which are organs directly in contact with drugs, and also the intestine, dorsal muscle, heart, and brain [22].

In this chapter we present methods to visualize the calpains in *Xenopus laevis* embryos and tadpoles. Immunohistochemistry has been used to determine the repartition of calpains throughout embryonic development (from one-cell stage to adult age). Immunofluorescence has been used to determine the presence of calpains in several organs of *Xenopus laevis* tadpole (stage 48).

2 Materials

2.1 Animals

1. The maintenance of the animals is carried out in accordance with ethical legislation.
2. Breeding: 40 L aquaria with dechlorinated tap water changed weekly, room temperature (22 °C ± 1 °C), and light/dark photoperiod (12 h:12 h).
3. Feeding of adults: twice a week with trout feed chopped pellets.
4. Feeding of tadpoles: twice a day with watercress soup.
5. Embryos obtained by in vitro fertilization [22].

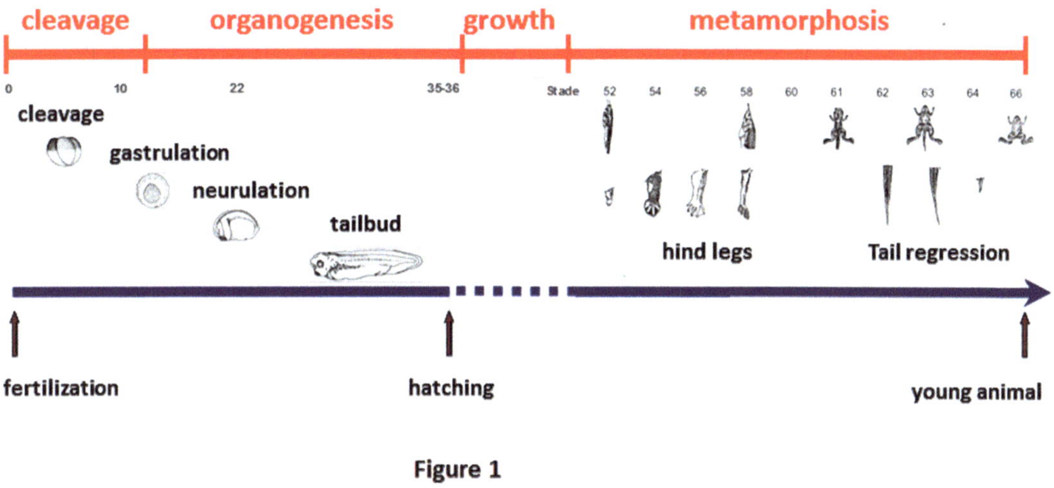

Fig. 1 The main development stages of *Xenopus laevis* according to Nieuwkoop and Faber's normal table of development [26]

6. The stages of development should be determined according to the *Xenopus laevis* normal table of development [26] (Fig. 1).

2.2 Histological Preparation of Tissues

1. Paraformaldehyde 4% in phosphate buffer 0.1 M:
   (a) Prepare phosphate buffer:
   - $Na_2HPO_4.2H_2O$, 0.1 M solution (17.8 g + 1 L H_2O).
   - NaH_2PO_4, $2H_2O$, 0.1 M (15.6 g + 1 L H_2O).
   - Adjust to pH 7.4 by adding approximatively 300 mL of NaH_2PO_4 in Na_2HPO_4.

   (b) Dissolve 40 g of paraformaldehyde in 1 L of 0.1 M phosphate buffer, and heat the solution 20 min at 80 °C. Add two drops of 12 N sodium hydroxide to clarify the fixative solution.

   This preparation must be done under a ventilated hood.
2. Bouin's fluid: 150 mL water saturated solution of picric acid, 50 mL 4% formalin, and 10 mL glacial acetic acid.
3. Ethanol: 100%, 96%, and 70%.
4. Cyclohexane.
5. Paraffin, Leuckart's bars, or embedding cases.
6. Minot microtome (vertical microtome).
7. Superfrost® plus slides.

2.3 Calpain Detection with Indirect Immunohist ochemistry

1. Dulbecco's Phosphate Buffered Saline (PBS): 1×, 0.1 M, pH 7.4.
2. Hydrogen peroxide: 3% solution diluted in PBS.

3. Mouse monoclonal anti-calpain 1 large subunit MAB3082 (Chemicon, Temecula, CA, USA). Specificity of the antibody against *X. laevis* enzymes must be verified prior to its use (*see* **Note 1**).
4. Rabbit polyclonal anti-calpain 2 large subunit AB1625 (Chemicon, Temecula, CA, USA). Specificity of the antibody against *X. laevis* enzymes must be verified prior to its use (*see* **Note 1**).
5. Rabbit polyclonal anti-p94 large subunit AB81018 (calpain 3) (Chemicon, Temecula, CA, USA). Specificity of the antibody against *X. laevis* enzymes must be verified prior to its use (*see* **Note 1**).
6. Horse normal serum, secondary antibody, and avidin-biotin-peroxidase complex: Vectastain ABC Universal Elite kit (mouse–rabbit, Vector Laboratories, Burlingame, USA).
7. Amino-ethylcarbazole (AEC): ready-to-use solution (Dako, Glostrup, Denmark).
8. Hematoxylin QS (Vector laboratories).
9. Hydrophilic medium (VectaMount AQ Mounting Medium, Vector laboratories).
10. Light microscope.

2.4 Calpain Detection with Indirect Immunofluorescence

1. PBS: 1×, 0.1 M, pH 7.4.
2. Bovine serum albumin (BSA): 1% in PBS (Sigma, St. Louis, MO).
3. Mouse monoclonal anti-calpain 1 large subunit MAB3082 (Chemicon, Temecula, CA, USA). Specificity of the antibody against *X. laevis* enzymes must be verified prior to its use (*see* **Note 1**).
4. Rabbit polyclonal anti-calpain 2 large subunit AB1625 (Chemicon, Temecula, CA, USA). Specificity of the antibody against *X. laevis* enzymes must be verified prior to its use (*see* **Note 1**).
5. Rabbit polyclonal anti-p94 large subunit AB81018 (calpain 3) (Chemicon, Temecula, CA, USA). Specificity of the antibody against *X. laevis* enzymes must be verified prior to its use (*see* **Note 1**).
6. Alexa 488-conjugated anti-mouse IgG (1:1000) (green fluorescent) (Invitrogen-Life Technologies, Carlsbad, CA, USA) or $Cy^{T}3$-labeled anti-rabbit IgG (1:200, orange-red fluorescent) (Chemicon, Temecula, CA, USA).
7. Fluorescence mounting medium (Dako) or ProLong Gold (molecular probes, Thermo Fisher Scientific, Waltham, MA, USA). Specific mounting medium is necessary to prevent

photobleaching. Alternatively, an antifading solution can be used in order to avoid the disappearance of fluorescence. Antifade solution: 9 volumes glycerol to 1 volume DABCO. Dissolve 0.233 g DABCO in 10 mL Tris–HCL 0.2 M (1 V), add 0.216 g $NaNO_3$ and 90 mL glycerol (9 V), and increase volume to 100 mL with purified water. Stir with a magnetic stirrer to homogenize the solution. Store at −20 °C protected from light.

8. Tris–HCl 0.2 M, pH 7.4.
9. Light microscope (Nikon Eclipse E400) equipped with epi-fluorescence optics (Nikon, Tokyo, Japan).

3 Methods

3.1 Histological Preparation of Tissues

1. Fix eggs, tadpoles, and organs dissected from adults in 4% paraformaldehyde for at least 24 h. Embryos (enriched in vitellus) and adult organs may be also fixed with Bouin's fluid (*see* **Note 2**). After 24 h, move the tissues into 70% ethanol for long-term storage (several weeks).
2. Dehydration: For eggs and tadpoles, ethanol 70% × 2 h, ethanol 96% 2 × 2 h, ethanol 100% 2 × 2 h. For organs: ethanol 70% × 4 h, ethanol 96% 2 × 12 h, ethanol 100% 2 × 4 h. The duration of each bath can be modified (*see* **Note 3**).
3. Clarification: This step is required to replace the solvent (ethanol) which cannot mix with paraffin. Incubate sections in butanol or cyclohexane 2 × 2 h for eggs and tadpoles or 2 × 12 h for organs.
4. Impregnation with paraffin: Leave the tissue in melted paraffin for 4–12 h depending on the tissues. For the liver, kidney, spleen, and lung, incubate for 4 h. for other tissues, incubate for 12 h (*see* **Note 4**). Embryos should be impregnated for 2 h. The temperature should be slightly higher than the melting point of paraffin.
5. Embedding in paraffin: The impregnated tissue or tadpoles are embedded in a paraffin block that is made by means of a mold using Leuckart's bars or embedding cases (*see* **Note 5**).
6. Cutting: 5 or 7 μm thickness transversal or sagittal sections should be obtained with the use of a vertical microtome. Sections should be stuck on slides Superfrost® plus (*see* **Note 6**). The slides can be kept indefinitely in boxes protected from dust.

3.2 Calpain Detection with Indirect Immunohist ochemistry

1. Dewax the slides in paraffin solvent.
2. Rehydrate the slides by dipping them for 5 min in each of 100% ethanol, 95% ethanol, 70% ethanol, and then distilled water. Rinse with PBS.
3. Place the slides in a humidified box.
4. Incubate sections for 5 min in 3% hydrogen peroxide diluted in PBS.
5. Wash in PBS.
6. Block with HSA for 20 min.
7. Prepare the calpain antibody solution at appropriate dilution: 1:50 dilution in PBS for the mouse monoclonal anti-calpain 1 large subunit MAB3082 and rabbit polyclonal anti-calpain 2 large subunit AB1625, 1:25 dilution in PBS for the rabbit polyclonal anti-p94 large subunit AB81018.
8. Incubate the slides with primary antibody solution for 1 h at laboratory temperature. Perform a negative control (*see* **step 22**).
9. Rinse three times for 5 min with PBS under gentle shaking.
10. Incubate the slides with secondary antibody solution for 1 h at laboratory temperature. Perform a negative control (*see* **step 23**).
11. Rinse three times for 5 min with PBS.
12. Apply avidin-biotin-peroxidase complex for 1 h.
13. Rinse three times for 5 min with PBS under gentle shaking.
14. Apply AEC solution at laboratory temperature.
15. Monitor the appearance of a red precipitate under a light microscope (about 5 min).
16. Rinse with distilled water.
17. Counterstain the sections with hematoxylin for 1 min at laboratory temperature, and then rinse with tap water.
18. Mount with hydrophilic medium.
19. Cover with a coverslip.
20. Let the medium polymerize overnight at laboratory temperature.
21. Visualize using a light microscope (*see* **Note** 7 and Figs. 2, 3, and 4).
22. Negative control: replace **step 8** with this one: incubate the slides for 1 h in PBS (0.1 M, pH 7.4). Return to **step 9** (Figs. 2 and 3).

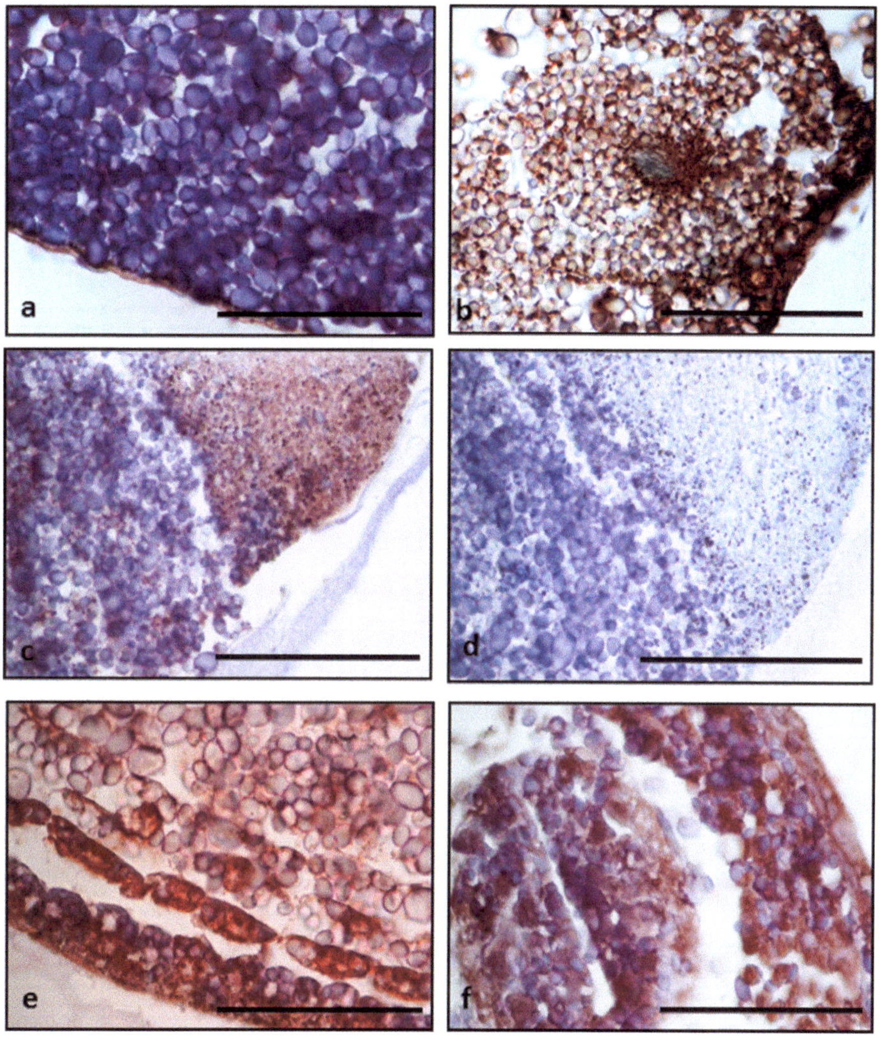

Fig. 2 Localization of calpains in *Xenopus laevis* during early stages of development, with indirect immunohistochemistry. (**a**) and (**b**) Calpain 2 in 4 cells embryo. (**c**) Calpain 1 in gastrula. (**d**) Negative control on gastrula (omission of primary antibody). (**e**) Calpain 1 in neurula. (**f**) Calpain 1 in tail bud. Scale bar = 100 μm

23. Negative control: replace **step 10** with this one: incubate the slides for 1 h in PBS (0.1 M, pH 7.4). Return to **step 11** (Fig. 4).

3.3 Calpain Detection with Indirect Immunofluorescence

1. Dewax the slides through cyclohexane, 2 × 15 min.
2. Rehydrate the slides by dipping them for 5 min in each of the following: 100% ethanol, 95% ethanol, 70% ethanol, and distilled water. Rinse with PBS.
3. Place the slides in a humidified box.
4. Block the slides with 1% BSA for 15 min.

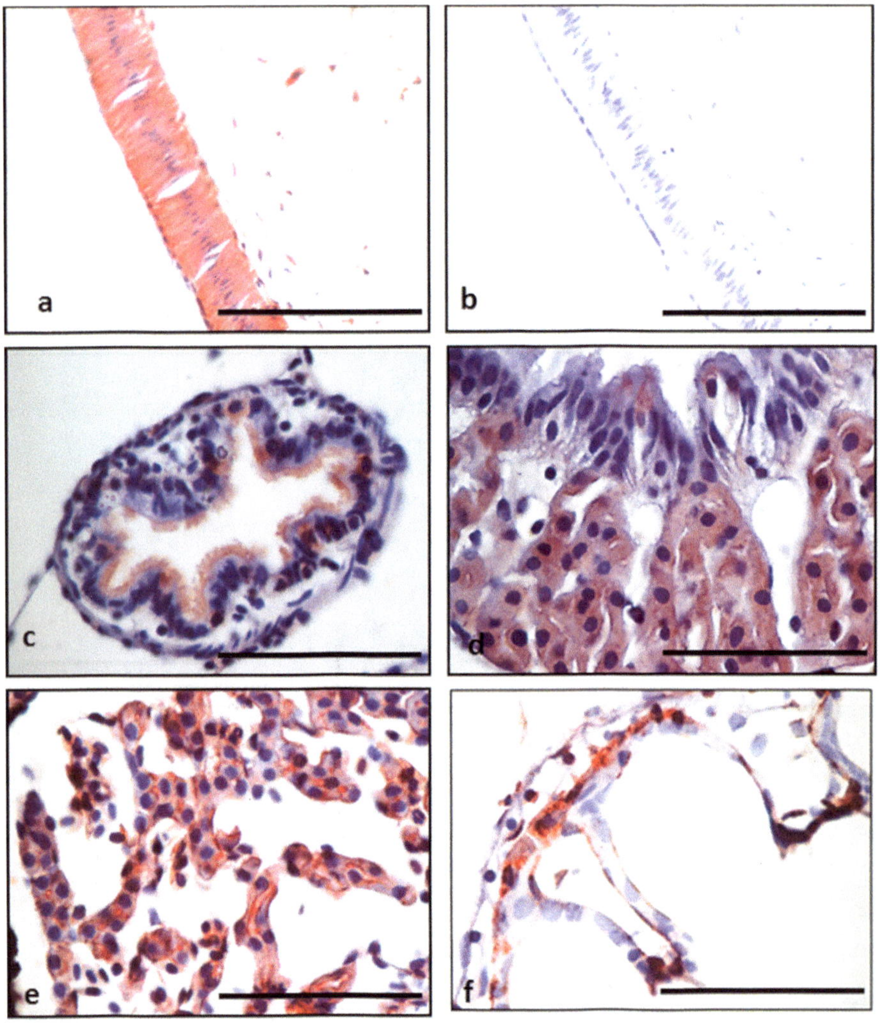

Fig. 3 Localization of calpains in *Xenopus laevis* tadpoles (stage 48), with indirect immunohistochemistry. (**a**) Calpain 3 on the velum. (**b**) Negative control on velum (primary antibody omission). (**c**) Calpain 2 in the esophagus. (**d**) Calpain 3 in the stomach. (**e**) Calpain 3 in the pancreas. (**f**) Calpain 3 in gills. Scale bar = 100 μm

5. Prepare the calpain antibody solution at appropriate dilution: 1:50 dilution in PBS for the mouse monoclonal anti-calpain 1 large subunit MAB3082 and rabbit polyclonal anti-calpain 2 large subunit AB1625, 1:25 dilution in PBS for the rabbit polyclonal anti-p94 large subunit AB81018.
6. Remove block, add calpain antibody solution to slides, and incubate for 1 h. Perform a negative control (*see* **step 13**).
7. Wash three times in PBS.

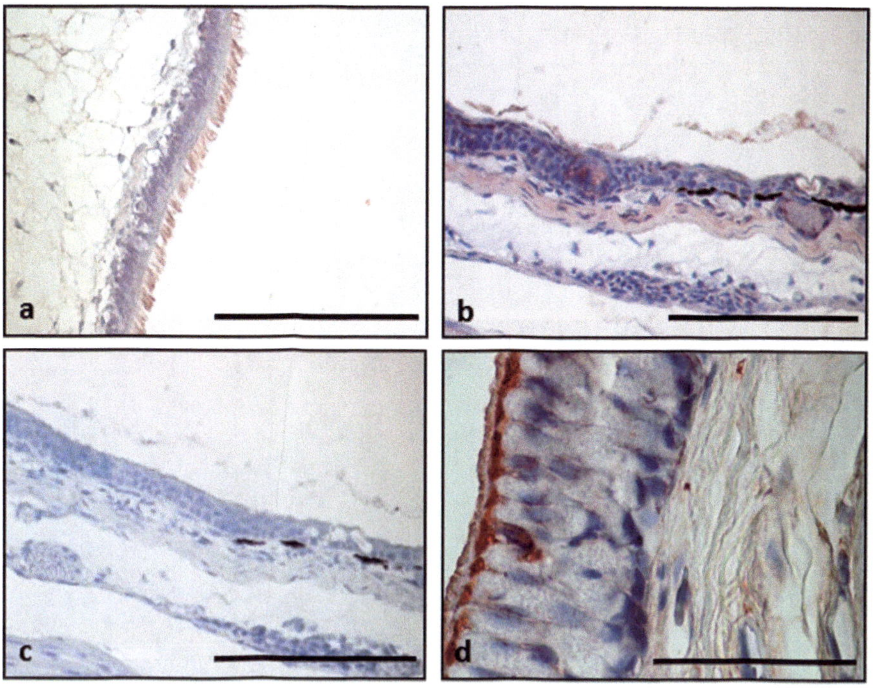

Fig. 4 Localization of calpains in *Xenopus laevis* at metamorphosis and in adult, with indirect immunohistochemistry. (**a**) Calpain 3 in the intestine (stage 59). (**b**) Calpain 2 on the skin (stage 62). (**c**) Negative control on the skin (omission of both primary and secondary antibodies). (**d**) Calpain 3 in the adult intestine. Scale bar = 100 μm

8. Incubate the sections for 1 h at room temperature (light protected) with Alexa 488-conjugated anti-mouse IgG or Cy^T3-labeled anti-rabbit IgG.
9. Wash 2–3 times with PBS.
10. Mount with antifade mounting medium.
11. Cover with a coverslip.
12. Visualize using a light microscope, equipped with epifluorescence optics (Fig. 5).
13. Negative control. Replace **step 6** only with this one: remove block, add PBS (0.1 M, pH 7.4) to slides, and incubate for 1 h. Return to **step 7** (Fig. 5).

4 Notes

1. The antibodies used to identify μ- and m-calpain (calpain 1 and calpain 2, respectively) are directed against human isoforms. Therefore, it is necessary to verify that these antibodies specifically recognize *Xenopus laevis* calpains. For that, Western blots should be performed with total extracts obtained from *Xenopus*

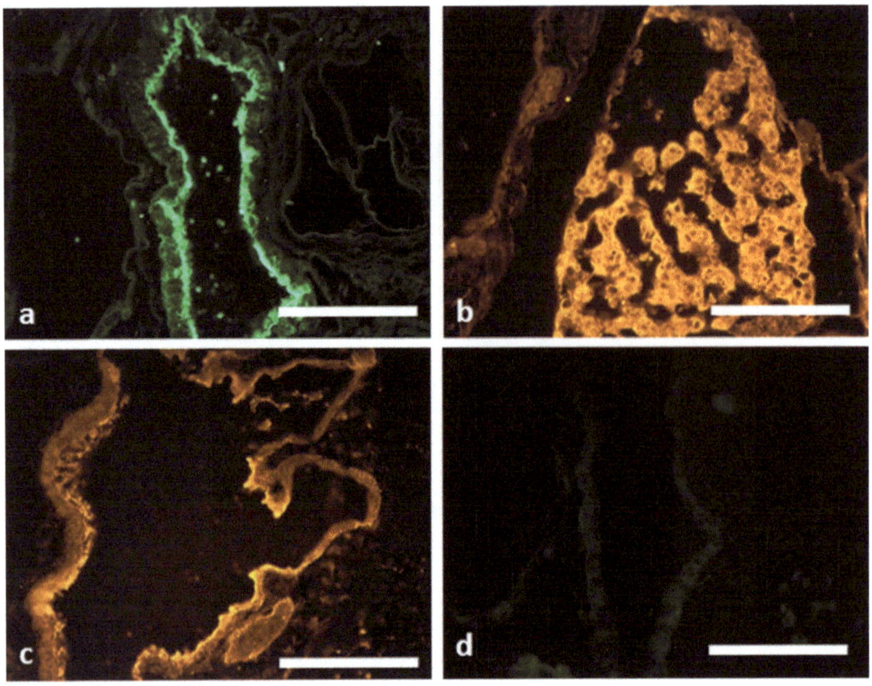

Fig. 5 Localization of calpains in *Xenopus laevis* tadpoles (stage 48), with immunofluorescence. (**a**) Calpain 1 in the esophagus, Alexa 488-conjugated anti-mouse IgG. (**b**) Calpain 2 in the liver, Cy^T3-labeled anti-rabbit IgG. (**c**) Calpain 2 in gills, Cy3-labeled anti-rabbit IgG. (**d**) Negative control in the esophagus (without primary antibody). Scale bar = 100 μm

laevis embryos [13]. The extracts should be electrophoresed on 8% SDS-PAGE, blotted onto a PVDF membrane, and blocked in nonfat dry milk before incubation with antihuman calpain antibodies. Purified calpain 1 (human erythrocyte) and calpain 2 (rat, recombinant, *Escherichia coli*) are used as positive controls. Appearance of bands running as the purified human calpain 1 or rat calpain 2 (80 kDa) can confirm the specificity of these antibodies against *Xenopus laevis* proteases. Likewise, the anti-calpain 3 antibody recognized two bands at 80 kDa and 78 kDa.

2. Immerse the sample rapidly after sampling in the fixer bath. Do not forget to label each sample.
3. The duration of each bath for the tadpole samples may be decreased to 30 min in ethanol 70% and 2 h in each subsequent bath.
4. The duration of fixation should be optimized for each new tissue type studied.
5. In the case of a manipulation mistake, it is sometimes useful to embed the pieces again. For that, immerse the cut block in melting paraffin. When the piece is melted, make the block again.

6. The sections are affixed to slides with ionized water maintained at 37 °C (use a heating plate). The slides are then allowed to dry for several hours in an oven at 37 °C.
7. Nonspecific signal can sometimes appear in the negative control sample due to the presence of endogenous peroxidases.

Acknowledgments

This work was supported by the Lyon Catholic University.

References

1. Ono Y, Sorimachi H (2012) Calpains: an elaborate proteolytic system. Biochim Biophys Acta 1824:224–236
2. Toyota H, Yanase N, Yoshimoto T, Moriyama M, Sudo T, Mizuguchi J (2003) Calpain-induced Bax-cleavage product is a more potent inducer of apoptotic cell death than wild-type Bax. Cancer Lett 189:221–230
3. Yajima Y, Kawashima S (2002) Calpain function in the differentiation of mesenchymal stem cells. Biol Chem 383:757–764
4. Arthur JS, Elce JS, Hegadorn C, Williams K, Greer PA (2000) Disruption of the murine calpain small subunit gene, CAPN4: calpain is essential for embryonic development, but not for cell growth and division. Mol Cell Biol 20:4474–4481
5. Zimmerman UJ, Boring L, Pak JH, Mukerjee N, Wang KK (2000) The calpain small subunit gene is essential: its inactivation results in embryonic lethality. IUBMB Life 50:63–68
6. Dutt P, Croall DE, Arthur JS, Veyra TD, Williams K, Elce JS, Greer PA (2006) M-Calpain is required for preimplantation embryonic development in mice. BMC Dev Biol 6:3
7. Lepage SE, Bruce AE (2008) Characterization and comparative expression of zebrafish calpain system genes during early development. Dev Dyn 237:819–829
8. Emori Y, Saigo K (1995) Calpain localization changes in coordination with actin-related cytoskeletal changes during early embryonic development of *Drosophila*. J Biol Chem 270:22652
9. Franco SJ, Huttenlocher A (2005) Regulating cell migration: calpains make the cut. J Cell Sci 118:3829–3838
10. Macqueen DJ, Wilcox AH (2014) Characterization of the definitive classical calpain family of vertebrates using phylogenetic, evolutionary and expression analyses. Open Biol 4:130219. https://doi.org/10.1098/rsob.130219
11. Sargianos N, Gaitanaki C, Beiste I (1994) Purification and characterization of m-calpain from the skeletal muscle of the amphibian Rana ridibunda. J Exp Zool 269:95–105
12. Sargianos N, Gaitanaki C, Beiste I (1995) Studies on the autolysis of m-calpain from the skeletal muscle of the amphibian Rana ridibunda. J Exp Zool 271:82–94
13. Cao Y, Zhao H, Grunz H (2001) XCL-2 is a novel m-type calpain and disrupts morphogenetic movements during embryogenesis in *Xenopus laevis*. Develop Growth Differ 43:563–571
14. Abrouk-Vérot L, Brun C, Exbrayat JM (2013) Expression patterns of CAPN1 and CAPN8b genes during embryogenesis in *Xenopus laevis*. CellBio 2:211–216
15. Di Primio C, Marracci S, Cecchettini A, Nardi I, Giorgi F, Fausto AM, Gambellini G, Mazzini M (2007) Differential tissue expression of a calpastatin isoform in *Xenopus embryos*. Micron 38:268–277
16. Moudilou EN, Mouterfi N, Exbrayat JM, Brun C (2010) Calpains expression during *Xenopus laevis* development. Tissue Cell 42:275–281
17. Exbrayat JM, Moudilou EA, Abrouk L, Brun C (2012) Apoptosis in amphibian development. Adv Biosci Biotechnol 3:669–678
18. Estabel J, Exbrayat JM (2002) Localisation des récepteurs AMPA/kaïnate dans les organes périphériques chez *Xenopus laevis* par immunohistochimie. Rev Fr Histotechnol 15:9–14
19. Estabel J, König N, Shiokawa K, Exbrayat JM (2005) Apoptosis in *Xenopus* genus. In: Scovassi I (ed) Apoptosis. Research Signpost, Trivandrum, pp 147–156
20. Moudilou M, Poirier AL, Brun C, Exbrayat JM (2009) Calpains expression during *Xenopus laevis* development. Mech Dev:S167

21. Hensey C, Gautier J (1998) Programmed cell death during *Xenopus* development: a spatio-temporal analysis. Dev Biol 203:36–48. https://doi.org/10.1006/dbio.1998.9028
22. Brun C, Moudilou EN, Bouchot C, Abrouk-Vérot L, Exbrayat JM (2013) Relationships between calpains and glutamate or kainate-induced apoptosis in *Xenopus laevis* tadpoles. Folia Histochem Cytobiol 51:300–311
23. König N, Poluch S, Estabel J, Durand M, Drian MJ, Exbrayat JM (2001) Synaptic and non-synaptic AMPA receptors permeable to calcium. Jpn J Pharmacol 86:1–17
24. Estabel J, Mercer A, Koenig N, Exbrayat JM (2003) Programmed cell death in *Xenopus laevis* metamorphosis development prior to, and during, metamorphosis. Life Sci 73:3298–3306
25. Estabel J, Exbrayat JM (2005) AMPA receptors localization by immunohistochemistry in *Xenopus* tadpoles. In: Ananjeva N, Tsinenko O (eds) Herpetologia Petropolitana, pp 133–135
26. Nieuwkoop PD, Faber J (1967) Normal table of *Xenopus laevis* (Daudin). North-Holland Publishing Company, Amsterdam

Chapter 8

Expression and Activity of Calpain A in *Drosophila melanogaster*

Maira Cardoso, Danielle Oliveira, and Helena Araujo

Abstract

Detecting calpain activity in *Drosophila* tissues is a fundamental tool to study calpain function. We use differential centrifugation to prepare membrane- versus cytosol-enriched fractions for measuring calpain activity with the fluorogenic substrate N-LY-AMC. With this method one can measure calpain A activity in wild-type flies and in several mutant fly backgrounds, revealing a strong correlation between in situ membrane distribution and in vitro determined activity measurements. Here we describe the steps for tissue preparation and calpain activity measurement in the *Drosophila* embryo.

Key words Calpain A, *Drosophila*, Calpain activity, Fluorogenic substrate

1 Introduction

Four loci-encoding calpains have been described in *Drosophila melanogaster*: CalpA, CalpB, CalpC, and small optic lobes or CalpD [1–4]. The active and broadly expressed CalpA and CalpB are produced during oogenesis and observed in the early embryo [2, 5], here used as our primary tissue for calpain activity measurements. Our analysis concentrates on detecting CalpA function, but the methodology may be used for other calpains with few adaptations. CalpA presents submembranous distribution, consistent with a membrane-anchoring segment and increased activity in the presence of phospholipids [3]. Alterations in membrane distribution correlate with altered CalpA activity to cleave a fluorogenic substrate, as well as cleavage of endogenous target substrates such as cactus/IκB and cyclin B [6–8]. Therefore, studying CalpA function may provide insights toward understanding calpain subcellular dynamics. Here we present a method to measure calpain activity in the fly embryo, using the fluorogenic substrate N-LY-AMC. Allied to the wealth of genetic tools available for modifying calpain expression, the *Drosophila* system is a powerful organism to investigate function of these modulatory proteases.

Jeannette S. Messer (ed.), *Calpain: Methods and Protocols*, Methods in Molecular Biology, vol. 1915,
https://doi.org/10.1007/978-1-4939-8988-1_8, © Springer Science+Business Media, LLC, part of Springer Nature 2019

2 Materials

2.1 Genetic Crosses and Embryo Collections

Fly lines used are wild-type Canton Special (CS), loss-of-function *cact*[A2] and *cact*[011], and gain-of-function *cact*[E10] all obtained from the Bloomington Indiana Drosophila Stock Center (BDSC, http://flystocks.bio.indiana.edu/) (*see* **Note 1**).

1. Embryo collection bottles.
2. Grape juice plates.
3. Nylon mesh, 100 μm.
4. 50% bleach in water.
5. TXN buffer: 0.04% Triton X-100 and 12 mM NaCl.
6. Storage buffer: 20 mM Tris–HCl, pH 7.4, 150 mM NaCl, and 0.05% Tween 20.
7. Greenhouse at 25 °C.

2.2 Tissue Collection and Preparation

1. Extraction buffer: 25 mM 2-(4-(2-hydroxyethyl)piperazin-1-yl)ethanesulfonic acid (HEPES), pH 7.2, 5 mM ethylenediaminetetraacetic acid (EDTA), 5 mM ethylenebis(oxyethylenenitrilo)tetraacetic acid (EGTA), 10 mM β-mercaptoethanol, and 25 mM sucrose supplied with 1 mM phenylmethylsulfonyl fluoride (PMSF) and 0.1 μM aprotinin. Keep this buffer on ice.
2. Potter-Elvehjem PTFE pestle and glass tube for 1 or 3 mL.
3. Freezers −80 °C and − 20 °C.
4. Micro-ultracentrifuge (i.e., Hitachi—himac CS100).

2.3 Calpain Activity Measurements

1. Calpain substrate N-succinyl-Leu-Tyr-7-amido-4-methyl-coumarin (N-LY-AMC) (Sigma, USA # S1153 or Calbiochem Merck, USA #208731).
2. Calpain 2: rat recombinant protein (Calbiochem Merck, USA #208718) (stock solution of 1.1 mg/mL or 10.9 mM). A control for the activity measurement protocol.
3. Calpastatin: human (Calbiochem Merck, USA #208902) (stock solution of 1.0 mg/mL or 314.7 μM). A control for the activity measurement protocol.
4. Reaction buffer (2×): 20 mM HEPES, pH 7.5, 300 mM NaCl, 2 mM EDTA, and 20 mM β-mercaptoethanol.
5. Fluorescence spectrophotometer or fluorescence microplate reader.
6. Fluorimeter cuvette or black plates with solid bottom.
7. Bovine serum albumin (BSA). This is a standard protein for quantification of the protein concentration in samples.
8. $CaCl_2$: stock solution of 1 M (analytical grade).

9. Trans-Epoxysuccinyl-L-leucylamido (4-guanidino) butane (E-64): 3 mM aqueous stock solution (Sigma, USA # $E3_132$).
10. EGTA: 0.1 M stock solution (analytical grade).
11. Dimethyl sulfoxide (DMSO): (anhydrous –99.9%–analytical grade). This is used to solubilize the calpain substrate.

3 Methods

3.1 Genetic Crosses and Embryo Collections

1. All genetic crosses and embryo collections are performed at 25 °C (schematized in Fig. 1). To analyze maternal effects on CalpA activity and distribution, we have used wild-type or mutant virgin female flies crossed to wild-type male flies (Fig. 1).

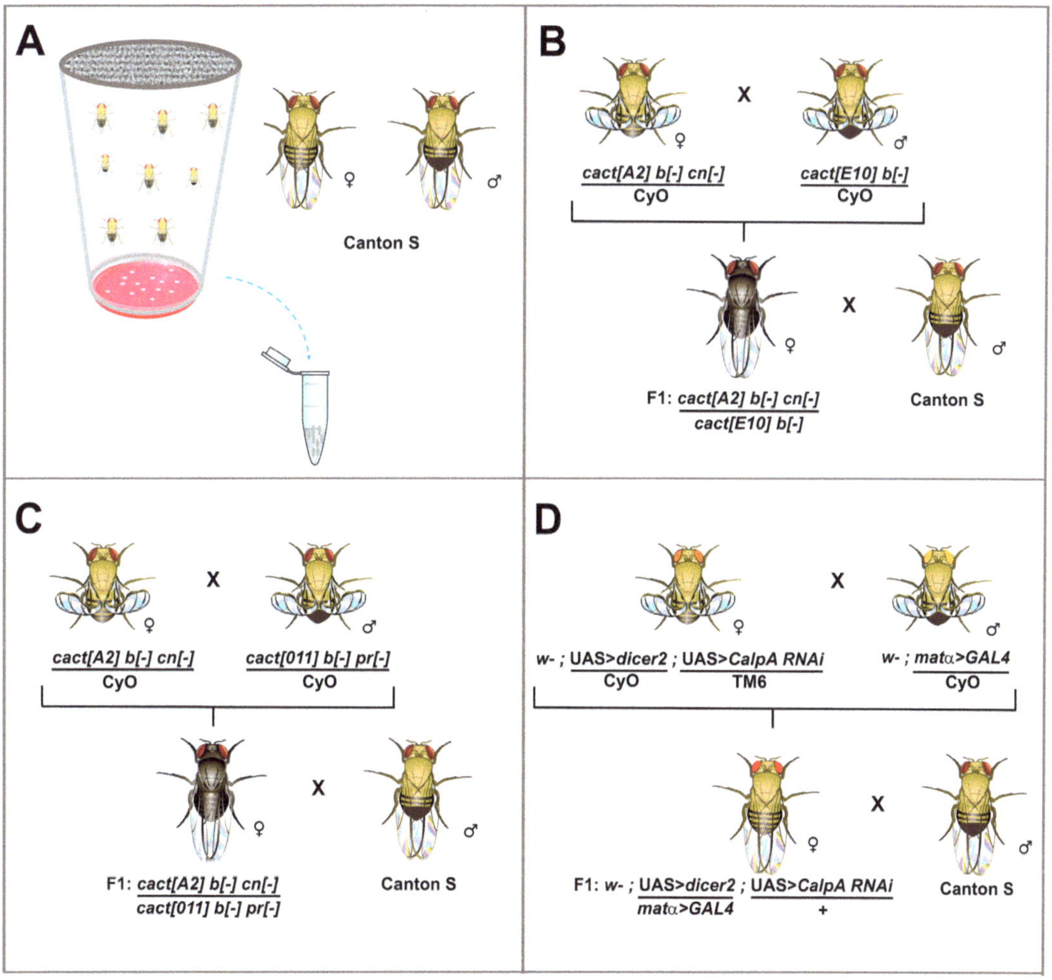

Fig. 1 Crossing schemes for analysis of CalpA maternal effect presented in Figs 2 and 3, used to generate cact [E10]/cact[A2] (**b**), cact[A2]/cact[011] (**c**), or knockdown of *CalpA* induced by UAS-GAL4 system mothers (**d**) crossed to wild-type Canton S males in collection bottles (**a**)

2. Mated flies (20–30 females +10–15 males) are transferred to embryo collection bottles, where eggs are laid on grape juice plates, maintained for 2 days before collection. Grape plates are changed once a day. For collections, we change plates two times before starting 2 h embryo collections (*see* **Note 2**).
3. Eggs are brushed off grape plates for fixation (as in Fontenele et al. [6]) or activity measurements. The embryos are transferred to nylon mesh, dechorionated for 1 to 2 min (*see* **Note 3**) in 50% bleach and washed in cold 1× TXN buffer. At least 20 μL of embryos are washed in cold storage buffer and stored at −80 °C. The sample groups (controls or mutant embryos/tissues) can be collected and stored at −80 °C until sufficient amounts of all desired samples are available for the next step (tissue collection and preparation) (*see* **Note 4**).

3.2 Tissue Collection and Preparation

1. Frozen samples containing mutant or wild-type 0–2-h-old embryos are thawed on ice and washed in extraction buffer. One volume of extraction buffer is added to each sample at a proportion of five- to tenfold embryo volume (total volume for 200 embryos makes approximately 20 μL).
2. Embryos are gently homogenized with a Potter-Elvehjem glass and pestle with 15–20 strokes on ice.
3. Embryo homogenates are subsequently centrifuged at 100,000 × *g* for 1 h at 4 °C. Supernatants are centrifuged a second time, at 15,000 × *g* for 10 min at 4 °C, to separate a fat portion on the top (less dense layer composed of lipid debris after centrifugation), which is discarded. The remaining supernatant portion is gently collected from each sample as the cytosolic fraction. Precipitates from the ultracentrifugation are resuspended in extraction buffer, using half of the volume used for the initial homogenate. This sample is termed the membrane-enriched fraction.
4. Protein content in each fraction is determined by the Folin-Lowry method [9]. Bovine serum albumin is used as protein standard. A no-protein control using extraction buffer is performed in all experiments. In general, extractions from mutant embryos produce lower protein concentrations compared to controls. A typical preparation of wild-type embryos produces a protein content of around 2 μg/μL to cytosolic fraction and around 4 μg/μL of membrane-enriched fraction. All samples are maintained at −20 °C until use (*see* **Notes 5** and **6**).

3.3 Calpain Activity Measurements

1. Calpain assays are performed using the fluorogenic substrate N-succinyl-Leu-Tyr-7-amido-4-methyl-coumarin (N-LY-AMC). A 2.5 mg/mL stock solution is prepared in 50% DMSO and stored at −20 °C. Components of the reaction are mixed immediately before measurements as below. Substrate volume addition starts the assays.

2. The total reaction is 100 μL, consisting of:
   - 50 μL 2× reaction buffer.
   - 0.03 mg membrane-enriched sample (final protein concentration 0.3 mg/mL).
   - 15 μL substrate (0.65 mM).
   - Water to complete 100 μL.

   Water volume should be adjusted if inhibitors or additives are to be used. Inhibitors or additives are used with the following concentrations:
   - E-64 60 μM (general cysteine protease inhibitor used as negative control).
   - EGTA 20 mM (calcium chelator, a negative control).
   - Ca^{2+} at 10 or 50 mM (calpain activity activator).
   - Rat calpain (recombinant) 0.04 mg/mL (positive control).
   - Human calpastatin 0.05 mg/mL (inhibitor, negative control).
3. Add all components but the substrate and incubate at room temperature for 1 min. Add N-LY-AMC and initiate measurements. Substrate hydrolysis is monitored in the kinetic mode in a fluorimeter at 380 nm excitation and 460 nm emission wavelengths. Analyses can be performed either in a fluorescence spectrophotometer or fluorescence microplate reader, with kinetic time interval of 3–5 min, depending on the progress curve. Steady-state velocities are obtained by linear regression of the substrate hydrolysis curve (*see* **Note 7**).
4. To characterize calpain activity, negative control assays are performed in the presence of the following modulators: E-64 (a general cysteine protease inhibitor), EGTA (a calcium chelator), and different calcium concentrations (10 to 50 mM of $CaCl_2$). Embryonic calpain activity is significantly inhibited by E-64 and EGTA (Fig. 2). Furthermore, calpain activity is only detected in the presence of added calcium. The membrane-enriched fraction has greater calpain activity compared to cytosolic fraction (Fig. 3), using 50 mM $CaCl_2$ (*see* **Note 8**).
5. It is also advisable to use external positive and negative control samples for the activity measurement protocol, by using recombinant calpain and the mammalian endogenous calpain inhibitor calpastatin. In our in vitro experiments, 90% inhibition of endogenous *Drosophila* calpain activity is observed in the presence of exogenous calpastatin (data not published), although no *calpastatin* gene has been described in the *Drosophila* genome [10].

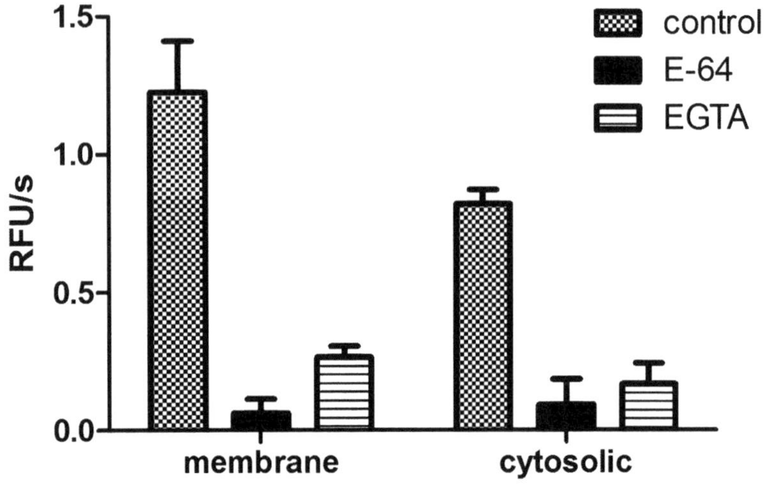

Fig. 2 Fluorometric determination of calpain activity. Samples prepared from wild-type 0–2-h-old embryos (membrane and cytosolic fractions) are assayed against calpain substrate in the presence of $CaCl_2$ (50 mM). Controls are performed using E-64 (60 μM) or EGTA (20 mM). Substrate hydrolysis is monitored continuously for 3 min. Each point represents the mean activities ± SEM of two independent experiments, performed in duplicate. Steady-state velocities are obtained by linear regression of the hydrolysis curve as followed in a fluorescence microplate reader using 380/460 nm as excitation/emission wavelengths

4 Notes

1. Available genetic resources for CalpA: Available stocks to study CalpA function include RNA interference lines under the control of a UAS promoter (y^1 v^1; P{TRiP.JF03391}attP2 at BDSC; w^{1118}; P{GD12275}v35261 and w^{1118}; P{GD12275} v35262 at Vienna Drosophila RNAi Center). We have used the BDSC line to good effect to knockdown CalpA function in the early embryo, immune system, and wings by driving double-stranded RNA expression with the Gal4 drivers maternal α-tubulin-Gal4/VP16 (matα), Cg-Gal4, and MS-1096 Gal4, respectively. Concomitant use of UAS-Dicer2 is necessary for a strong effect. All lines are available from BDSC. For overexpression, we have developed pTiger-CalpAeGFP (under UASp promoter for germline and somatic expression; [7]). A non-tagged UAS-CalpA is also available [11]. The following loss-of-function lines for CalpA have been developed, CalpA [72] and CalpA[75] [11], that delete the first CalpA exon.
2. Collecting a wide variety of unwanted embryonic development stages is avoided when, before starting embryo collections, the

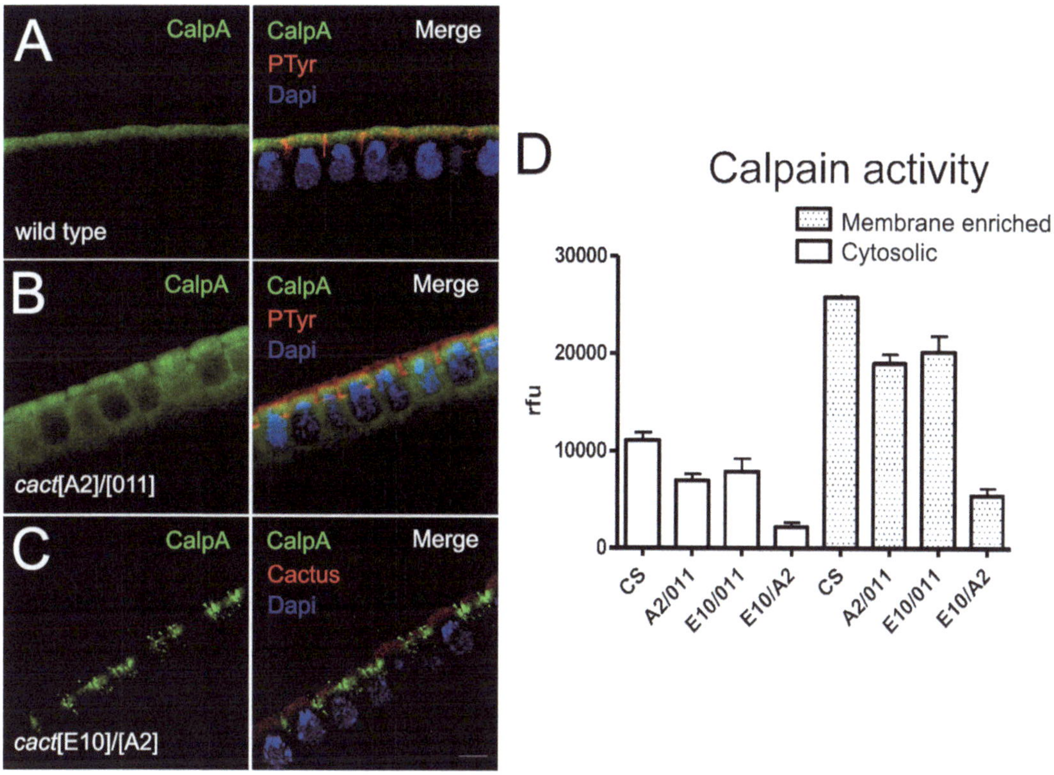

Fig. 3 Cactus mutants are used to show the correlation between CalpA membrane distribution and calpain activity in the embryonic blastoderm. Immunolabeling with rabbit anti-CalpA antibody (kindly provided by Dr. Emori Saigo) and anti-rabbit Alexa Fluor 488 reveals CalpA membrane distribution in wild-type embryos (**a**). In embryos from loss-of-function *cact* mothers (*cact*[A2]/*cact*[011]), CalpA redistributes to the cytoplasm (**b**), while in embryos from gain-of-function *cact* mothers (*cact*[E10]/*cact*[A2]), CalpA is distributed as patches close to the membrane (**c**). Anti-phosphotyrosine antibody (red in **a**, **b**) is used as membrane marker, or anti-cactus antibody is used for submembranous distribution (red in **c**). Nuclei are shown in blue (DAPI). These differences in CalpA distribution result in a decrease in calpain activity, as measured with the fluorescent substrate N-LY-AMC (**d**) (extracted and modified from Fontenele et al. [7], standard fixation and immunolabeling protocols within)

grape juice plate is discarded two times along with the embryos collected for 1 h. In these first two collections, the female flies can lay older eggs since some are stored in the oviduct. Only the third collection will contain embryos of the exact stage (0–2-h-old embryos) that will be prepared for analysis of CalpA activity as described.

3. Perfect dechorionation is observed under a dissecting scope by loss of the anterior dorsal appendages. Care should be used to avoid excessive time in bleach, as well as thorough washout of bleach for the following tissue preparation steps.
4. We have kept embryos at −80 °C for activity measurements up to 6 months in extraction buffer, without homogenization,

with no significant decrease in calpain activity compared to freshly collected and prepared embryos. Other tissues, especially those rich in lipids, will need to be evaluated to ensure that activity is not lost under long-term storage.

5. Calpain activity is detected in these membrane-enriched samples for up to 6 months in storage at −20 °C.
6. It has been reported that CalpA and CalpB are differentially distributed in cells, with CalpA concentrated below the plasma membrane and CalpB associated with intracellular membranes [5, 3]. Since we observe that CalpA activity correlates well with membrane distribution, we have used plasma membrane enrichment instead of differential phospholipids to preferentially access CalpA activity. Use of a CalpA knockdown can be used to confirm CalpA activity in this fraction. For those interested in measuring CalpB activity, appropriate phospholipids might need to be added to cytosolic fractions. CalpD (sol) is not expressed during the stage here tested and therefore does not contribute to overall calpain activity in the early embryo [1]. Since CalpC is not an active calpain, it does not contribute to calpain activity either [12].
7. Most fluorescent microplate readers automatically provide a linear regression curve and relative fluorescence unit by time (RFU/s). When substrate hydrolysis is monitored in a fluorescence spectrophotometer, a progress curve is followed during 3 min, with 0.5 s of time intervals. Data is transferred to Graph Pad Prism 4.0 software, and the initial velocities are determined by linear regression. For more information on linear regression methodology applied to this context, please consult [13].
8. The phospholipid requirement for recombinant CalpA versus CalpB activity measurements is different [3], providing important means to differentiate between their activities using fluorescent substrates. We have assayed endogenous calpain activity measurements using phospholipids (phosphatidic acid or phosphatidylinositol, 50–100 μg/mL)) using membrane-enriched and cytosolic fractions (not shown). However, results were variable likely due to endogenous phospholipids in these early embryo samples.

Acknowledgments

This work was supported by CNPq/Brazil grant to HA and fellowship to MAC.

References

1. Delaney SJ, Hayward DC, Barleben F, Fischbach KF, Miklos GL (1991) Molecular cloning and analysis of small optic lobes, a structural brain gene of Drosophila melanogaster. Proc Natl Acad Sci U S A 88(16):7214–7218
2. Emori Y, Saigo K (1994) Calpain localization changes in coordination with actin-related cytoskeletal changes during early embryonic development of Drosophila. J Biol Chem 269 (40):25137–25142
3. Jekely G, Friedrich P (1999) Characterization of two recombinant Drosophila calpains. CALPA and a novel homolog, CALPB. J Biol Chem 274(34):23893–23900
4. Theopold U, Pinter M, Daffre S, Tryselius Y, Friedrich P, Nassel DR, Hultmark D (1995) CalpA, a Drosophila calpain homolog specifically expressed in a small set of nerve, midgut, and blood cells. Mol Cell Biol 15(2):824–834
5. Farkas A, Tompa P, Schad E, Sinka R, Jekely G, Friedrich P (2004) Autolytic activation and localization in Schneider cells (S2) of calpain B from Drosophila. Biochem J 378 (Pt 2):299–305. https://doi.org/10.1042/BJ20031310
6. Fontenele M, Carneiro K, Agrellos R, Oliveira D, Oliveira-Silva A, Vieira V, Negreiros E, Machado E, Araujo H (2009) The Ca2+–dependent protease calpain A regulates Cactus/I kappaB levels during Drosophila development in response to maternal Dpp signals. Mech Dev 126(8-9):737–751. https://doi.org/10.1016/j.mod.2009.04.005
7. Fontenele M, Lim B, Oliveira D, Buffolo M, Perlman DH, Schupbach T, Araujo H (2013) Calpain A modulates toll responses by limited Cactus/IkappaB proteolysis. Mol Biol Cell 24 (18):2966–2980. https://doi.org/10.1091/mbc.E13-02-0113
8. Vieira V, Cardoso MA, Araujo H (2017) Calpain A controls mitotic synchrony in the Drosophila blastoderm embryo. Mech Dev 144 (Pt B):141–149. https://doi.org/10.1016/j.mod.2016.05.005
9. Lowry OH, Rosebrough NJ, Farr AL, Randall RJ (1951) Protein measurement with the Folin phenol reagent. J Biol Chem 193(1):265–275
10. Friedrich P, Tompa P, Farkas A (2004) The calpain-system of Drosophila melanogaster: coming of age. BioEssays 26(10):1088–1096. https://doi.org/10.1002/bies.20106
11. Kanamori T, Kanai MI, Dairyo Y, Yasunaga K, Morikawa RK, Emoto K (2013) Compartmentalized calcium transients trigger dendrite pruning in Drosophila sensory neurons. Science 340(6139):1475–1478. https://doi.org/10.1126/science.1234879
12. Spadoni C, Farkas A, Sinka R, Tompa P, Friedrich P (2003) Molecular cloning and RNA expression of a novel Drosophila calpain, Calpain C. Biochem Biophys Res Commun 303 (1):343–349
13. Lima AP, dos Reis FC, Serveau C, Lalmanach G, Juliano L, Menard R, Vernet T, Thomas DY, Storer AC, Scharfstein J (2001) Cysteine protease isoforms from Trypanosoma cruzi, cruzipain 2 and cruzain, present different substrate preference and susceptibility to inhibitors. Mol Biochem Parasitol 114(1):41–52

Chapter 9

Use of the β-Glucuronidase (GUS) Reporter System to Localize Promoter Activities of the Endogenous Plant Calpain *DEFECTIVE KERNEL1 (DEK1)*

Zhe Liang and Hilde-Gunn Opsahl-Sorteberg

Abstract

The *DEFECTIVE KERNEL1* (*DEK1*) gene encodes the unique plant calpain protein, which is fundamental for cell specification, development, and growth of land plants. The β-glucuronidase (GUS) reporter gene system has been used to characterize and localize gene expression over several decades. When cloning a promoter upstream of the *uidA/GUS* reporter gene, you visualize when the promoter is activating expression by monitoring enzymatic activity of GUS, by detecting β-glucuronidase cleavage products. Here we describe a protocol for monitoring the *DEK1* promoter activity in different tissues in *Arabidopsis* by GUS staining.

Key words DEFECTIVE KERNEL1, DEK1, Calpain, GUS, *uidA*, *Arabidopsis*

1 Introduction

Reporter genes are useful to unravel gene regulation, by monitoring where and when promoters are transcriptionally active or by detecting a gene's activity or product. A common reporter gene, β-glucuronidase (GUS), is a hydrolase encoded by the *uidA* gene that catalyzes the cleavage of a wide variety of β-glucuronides. The technique of GUS staining is based on conversion of a colorless chemical like 5-bromo-4-chloro-3-indolyl glucuronide (X-Gluc) to a visual blue product by the *Escherichia coli* enzyme β-glucuronidase [1].

DEK1 is a single copy gene in plants that is a member of the multigene family of cysteine proteases, a gene family that is evolutionarily highly conserved across kingdoms [2–6]. Evolutionary studies show that DEK1 is a TML calpain that evolved about 1.5 billion years ago through a fusion between the calpain catalytic core domain CysPc and other domains [5]. As a single copy calpain found in the majority of land plants, the function was expected to

Jeannette S. Messer (ed.), *Calpain: Methods and Protocols*, Methods in Molecular Biology, vol. 1915,
https://doi.org/10.1007/978-1-4939-8988-1_9, © Springer Science+Business Media, LLC, part of Springer Nature 2019

be conserved and was demonstrated by rescuing *Arabidopsis dek1* mutants with the moss *Physcomitrella patens* calpain *DEK1* sequence [2]. In *Arabidopsis*, DEK1 is essential to confer and maintain epidermal cell identity that allows development beyond the globular embryo stage through regulating microtubule-orchestrated cell wall orientation during cell divisions [7].

The plant calpain *DEK1* has been studied by various techniques and was generally regarded to be pretty evenly expressed in most plant parts. To study *DEK1* regulation more closely and link it to a putative cellular function, we visualized the promoter's activity in seeds, flowers, and vegetative plant parts by GUS staining [7]. The *DEK1* promoter was cloned directing the *uidA GUS* marker gene (*pDEK1::uidA*) and subsequently monitored. GUS localization showed the plant calpain to be localized to dividing cells of meristems, vasculature, and young embryos and in the basal endosperm transfer (BET) cell region not previously reported, of young *Arabidopsis* seeds. In this chapter, we will describe the protocol used for the GUS staining assay and the measurement of the *DEK1* promoter activity in *Arabidopsis*.

2 Materials

The *uidA* gene [8, 9] amplified by PCR and cloned to generate *pDEK1::uidA* and transformed to generate *Arabidopsis* plants expressing the promoter::*uidA* construct.

1. 1.5 ml Eppendorf tubes.
2. 37 °C, 42 °C, and 55 °C incubators.
3. Vacuum manifold and vacuum pump.
4. Glass slides.
5. Shaker.
6. Microtome.
7. Microscope.
8. Dimethylformamide (DMF).
9. Acetone.
10. Ethanol.
11. Acetic acid.
12. Formaldehyde.
13. Phosphate-buffered saline (PBS) buffer.
14. Histoclear.
15. Eosin.
16. Paraplast.
17. Wax.

18. Rinse buffer (*see* **Note 1**): 0.1% Triton X-100, 50 mM $NaPO_4$ (pH 7.2), 0.5 mM $K_3Fe(CN)_6$, and 0.5 mM $K_4Fe(CN)_6$.
19. Staining buffer (*see* **Note 1**): 50 mM $NaPO_4$ (pH 7.2), 0.5 mM $K_3Fe(CN)_6$, 0.5 mM $K_4Fe(CN)_6$, and 2 mM 5-bromo-4-chloro-3-indolyl beta-D-glucuronide cyclohexylamine salt (X-Gluc) (*see* **Note 2**).

3 Methods

1. Clone your promoter of interest in front of the *uidA* or *GUS* marker gene.
2. Transfer this gene construct (gene expression cassette) to your plant of interest, e.g., *Arabidopsis* flowers, to generate transgenic seeds containing the reporter gene construct.
3. Select transgenic *Arabidopsis* seeds, containing the expression cassette.
4. Harvest and place the samples in Eppendorf tube(s) containing cold 90% acetone on ice for 20 min.
5. When all samples are harvested, incubate them at room temperature for 20 min.
6. Remove acetone from the samples using a pipette, and wash samples 1–2 times with rinse buffer.
7. Remove rinse buffer, and add staining buffer (submerge the samples).
8. Leave the tube(s) open on ice, and apply vacuum until the samples begin to sink (*see* **Note 3**).
9. Incubate samples in the tubes at 37 °C overnight (*see* **Note 4**).
10. Remove staining buffer.
11. Serially wash the samples in 10%, 30%, 50%, and 70% ethanol (*see* **Note 5**) at room temperature, allowing 1 h for each step.
12. Remove ethanol.
13. Proceed to wax embedding procedure, or observe directly under a dissecting or light microscope.
14. Sample dehydration and wax embedding, this will take 7 days.
    (a) Day 1 (All steps at 4 °C and shaking):
    - 85% ethanol wash 1 h.
    - 95% ethanol + eosin (typically 1~5% (w/v) if needed, *see* **Note 6**) (until light pink; to visualize tissue) overnight.

    (b) Day 2 (All steps should be performed at room temperature with shaking):

- 100% ethanol + eosin wash 30 min.
- 100% ethanol + eosin wash 30 min.
- 100% ethanol + eosin wash 1 h.
- 100% ethanol + eosin wash 1 h.
- 25% Histoclear, 75% ethanol wash 1 h.
- 50% Histoclear, 50% ethanol wash 1 h.
- 75% Histoclear, 25% ethanol wash 1 h.
- 100% Histoclear wash 1 h.
- 100% Histoclear wash 1 h.
- 100% Histoclear + 1/4 volume Paraplast overnight (no shaking).

(c) Day 3:

- Place at 42 °C until the Paraplast melts completely.
- Add 1/4 volume of Paraplast and place at 42 °C until the Paraplast melts completely.
- Move to 55 °C for several hours.
- Replace wax/Histoclear with freshly melted wax overnight (55 °C).

(d) Day 4–Day 6:

- Change with fresh melted wax every morning and every evening.

(e) Day 7:

- Arrange the samples so that they are easily separated when cutting the wax (at least 3 mm apart) and cool down in appropriate molds. Store at 4 °C.

15. Section the paraffin-embedded samples into 8 μm thick sections and place the samples onto slides. Incubate the slides at 42 °C overnight so that samples adhere to slides (*see* **Note 7**).
16. Deparaffinize sections: (All steps should be performed at room temperature unless specified differently).
    - Histoclear wash 2 × 10 min.
    - 100% ethanol wash 2 × 1 min.
    - 95% ethanol wash 1 min.
    - 90% ethanol wash 1 min.
    - 80% ethanol wash 1 min.
    - 60% ethanol wash 1 min.
    - 30% ethanol wash 1 min.
    - H_2O wash 1 min.

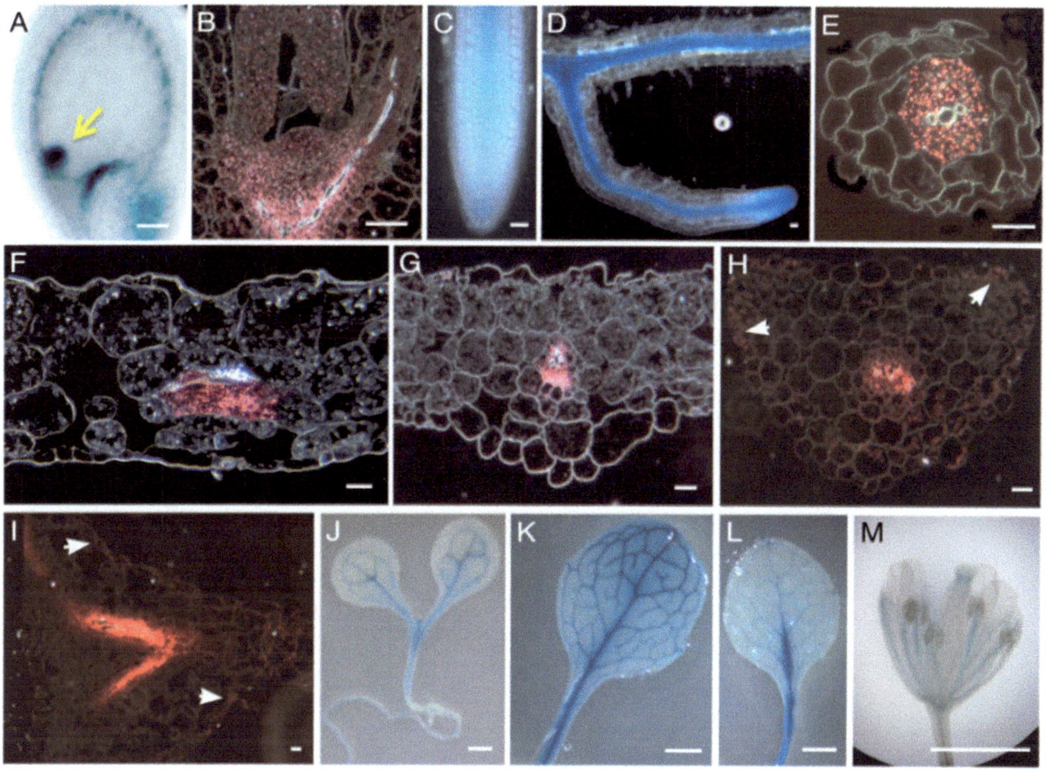

Fig. 1 *DEK1* (*pDEK1::uidA*) promoter expression in (**a**) seed especially in young embryo (yellow arrow) and BET-like cells, (**b**) SAM, (**c**) RAM, (**d**) lateral root meristem (RM) and (**e**) root vasculature, (**f**) phloem/cambium in cotyledon, (**g**) rosette leaf, (**h**) leaf petiole and (**i**) vascular branch, (**j**) whole seedling vasculature, (**k**) adaxial and (**l**) abaxial side of rosette leaf, and (**m**) all flower organs except petals. White arrows indicate *pDEK1::uidA* activity in the epidermal layer. Scale bars = 20 mm (**a–i**), 500 mm (**j–m**). This figure was originally published in Plant Cell and Physiology [7]

17. Dehydrate sections.
    - 70% ethanol wash 30 s.
    - 100% ethanol wash 2 × 30 s.
18. Air-dry the slides and mount slides with 50% glycerol. View the sample under a microscope; the blue color or red (with a color filter) demonstrates promoter activity via the visual GUS localization (Fig. 1).

4 Notes

1. Rinse buffer and staining buffer must be kept at 4 °C.
2. We recommend making aliquots of X-Gluc in DMF and store it at −20 °C.

3. Make sure that all samples sink to the bottom of the tubes; however, do not leave samples under vacuum conditions for longer than required. Also note that flowers will never sink using this method because of air spaces in the tissue.
4. Do not incubate for too long, since time will lead to over-staining or damaging tissues.
5. Tissues can be kept for several months in 70% ethanol.
6. Eosin normally use 1~5% (w/v). Eosin is not necessary if the GUS signal is strong.
7. Sectioned samples can be stored with desiccant for several weeks at 4 °C.

Acknowledgments

We thank Stein Erik Lid for cloning the *DEK1::uidA* construct [7] and Dr. Lisha Shen for critical reading of the manuscript. This work was supported by the Norwegian Research Council and NMBU to H-G.O-S.

References

1. Jefferson RA, Kavanagh TA, Bevan MW (1987) GUS fusions: beta-glucuronidase as a sensitive and versatile gene fusion marker in higher plants. EMBO J 6:3901–3907
2. Liang Z, Demko V, Wilson RC, Johnson KA, Ahmad R, Perroud PF, Quatrano R, Zhao S, Shalchian-Tabrizi K, Otegui MS, Olsen OA, Johansen W (2013) The catalytic domain CysPc of the DEK1 calpain is functionally conserved in land plants. Plant J 75:742–754
3. Lid SE, Gruis D, Jung R, Lorentzen JA, Ananiev E, Chamberlin M, Niu X, Meeley R, Nichols S, Olsen OA (2002) The defective kernel 1 (dek1) gene required for aleurone cell development in the endosperm of maize grains encodes a membrane protein of the calpain gene superfamily. Proc Natl Acad Sci U S A 99:5460–5465
4. Sorimachi H, Hata S, Ono Y (2011) Impact of genetic insights into calpain biology. J Biochem 150:23–37
5. Zhao S, Liang Z, Demko V, Wilson R, Johansen W, Olsen OA, Shalchian-Tabrizi K (2012) Massive expansion of the calpain gene family in unicellular eukaryotes. BMC Evol Biol 12:193
6. Johansen W, Ako AE, Demko V, Perroud PF, Rensing SA, Mekhlif AK, Olsen OA (2016) The DEK1 calpain linker functions in three-dimensional body patterning in Physcomitrella patens. Plant Physiol 172:1089–1104
7. Liang Z, Brown RC, Fletcher JC, Opsahl-Sorteberg HG (2015) Calpain-mediated positional information directs cell wall orientation to sustain plant stem cell activity, growth and development. Plant Cell Physiol 56:1855–1866
8. Jefferson RA, Burgess SM, Hirsh D (1986) Beta-Glucuronidase from Escherichia coli as a gene-fusion marker. Proc Natl Acad Sci U S A 83:8447–8451
9. Opsahl-Sorteberg HG, Divon HH, Nielsen PS, Kalla R, Hammond-Kosack M, Shimamoto K, Kohli A (2004) Identification of a 49-bp fragment of the HvLTP2 promoter directing aleurone cell specific expression. Gene 341:49–58

Part III

Identification of Calpain Substrates and Calpain-Activated Protein Functions

Chapter 10

LabCaS for Ranking Potential Calpain Substrate Cleavage Sites from Amino Acid Sequence

Yong-Xian Fan, Xiaoyong Pan, Yang Zhang, and Hong-Bin Shen

Abstract

Calpains are a family of Ca^{2+}-dependent cysteine proteases involved in many important biological processes, where they selectively cleave relevant substrates at specific cleavage sites to regulate the function of the substrate proteins. Presently, our knowledge about the function of calpains and the mechanism of substrate cleavage is still limited due to the fact that the experimental determination and validation on calpain bindings are usually laborious and expensive. This chapter describes LabCaS, an algorithm that is designed for predicting the calpain substrate cleavage sites from amino acid sequences. LabCaS is built on a conditional random field (CRF) statistic model, which trains the cleavage site prediction on multiple features of amino acid residue preference, solvent accessibility information, pair-wise alignment similarity score, secondary structure propensity, and physical-chemistry properties. Large-scale benchmark tests have shown that LabCaS can achieve a reliable recognition of the cleavage sites for most calpain proteins with an average AUC score of 0.862. Due to the fast speed and convenience of use, the protocol should find its usefulness in large-scale calpain-based function annotations of the newly sequenced proteins. The online web server of LabCaS is freely available at http://www.csbio.sjtu.edu.cn/bioinf/LabCaS.

Key words Protease substrate recognition, Cleavage site prediction, Sequence labeling, Ensemble learning, Calpain, Conditional random fields

1 Introduction

1.1 Background

Calpains are an important family of Ca^{2+}-regulated cysteine proteases that serve to regulate gene expression, signal transduction, cell death, and apoptosis [1, 2] through interacting and cleaving their substrates [3]. Many previous studies have shown that calpains are involved in pathophysiological mechanisms, and as a result, their malfunctions may lead to the occurrence of various diseases, such as muscular dystrophies, diabetes, and tumorigenesis [4, 5]. Although several promising calpain-targeted therapeutic strategies have been reported, there are still remaining challenges for drug discovery [4] due to the fact that the mechanisms of calpain modulation and the cleavage specificity of calpain substrates are still unclear.

Jeannette S. Messer (ed.), *Calpain: Methods and Protocols*, Methods in Molecular Biology, vol. 1915,
https://doi.org/10.1007/978-1-4939-8988-1_10,

As one of the most important functions of calpain is recognizing and cleaving their substrates, knowing the exact positions of the corresponding substrates' cleavage sites is very important for understanding their working mechanisms because the locations of the cleavage sites are closely related to how calpains will modulate substrate functions [6]. Although the cleavage sites can be determined based on various wet-lab experimental approaches, the process is very laborious and time-consuming to test all the residues throughout the substrate sequence. Considering the fact that the number of known protein sequences has been exponentially increasing with the development of modern sequencing technologies, automated computational tools are highly desired to screen out the most probable cleavage sites from the whole sequence space; this would reduce the search space and provide a small subset of high confidence for further experimental verification. Motivated by this, the development of calpain-related bioinformatics tools has been an important area of research, and many databases [7–11] and computational methods [12–18] have been reported in the literature.

1.2 Public Databases Related with Calpain

With the efforts of past decades, many calpain-based data resources have been constructed that are publicly available and can be accessed through the Internet. Table 1 lists some of these resources. For instance, the MEROPS database is a manually curated resource for proteolytic enzymes, their inhibitors, and substrates [7]. The CaMPDB database consists of three resources: calpain, substrates, and calpastatin [8]. The ENZYME is another repository of the nomenclature of enzymes where a large amount of calpain-related data exists. This database primarily describes each type of characterized enzyme for which an EC (Enzyme Commission) number has been attached [9]. The CutDB focuses on the annotations of individual proteolytic events, both actual and computationally predicted [10], while PMAP provides physiological substrate cleavage data and pathways and networks for proteolytic-related systems [11].

1.3 Publicly Available Bioinformatics Predictors Related with Calpain

Apart from the efforts of collecting calpain-related data from various large biological pools, another branch of bioinformatics has dealt with the development of intelligent predictors, which have been employed for the prediction of the most probable targets utilized in wet-lab experiments; this helps in the reduction of both time and expense. For instance, they can be applied to rank the most potential cleavage sites from the entire substrate amino acid sequence. Most of these substrate cleavage site predictors will need an annotated dataset, which is referred to as training set and consists of the substrate sequences with well-annotated true cleavage sites. Subsequently, four general categories of techniques can be used to construct the cleavage site ranking predictors: (1) statistical

Table 1
Some publicly available calpain-related databases

Database name	Description	Website	Reference
MEROPS	Database of proteolytic enzymes, their substrates and inhibitors	https://www.ebi.ac.uk/merops	[7]
CaMPDB	Resource for calpain and modulatory proteolysis	http://calpain.org	[8]
ENZYME	Enzyme nomenclature database	http://enzyme.expasy.org	[9]
CutDB	Proteolytic event database	http://cutdb.burnham.org	[10]
PMAP	Databases for analyzing proteolytic events and pathways	http://pmap.burnham.org/proteases	[11]

Table 2
The state-of-the-art predictors for calpain substrate cleavage sites

Predictors	Description	Address	Reference
Preference matrix	Sequential determinants of calpain cleavage	/	[12]
PoPS	Modeling and predicting protease specificity	http://pops.csse.monash.edu.au	[13]
Site prediction	Predicting the cleavage of proteinase substrates	http://www.dmbr.ugent.be/prx/bioit2-public/SitePrediction	[14]
GPS-CCD	Predicting calpain cleavage sites	http://ccd.biocuckoo.org	[15]
SVM (MKL)	Calpain cleavage site prediction	http://calpain.org	[16]
LabCaS	Labeling calpain substrate cleavage sites	http://www.csbio.sjtu.edu.cn/bioinf/LabCaS	[17]
Binary QSAR model	Prediction of cleavability of calpain proteolysis	/	[18]

propensity score-based method, (2) machine learning-based cleavage/non-cleavage two-class classifier, (3) the sequential learning-based labeling method, and (4) the quantitative structure activity-based method.

In the propensity score method, the amino acid propensities around the true cleavage sites are first calculated according to the training dataset. In the following step, a total score of a peptide segment in a pre-defined fixed-size sliding window is obtained that is judged with a derived optimal threshold to determine whether the central residue located in the segment is cleavable or not. The advantage of such a propensity score-based approach is that it is very straightforward and fast. However, it is very sensitive to the size of the dataset. In general, a larger training set generates more reliable propensity statistics, while an optimal threshold often varies with the different dataset. In the machine learning-based two-class classification approach, the positive and negative (cleavable and non-cleavable, respectively) subsets are first structured by partitioning the whole dataset. The two subsets are then accessed by machine learning-based models, i.e., artificial neural networks (ANN) and support vector machine (SVM), to learn the classification rules. After this training procedure, a trained classifier is used for the prediction for a new query, which is not contained in the training set. The merit of the two-class classifiers is that they can partially reduce the small sample size effects. However, the classification performance can be significantly affected by the imbalance between positive and negative samples. The sequential learning-based method, which is suitable for sequential labeling and classification problems, is often insensitive to the ratio between positive and negative training subsets, and hence all the negative samples can be used to establish the labeling model that can avoid information loss. When the substrate proteins have solved structures, the method based on quantitative structure-based activity analysis can be applied. Different state-of-the-art computational methods that are available for calpains analysis are tabulated in Table 2.

Here, we have adopted the typical conditional random field (CRF) sequential learning model and developed a label calpain substrate cleavage site predictor (LabCaS). The LabCaS tool has integrated multiple sequence-derived features to enhance the prediction performance. LabCaS is freely accessible at http://www.csbio.sjtu.edu.cn/bioinf/LabCaS. Through web browsers such as Safari, Internet Explorer, and Firefox, users can use the web server by providing the sequence of query proteins in FASTA format as input. For each query sequence, LabCaS ranks the potential cleavage sites and gives the confidence scores belonging to the cleavage sites.

2 Methods for Constructing LabCaS

The LabCaS predictor can directly label the cleavable residues from the entire sequence using CRF algorithm, where five different sequential features are integrated. When a sequence is submitted to the LabCaS tool, it first extracts the features from the sequence that are fed into the CRF model for prediction.

2.1 Extracted Features for Prediction

Five different sequential features are extracted from the sequences:

1. Amino acid preference feature

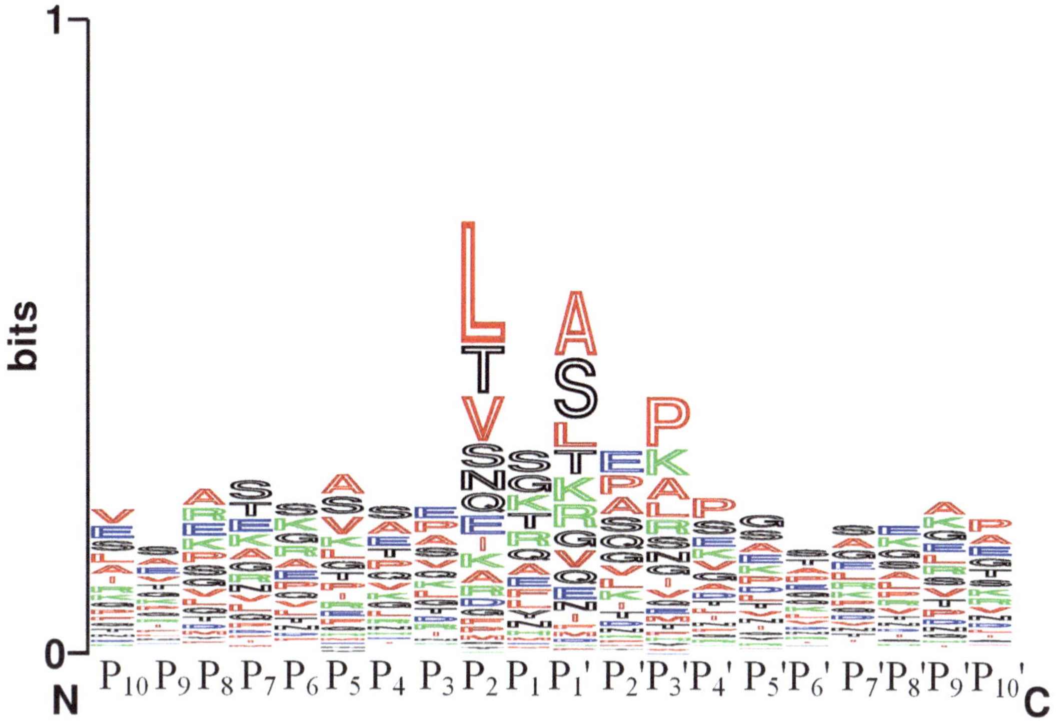

Fig. 1 A statistical view of the amino acid occurrences in a sliding window

Considering a sliding window of 20 residues (P_{10} to P_{10}' as shown in Fig. 1) in a peptide, we determine the probability of the occurrences of amino acids at each position. Such a statistics reflects the residue conservations around the cleavage sites [17]. We did find that such a preference feature is useful for the final predictions [17].

2. Sequence-based solvent accessibility information

From a structural point of view, the solvent accessibility of a residue is closely related to its interaction with other molecules. Hence, the solvent accessibility is also extracted from the sequence. In LabCaS, we apply I-TASSER package [19, 20] to extract solvent accessibility score for each residue that ranges from 0 (buried residue) to 9 (highly exposed residue).

3. Pair-wise alignment similarity score

We also construct a pool of cleavable peptides from the training dataset. Given an amino acid sequence, we split it into many peptides by using the sliding window approach. Each of these segments is aligned against the pre-collected cleavable segments from the pool to generate the pair-wise alignment similarity score, which is considered as a feature.

4. Sequence-based secondary structure information

The cleavable residues are often related with their secondary structures. Therefore, a feature with secondary structures predicted by PSIPRED [21] is consisted in LabCaS.

5. Physical-chemistry property information

In order to incorporate the amino acid physical-chemistry properties [22–24], we grouped the 20 amino acids into five groups: [(V, A, F, I, L, and M), (C, G, P, H, N, Q, S, and T), (W and Y), (D and E), and (R and K)]. The common physical-chemistry properties in each group are fed into the LabCaS program.

2.2 Prediction Models

LabCaS assigns a cleavable label to each residue of a calpain substrate sequence by using CRFs. According to the Hammersley-Clifford theorem of random field [25], the conditional distribution over a labeled sequence $\vec{y}$ given a calpain substrate corresponding sequence $\vec{x}$ can be calculated as:

$$P(y|x) = \frac{1}{Z(x)} \Pi_{t=1}^{T} \exp\left(\sum_i \sum_j \lambda_{ij} f_{ij}(y_{t-1}, y_t) + \sum_j \sum_k \mu_{jk} g_{jk}(y_t, x)\right) \quad (1)$$

where $Z(x)$ is a normalization factor; $f_{ij}(y_{t-1}, y_t)$ is a transition feature function of the labels at position t and t-1 in the labeled sequence; $g_{jk}(y_t, x)$ is a state feature function of the label at position t and the observation sequence; λ_{ij} and μ_{jk} are model parameters corresponding to feature functions $f_{ij}(\cdot)$ and $g_{jk}(\cdot)$; i and j denote the ith and jth kind labels, respectively; and k represents the kth kind sequence pattern.

Five CRF models can thus be constructed from the five features, and the models are further combined using the product rule to obtain the final prediction by LabCaS:

$$\mathrm{LabCaS}(x) = \sqrt[5]{\Pi_{j=1}^{5} p_j^*(y|x)} \quad (2)$$

3 How to Use LabCaS

3.1 Query Input Page of LabCaS

Figure 2 shows the main page with the sequence entry section, where the sequences for calpain substrate are manually inputted in FASTA format by the users to predict cleavage sites. Upon the submission of each job, a link is provided to the user for tracking the progress of the job and for accessing the final prediction results. The following points are important when using LabCaS. First, LabCaS focuses on the prediction of calpain cleavage sites only. Second, the input sequence length should be at least 50 amino acid

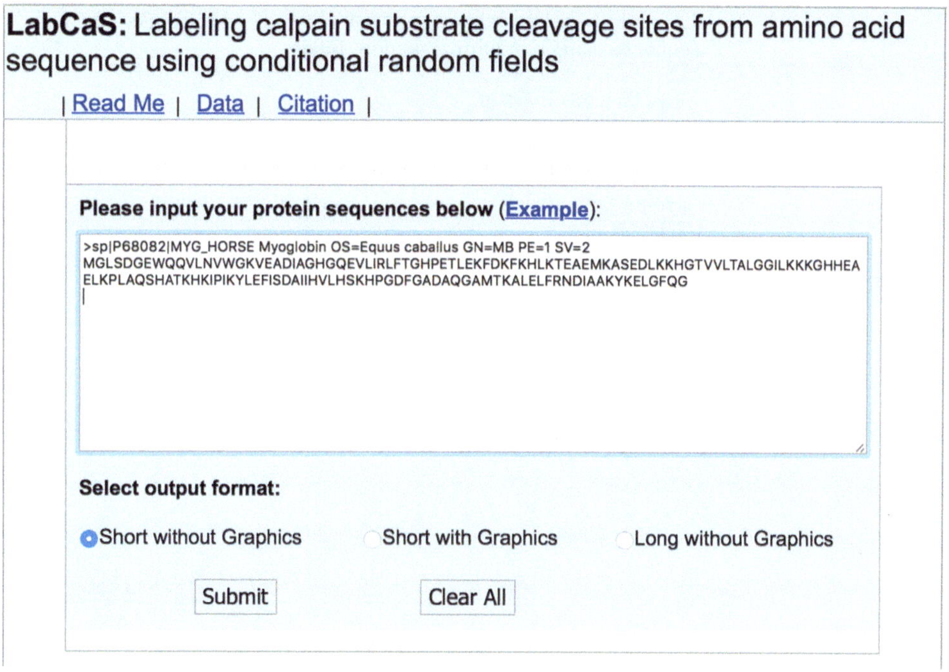

Fig. 2 Screenshot to show the query input page of LabCaS

residues and no more than 6000 residues. Additionally, the number of submitted sequences should not be more than three at a time. Third, there are three output formats of the predicted results, where "short without graphics" format outputs the predicted results at the high threshold; "long without graphics" outputs the results at the high, middle, and low thresholds without the corresponding graphics; and "short with graphics" outputs the results at the high threshold with the corresponding graphics.

3.2 Output Page of LabCaS

Once the submitted job is completed, the results are presented through both webpage and as a downloadable file. An illustration of a typical output is shown in Fig. 3, with a tab for display. The results show predicted calpain substrate cleavage sites, which are presented in a descending order of the confidence score. Additionally, a link is provided to view and download a TXT file comprising the results from LabCaS.

3.3 Case Study for Using LabCaS

Previously, a binary QSAR model was developed to predict the cleavability of calpain proteolysis [18]. Based on this binary QSAR model, the authors obtained 12 predicted cleavage sites of the sequence of horse myoglobin (MYO). Two of these sites, $Lys^{51}{\downarrow}Thr^{52}$ and $Gly^{151}{\downarrow}Phe^{152}$, were cleaved by calpains, as demonstrated by experiments utilizing LC-MS/MS. Table 3 shows the results when we submit the sequence to the LabCaS server. The first $Gly^{151}{\downarrow}Phe^{152}$ cleavage site is ranked as first, and

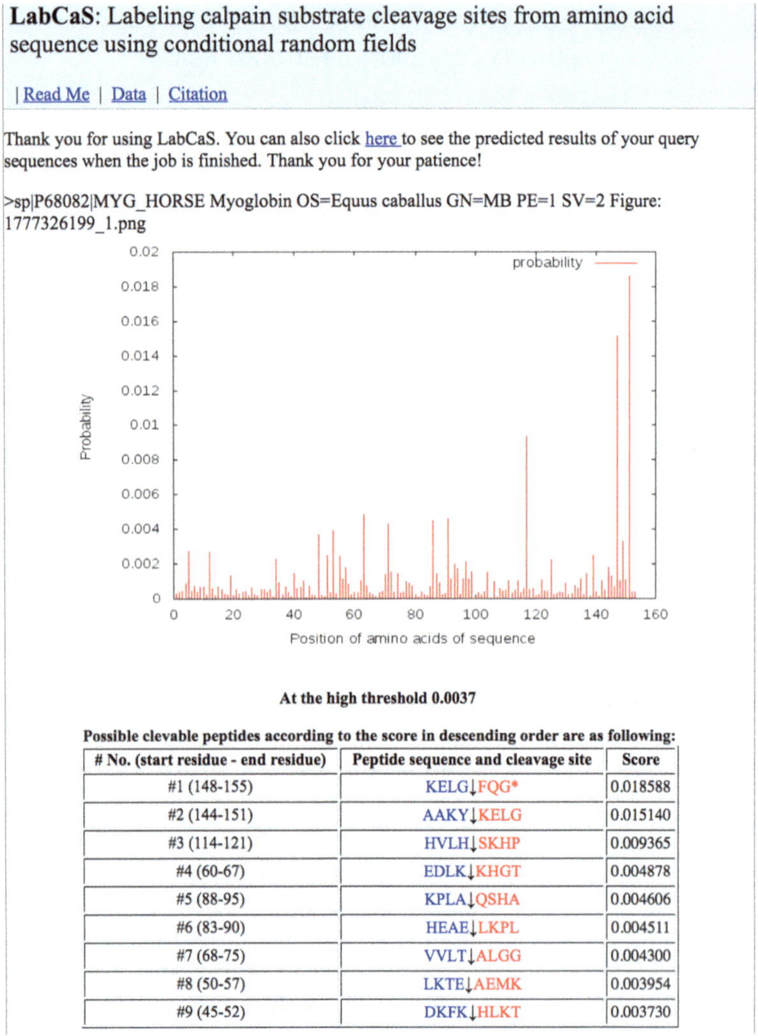

Fig. 3 Screenshot to show an example page of predicted cleavage sites by LabCaS

Table 3
Ranking results for horse myoglobin (accession number: P68082) by LabCaS

Rank	LabCaS
From first to fifth	$Gly^{151}\downarrow Phe^{152}$ (first)
From sixth to tenth	/
From 11th to 15th	$Lys^{51}\downarrow Thr^{52}$ (13th)

the second cleavage site of $Lys^{51}\downarrow Thr^{52}$ is ranked as 13th, as shown in Table 3. These results demonstrate that LabCaS can confidently rank the true cleavage sites and provide a hypothesis for wet-lab verifications.

4 Conclusions

In this chapter, we first briefly reviewed several bioinformatics tools, which include many databases and computational methods related to calpains. Then, the prediction of calpain substrate cleavage sites was formulated as a sequence labeling problem that was achieved by a novel ensemble method called LabCaS, which combines the models based on CRF algorithm. As a web server implementation of our approach, LabCaS is freely available for academic use at http://www.csbio.sjtu.edu.cn/bioinf/LabCaS, which is expected to become a powerful tool for in silico recognition of calpain substrate cleavage sites.

Acknowledgment

We are grateful to Mr. Wallace Chan and Dr. S M Golam Mortuza for proofreading the manuscript. This work was supported in part by the National Natural Science Foundation of China (No. 61462018, 61762026, 61671288, 91530321, 61725302, and 61603161), Guangxi Natural Science Foundation (No. 2017GXNSFAA198278), Guangxi Key Laboratory of Trusted Software (No. kx201403), Guangxi Colleges and Universities Key Laboratory of Intelligent Processing of Computer Images and Graphics (No. GIIP201502), Science and Technology Commission of Shanghai Municipality (No. 16JC1404300, 17JC1403500), and the National Science Foundation (ABI 1564756).

References

1. Campbell RL, Davies PL (2012) Structure-function relationships in calpains. Biochem J 447:335–351
2. Franco SJ, Huttenlocher A (2005) Regulating cell migration: calpains make the cut. J Cell Sci 118:3829–3838
3. Storr SJ, Carragher NO, Frame MC et al (2011) The calpain system and cancer. Nat Rev Cancer 11:364–374
4. Bertipaglia I, Carafoli E (2007) Calpains and human disease. Subcell Biochem 45:29–53
5. Croall DE, Ersfeld K (2007) The calpains: modular designs and functional diversity. Genome Biol 8:218
6. Friedrich P, Bozoky Z (2005) Digestive versus regulatory proteases: on calpain action in vivo. Biol Chem 386:609–612
7. Rawlings ND, Barrett AJ, Finn R (2016) Twenty years of the MEROPS database of proteolytic enzymes, their substrates and inhibitors. Nucleic Acids Res 44:D343–D350
8. Duverle D, Takigawa I, Ono Y et al (2010) CaMPDB: a resource for calpain and modulatory proteolysis. Genome Inform 22:202–213
9. Bairoch A (2000) The ENZYME database in 2000. Nucleic Acids Res 28:304–305
10. Igarashi Y, Eroshkin A, Gramatikova S et al (2007) CutDB: a proteolytic event database. Nucleic Acids Res 35:D546–D549
11. Igarashi Y, Heureux E, Doctor KS et al (2009) PMAP: databases for analyzing proteolytic events and pathways. Nucleic Acids Res 37: D611–D618
12. Tompa P, Buzder-Lantos P, Tantos A et al (2004) On the sequential determinants of calpain cleavage. J Biol Chem 279:20775–20785
13. Boyd SE, Pike RN, Rudy GB et al (2005) PoPS: a computational tool for modeling and

predicting protease specificity. J Bioinforma Comput Biol 3:551–585

14. Verspurten J, Gevaert K, Declercq W et al (2009) SitePredicting the cleavage of proteinase substrates. Trends Biochem Sci 34:319–323
15. Liu Z, Cao J, Gao X et al (2011) GPS-CCD: a novel computational program for the prediction of calpain cleavage sites. PLoS One 6: e19001
16. Duverle DA, Ono Y, Sorimachi H et al (2011) Calpain cleavage prediction using multiple kernel learning. PLoS One 6(5):e19035
17. Fan YX, Zhang Y, Shen HB (2013) LabCaS: labeling calpain substrate cleavage sites from amino acid sequence using conditional random fields. Proteins 81:622–634
18. Shinkai-Ouchi F, Koyama S, Ono Y et al (2016) Predictions of cleavability of calpain proteolysis by quantitative structure-activity relationship analysis using newly determined cleavage sites and catalytic efficiencies of an oligopeptide array. Mol Cell Proteomics 15:1262–1280
19. Roy A, Kucukural A, Zhang Y (2010) I-TASSER: a unified platform for automated protein structure and function prediction. Nat Protoc 5:725–738
20. Xu D, Zhang J, Roy A et al (2011) Automated protein structure modeling in CASP9 by I-TASSER pipeline combined with QUARK-based ab initio folding and FG-MD-based structure refinement. Proteins 79 Suppl 10:147–160
21. Jones DT (1999) Protein secondary structure prediction based on position-specific scoring matrices. J Mol Biol 292:195–202
22. Granseth E, Von Heijne G, Elofsson A (2005) A study of the membrane-water interface region of membrane proteins. J Mol Biol 346:377–385
23. Mak MW, Wang W, Kung SY (2009) Fusion of conditional random field and signalp for protein cleavage site prediction. In: In acoustics, speech and signal processing. Taipei, pp 716–721
24. Fan YX, Song J, Shen HB et al (2011) PredCSF: an integrated feature-based approach for predicting conotoxin superfamily. Protein Pept Lett 18:261–267
25. Hammersley J, Clifford P (1971) Markov field on finite graphs and lattices. Unpublished manuscript

Chapter 11

CalCleaveMKL: a Tool for Calpain Cleavage Prediction

David A. duVerle and Hiroshi Mamitsuka

Abstract

Calpain, an intracellular Ca^{2+}-dependent cysteine protease, is known to play a role in a wide range of metabolic pathways through limited proteolysis of its substrates. However, only a limited number of these substrates are currently known, with the exact mechanism of substrate recognition and cleavage by calpain still largely unknown.

Current sequencing technologies have made it possible to compile large amounts of cleavage data and brought greater understanding of the underlying protein interactions. However, the practical impossibility of exhaustively retrieving substrate sequences through experimentation alone has created the need for efficient computational prediction methods. Such methods must be able to quickly mark substrate candidates and putative cleavage sites for further analysis. While many methods exist for both calpain and other types of proteolytic actions, the expected reliability of these methods depends heavily on the type and complexity of proteolytic action, as well as the availability of well-labeled experimental datasets, which both vary greatly across enzyme families.

This chapter introduces *CalCleaveMKL*: a tool for calpain cleavage prediction based on multiple kernel learning, an extension to the classic support vector machine framework that is able to train complex models based on rich, heterogeneous feature sets, leading to significantly improved prediction quality. Along with its improved accuracy, the method used by CalCleaveMKL provided numerous insights on the respective importance of sequence-related features, such as solvent accessibility and secondary structure. It notably demonstrated there existed significant specificity differences across calpain subtypes, despite previous assumption to the contrary.

An online implementation of this prediction tool is available at http://calpain.org.

Key words Calpain, Cleavage prediction, Support vector machines, Multiple kernel learning

1 Introduction

Proteolytic cleavage is omnipresent in systems biology: from protein interactions in gene regulation to intracellular processes such as cell apoptosis. Understanding the mode of proteolytic action of a particular protease and identifying its target substrates is a crucial step to understanding its cellular processes. This understanding can also lead to efficient design of viral protease inhibitors: a promising therapeutic vector for major diseases such as AIDS [13].

Jeannette S. Messer (ed.), *Calpain: Methods and Protocols*, Methods in Molecular Biology, vol. 1915,
https://doi.org/10.1007/978-1-4939-8988-1_11,

Calpain (EC 3.4.22.17, Clan CA, family C02) is an intracellular Ca^{2+}-dependent cysteine protease known to regulate substrate functions by limited proteolysis, i.e., proteolytic processing [6, 7, 15, 24, 35, 47, 54, 56], resulting in the modulation of a wide variety of biological phenomena. The many known homologues of calpain constitute a major protease family distributed over a wide range of organisms. Calpain has been associated with regulation of signal transduction systems, cell motility, and apoptosis, while malfunction has been observed in several serious diseases in humans [6], including muscular dystrophies [39, 42], diabetes [27, 28], and tumorigenesis [33, 64].

For precise modulation of substrate functions by calpains, the cleavage sites are anticipated to be strictly determined depending on substrates [22]. In other words, the positions of the cleavage sites are essential determinants for how calpains modulate substrate functions. Therefore, prediction of cleavage sites by calpains is crucial to gain insight into how calpain proteolysis modulates cellular functions through substrate proteolysis [47]. The prediction holds an advantage when available amounts of substrates are low and cleavage site determination by protein chemistry such as protein sequencing and mass-spectrometry is impossible. If cleavage sites are determined, antibodies specific to the sites [32, 45, 46] and inhibitors for specific substrate proteolysis [44, 55, 67] can be designed to analyze proteolytic events by calpain under various conditions. Many studies have been attempted to predict calpain cleavage sites [3, 17, 58]; however, precise prediction has never been successful so far.

Occurrence of proteolysis of a polypeptide, that is, hydrolysis of one or many peptide bonds by a proteolytic enzyme, is naturally tied to the dynamic spatial configuration of the two proteins: their respective tertiary structures, including possible changes incurred during docking (*see*, for example, Fig. 1). It has been shown, however, that cleavage can often be predicted with good accuracy, from knowledge of the substrate's amino acid sequence alone [59]. The predictive power of sequence information depends heavily on the protease considered and the strength of its affinities for particular amino acid at specific positions. Table 1 gives a good illustration of the broad range of specificity degrees across protease families, through a cursory analysis of sequence conservation, hinting at a class of proteases for which more advanced models must be used.

Mechanisms of substrate recognition by calpain are altogether poorly understood, compared to other types of proteases. For example, while PEST motifs (sequences rich in proline, glutamic acid, serine, and threonine) have been shown to play a role in calpain recognition for some substrates [61], numerous studies have also identified cases for which PEST motifs do not impact substrate recognition or cleavage [10, 38]. Attempts at predicting

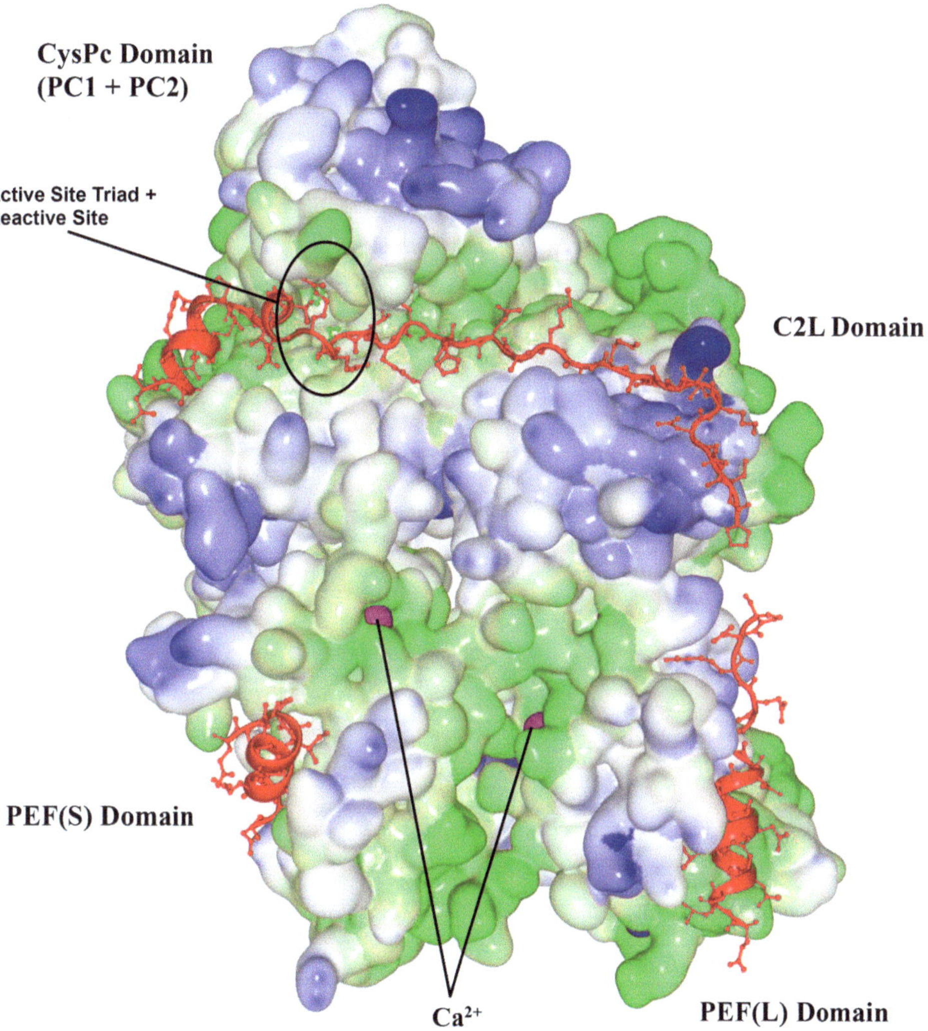

Fig. 1 Spatial representation of calpastatin–calpain complex. After activation by Ca^{2+} ions, the calpain molecule can dock (but not cleave) with calpastatin (in red), an endogenous calpain inhibitor

substrate cleavage by calpain have so far been entirely built on empirically derived rules for position-based residue preferences [55] and, more generally, position-specific scoring matrix methods [17], although the importance of higher-order structure information has long been established [48].

A number of different methods [2, 51, 63, 65] have been developed to predict substrate recognition and cleavage by proteases other than calpain, notably caspases: another family of cysteine protease involved mainly in apoptosis, as well as in various biological phenomena also involving calpain at times [60]. However, despite their similarity, calpain's particular mode of proteolytic action would appear to set it apart from caspases, and different methods seem needed in order to attain similar prediction results.

Table 1
Sequence entropies for main MEROPS protease substrates

ID	Protein name	Count	Avg. 8	Best 3	P4	P3	P2	P1	P1′	P2′	P3′	P4′
A01.001	Pepsin A	90	0.73	0.71	0.75	0.72	0.7	0.74	0.74	0.72	0.74	0.71
A01.009	Cathepsin D	728	0.71	0.67	0.74	0.73	0.73	0.6	0.7	0.71	0.72	0.73
A01.010	Cathepsin E	1286	0.69	0.64	0.74	0.75	0.71	0.62	0.62	0.69	0.71	0.71
A01.053	Necepsin-1	167	0.73	0.71	0.73	0.73	0.76	0.73	0.75	0.71	0.71	0.72
A02.001	HIV-1 retropepsin	479	0.68	0.61	0.73	0.75	0.6	0.7	0.64	0.58	0.75	0.71
C01.032	Cathepsin L	72	0.71	0.67	0.68	0.76	0.62	0.73	0.72	0.74	0.73	0.72
C01.033	Cathepsin L1 (*Fasciola* sp.)	177	0.74	0.73	0.75	0.75	0.72	0.74	0.73	0.75	0.74	0.73
C01.036	Cathepsin K	116	0.68	0.6	0.75	0.73	0.53	0.59	0.69	0.74	0.74	0.69
C01.046	Falcipain-2	126	0.74	0.73	0.74	0.74	0.73	0.74	0.72	0.74	0.76	0.75
C01.063	Falcipain-3	102	0.73	0.71	0.75	0.72	0.7	0.72	0.71	0.75	0.76	0.74
C02.001	Calpain-1	88	0.69	0.66	0.71	0.72	0.68	0.72	0.65	0.7	0.66	0.7
C02.002	Calpain-2	138	0.7	0.67	0.74	0.72	0.64	0.72	0.68	0.73	0.69	0.72
C14.001	Caspase-1	87	0.64	0.55	0.71	0.68	0.67	0.44	0.57	0.69	0.64	0.7
C14.003	Caspase-3	504	0.58	0.38	0.54	0.7	0.64	0.1	0.52	0.71	0.73	0.73
C14.004	Caspase-7	99	0.54	0.3	0.44	0.7	0.63	0.08	0.39	0.67	0.71	0.69
C14.005	Caspase-6	178	0.6	0.38	0.62	0.47	0.75	0.04	0.73	0.74	0.73	0.75
C14.009	Caspase-8	61	0.55	0.34	0.6	0.57	0.66	0.04	0.42	0.71	0.72	0.68
M04.001	Thermolysin	240	0.71	0.65	0.73	0.74	0.74	0.73	0.47	0.74	0.74	0.73
M10.001	Matrix metallopeptidase-1	63	0.68	0.63	0.71	0.65	0.72	0.7	0.6	0.72	0.7	0.65
M10.002	Matrix metallopeptidase-8	86	0.65	0.6	0.67	0.61	0.69	0.68	0.65	0.71	0.55	0.68
M10.003	Matrix metallopeptidase-2	1802	0.65	0.58	0.74	0.62	0.63	0.63	0.5	0.69	0.63	0.73
M10.004	Matrix metallopeptidase-9	320	0.65	0.59	0.62	0.61	0.69	0.61	0.66	0.73	0.55	0.72
M10.005	Matrix metallopeptidase-3	160	0.67	0.62	0.72	0.6	0.63	0.64	0.66	0.71	0.69	0.73
M10.008	Matrix metallopeptidase-7	165	0.64	0.58	0.71	0.61	0.67	0.68	0.46	0.68	0.66	0.67
M10.009	Matrix metallopeptidase-12	191	0.62	0.55	0.65	0.54	0.67	0.57	0.65	0.66	0.56	0.66
M10.013	Matrix metallopeptidase-13	98	0.62	0.55	0.6	0.55	0.63	0.56	0.64	0.72	0.53	0.7
M10.014	Membrane-type matrix metallopeptidase-1	113	0.69	0.63	0.72	0.64	0.7	0.64	0.6	0.76	0.73	0.75
M12.221	ADAMTS4 peptidase	57	0.73	0.7	0.72	0.75	0.74	0.72	0.76	0.69	0.69	0.74
M13.001	Neprilysin	74	0.71	0.66	0.73	0.73	0.74	0.71	0.53	0.75	0.73	0.74
M16.002	Insulysin	91	0.67	0.64	0.64	0.67	0.62	0.68	0.64	0.67	0.71	0.71
M16.003	Mitochondrial processing peptidase beta-subunit	75	0.56	0.44	0.64	0.61	0.35	0.65	0.49	0.49	0.62	0.66
M24.001	Methionyl aminopeptidase 1	539	0.51	0.35	-	-	-	0	0.34	0.74	0.72	0.74
M24.002	Methionyl aminopeptidase 2	131	0.53	0.39	-	-	-	0	0.43	0.73	0.77	0.74
M24.028	Mitochondrial methionyl aminopeptidase	131	0.47	0.33	-	-	-	0	0.32	0.69	0.66	0.7
M35.004	Peptidyl-Lys metallopeptidase	2104	0.65	0.5	0.75	0.74	0.75	0.74	0	0.75	0.74	0.75
S01.001	Chymotrypsin A (cattle-type)	929	0.69	0.62	0.72	0.73	0.72	0.45	0.7	0.75	0.74	0.74
S01.010	Granzyme B (*Homo sapiens*-type)	344	0.59	0.42	0.58	0.66	0.66	0.03	0.69	0.7	0.72	0.69
S01.029	Kallikrein-related peptidase 14	52	0.64	0.56	0.71	0.66	0.62	0.41	0.64	0.67	0.74	0.68
S01.131	Elastase-2	325	0.68	0.59	0.74	0.73	0.66	0.44	0.68	0.71	0.71	0.73
S01.133	Cathepsin G	276	0.69	0.64	0.74	0.72	0.67	0.58	0.67	0.73	0.72	0.7
S01.135	Granzyme A	270	0.61	0.47	0.66	0.67	0.73	0.15	0.62	0.64	0.7	0.73
S01.136	Granzyme B, rodent-type	163	0.59	0.42	0.6	0.7	0.68	0	0.71	0.65	0.7	0.68
S01.151	Trypsin 1	10191	0.66	0.51	0.74	0.74	0.74	0.05	0.75	0.74	0.75	0.74
S01.162	Kallikrein-related peptidase 3	58	0.66	0.62	0.59	0.63	0.69	0.63	0.67	0.68	0.67	0.71
S01.217	Thrombin	111	0.56	0.33	0.71	0.74	0.51	0	0.49	0.7	0.67	0.66
S01.233	Plasmin	100	0.63	0.49	0.71	0.73	0.71	0.09	0.66	0.72	0.74	0.72
S01.269	Glutamyl peptidase I	1073	0.66	0.52	0.75	0.75	0.76	0.13	0.7	0.73	0.73	0.73
S01.278	HtrA2 peptidase	55	0.67	0.56	0.61	0.75	0.72	0.45	0.7	0.72	0.62	0.76
S01.280	Lysyl peptidase (bacteria)	803	0.65	0.5	0.74	0.75	0.74	0	0.75	0.75	0.75	0.75
S08.070	Kexin	70	0.59	0.39	0.67	0.77	0.31	0.27	0.6	0.72	0.67	0.74
S08.071	Furin	98	0.51	0.24	0.35	0.7	0.31	0.07	0.56	0.63	0.74	0.72
S08.072	Proprotein convertase 1	74	0.54	0.28	0.72	0.7	0.21	0	0.69	0.69	0.64	0.71
S08.073	Proprotein convertase 2	99	0.58	0.34	0.73	0.73	0.27	0.05	0.71	0.71	0.71	0.75
S08.109	KPC2-type peptidase	60	0.49	0.24	0.61	0.52	0.21	0	0.55	0.72	0.68	0.63
S26.001	Signal peptidase I	266	0.53	0.31	0.57	0.3	0.69	0.04	0.59	0.61	0.71	0.72
S26.010	Signalase (animal) 21 kDa component	360	0.63	0.49	0.65	0.52	0.76	0.3	0.68	0.71	0.69	0.73
XM12-001	Meprin A complex peptidase	66	0.74	0.74	0.75	0.76	0.73	0.75	0.74	0.75	0.74	0.74
XS26-001	Signal peptidase complex (animal)	1907	0.63	0.44	0.65	0.45	0.76	0.21	0.74	0.74	0.73	0.73

Group-based information-theoretic entropy of positions P4 to P4′ for protease substrate sequences from the MEROPS database. Only proteases for which 50 substrate sequences or more were available are listed (column "count"). Entropy is computed along amino acid type (aliphatic, aromatic, acidic, basic, small and "other"), with an expected background entropy of 0.75 (based on UniProtKB statistics for protein composition). An entropy of 0 indicates perfect conservation across all substrate sequences. "Avg. 8" and "Best 3" give the average entropy for the whole octapeptide and its 3 most conserved positions respectively. Darker background color indicates that a position is more conserved (and therefore presumably easier to predict)

Difficulties of predicting calpain cleavage sites probably originate from the structure and functions of calpains: calpains can proteolyze various substrates in vitro and in vivo that are involved in a variety of cellular processes [47]. To achieve this, substrate binding sites of calpain molecules may have evolved to recognize their substrates in a wide range of peptide sequences, rather than binding strongly to a few specific amino acid residues around cleavage sites in a fashion similar to trypsin or caspases (which have predominant K/R and D residue preferences at the P1 site, respectively [5, 14, 18]). As a consequence, elucidating the mechanisms of substrate cleavage by calpains requires complex combinatorial analysis of a wide range of amino acid sequences around substrate cleavage sites.

Currently, CaMPDB, an online repository of calpain sequences [19], lists a little over a hundred confirmed substrate sequences, along with a computationally expanded set of many thousands potential substrate candidates, obtained through BLAST alignment search. While a crucial help to devise machine-learning cleavage prediction methods, the limited number of confirmed cleavage data, compounded by the presence of important selection biases in the set, further complicates the task of reaching prediction performances on calpain cleavage comparable to other types of cysteine protease (by contrast, in their recent work on caspase cleavage prediction, Song et al. [51] had access to data for 562 cleavage sites over 370 sequences).

Over the past 20 years, support vector machine algorithms have become a ubiquitous tool in machine learning and occupy a prominent position in bioinformatics research. In addition to belonging to the margin-maximizer group of classifiers (thus providing a bound on the generalization error), SVM distinguishes themselves by the use of so-called kernel functions to transform the input data before classification. Traditional SVM algorithms such as used by recent related work on protease substrate prediction [4, 51] require selecting a single-kernel function and using it on all input data throughout the algorithm. Such work emphasized the importance of using richer feature sets (such as secondary structure information in addition to sequence), however, because of the nature of standard single-kernel methods, had to compromise on the type and format of features that could be used.

The use of extensions to the SVM framework, commonly known as multiple kernel learning (MKL) algorithms, allowed us to combine heterogeneous feature sets, each with their own adapted kernel function, while optimizing the contribution of each sub-kernel to the resulting classifier.

Most interestingly, it has been shown [52] that multiple kernel learning can give a good understanding of which feature sets are important for discrimination. While standard SVM methods produce classification function that are notoriously difficult to interpret in terms of feature contribution compared to other classification techniques, MKL yields weights for each sub-kernel that, once properly scaled, provide a useful representation of the relative discriminative power of each set of features.

Along with an overview of cleavage prediction issues, this chapter aims to provide a complete introduction to our calpain cleavage prediction tool, *CalCleaveMKL*, by covering the essence of our two journal publications on the topic [20, 21].

2 Methods

2.1 Proteolytic Enzymes

Because mechanism of proteolytic action and levels of substrate specificity vary widely across protease families, often within subgroups of a family, the development of prediction methods has usually been tied to a single type of proteolytic enzyme at the exclusion of others.

Many early efforts in cleavage prediction models have focussed on the HIV-1 aspartic-type protease [13], with a view on developing better protease inhibitors for therapeutic purposes. For historical reasons, these research heavily favored models based on artificial neural networks (ANN) [8, 57], later to shift focus toward support vector machines (SVM) [9]. It was suggested that the HIV-1 cleavage dataset might be linearly separable [43] and that less complex models such as linear SVM or decision trees would therefore provide equal accuracy with better understanding. This is corroborated by overall high prediction performance [13, 50] for HIV proteases, regardless of method.

Among cysteine proteases, cleavage prediction (and the closely related problem of inhibitor prediction) has largely been dominated by caspases: a family of proteolytic enzymes associated with apoptosis (programmed cell death), necrosis, and inflammation. Due to their key role in major cell regulation mechanisms and considerable dataset of known substrates (with an even larger number of putative substrates yet to be identified or confirmed), caspases make prime candidates for the application of machine-learning algorithms to cleavage prediction. Indeed, a wide array of caspase cleavage prediction methods have been explored, with many implemented and available to the public [2, 62, 65].

In contrast to the previously mentioned proteolytic enzymes, calpains' mechanisms of proteolytic action are still poorly understood but hint at a more complex, less specific, mode of substrate recognition: in vitro and in vivo experiments have shown cleavage for a wide range of peptide sequences, with none of the strong position-specific amino acid residue preferences exhibited by most proteases (such as trypsin and caspases, predominantly binding to K/R and D residues at the P1 site, respectively [5, 14, 18]).

A superficial preliminary study of substrate sequence conservation in the MEROPS database, using information entropy (*see* Table 1), shows that a majority of proteases exhibit strong specificity in many positions around the cleavage site, while a few others exhibit little to no amino acid preferences, suggesting a more complex model of substrate recognition. Substrates of serine peptidases and caspases not only exhibit near-perfect conservation in position P1, but also have strong conservation (low entropy) for multiple other residues in the neighborhood of the cleavage site. On the other hand, HIV-1 or calpains show no absolute preference on any position (high entropy scores for all positions). As an illustration: thrombin substrates, with information entropy values of 0.51 and 0.49 in positions P2 and P1′, respectively, offer much higher predictability than calpain-2 substrates, for which none of the 8 positions have entropies lower than 0.64 (and most positions are close to the expected background entropy value of 0.75).

An important strength of such statistical algorithms based on supervised learning is their ability to adapt to a broad range of underlying physical models (such as biological differences between proteases). Combinatorial complexity (as represented by the sequence conservation across substrates) might therefore provide a better indicator of a method's affinity with a type of proteolytic action, than the biological taxonomy of its protease.

2.2 Problem Setting

The two sub-problems of predicting adequate protein docking and actual cleavage location prediction can be treated sequentially or as one single task. In theory, separating the two tasks could help reduce the complexity of each step and the scope of possible input features. However, issues such as unclear statistical dependencies make it a difficult task to combine two such classifiers into a full working predictor.

The problem can therefore be defined as the training of a scoring function $C(S, i, \theta_S)$, where $S = s_1 s_2 \ldots s_{n-1} s_n$ is a candidate sequence of n amino acids, i: the putative cleavage position, and θ_S: some optional input data associated with the sequence S (such as structural features or literature-curated information). The prediction is usually dependent on a limited neighborhood around the cleavage site and the function can therefore be simplified to $C(s_{i-k_1} \ldots s_{i-1} s_i s_{i+1} \ldots s_{i+k_2})$, where $k = k_1 + k_2 + 1$ is the length of substring considered by the predictor (window size) for each position.

In this standard approach, scores are generated for every position in the sequence by "sliding" the substring window along the whole sequence (*see* Fig. 2 for a simplified presentation of such a workflow). Assuming a fixed score threshold δ, the set of predicted cleavage positions is then defined as $\{i | C(s_{i-k_1} \ldots s_{i+k_2}) > \delta, i \in [1, n]\}$. Sequence edges ($i \leq k_1$ or $i \geq n - k_2$) need special care (*see* Subheading 2.3).

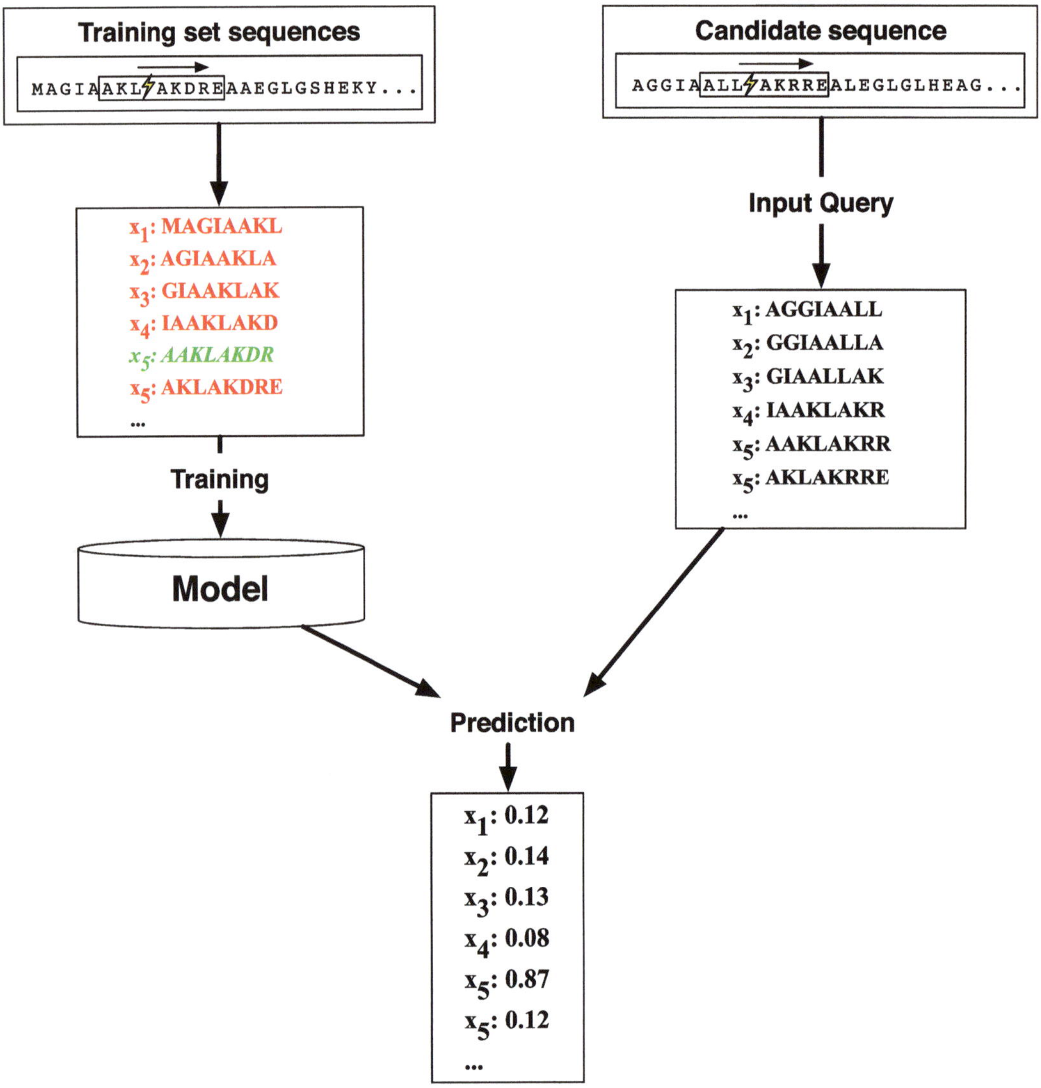

Fig. 2 Typical workflow of sequence-based cleavage prediction

2.3 Choice of Feature Types

A critical step in designing and applying a supervised-learning algorithm to the cleavage prediction problem is the selection of a set of appropriate input features. While primary structure (amino acid sequence) is the ubiquitous choice, biological insights and empirical results have shown the importance of higher-order feature types [48], such as secondary structure or tertiary structure (through solvent accessibility, for example).

1. **Vector Encoding**

   In order to use categorical or position-based sequence information with supervised learning frameworks, it is generally necessary to define an adequate transformation from a labeled sequence to a numeric vector. For discrete categories such as residues, it is common to use a canonical binary encoding where each amino acid is represented by a unique 20-long binary vector [41]: alanine: [1, 0, 0..., 0, 0], arginine: [0, 1, 0..., 0, 0], etc. Similarly, discretized secondary structure label (e.g., "alpha-helix," "beta-sheet," and "other") can be encoded on 3-bit long binary vectors. Any form of encoding that attempts to blindly convert class labels to integer values of a single variable would result in arbitrarily biased Euclidean distances between instances. It can be argued that binary encoding, which assumes equidistance between amino acids, has no theoretical basis in chemistry or biology; however, attempts to introduce tailored measures of similarity based on biological insights [66] have so far failed to show consistent improvements over standard techniques.

Sequence edges are a particular case and can require special treatment, depending on the type of protease (endopeptidases or exopeptidases), it can be preferable to use an asymmetric model or choose to restrict the substring window to the sequence bounds and discard any cleavage site which would occur within the $s_1 \dots s_{k_1}$ or $s_{n-k_2} \dots s_n$ edge sub-sequences. In this instance, given calpain's mode of proteolysis, we chose to pad the sequences with respectively k_1 and k_2 additional blank positions on each side.

2. **Dataset Imbalance**

Supervised learning requires labeling of training instances as either "positive" (for instance, cleaved sub-sequence) or "negative" (uncleaved sub-sequences). The nature of proteolysis means there is of course considerably more negative than positive instances in a normal dataset. Failing proper correction, this imbalance can greatly affect performances of algorithms such as SVM. A remedy can be to increase the penalty factor associated with misclassification of the positive class (cost parameter, in a standard SVM algorithm). Another universal solution is to preprocess the data by sampling a subset of negative instances, so as to bring back approximate parity between the two classes. Such sampling can follow a more or less complex distribution: from uniform sampling to a mixture of Gaussians centered on cleavage sites (so as to give more focus on regions where the cleavage occurs). Based on empirical results showing that best results can be achieved with a combination of both approaches [1], we used both sampling and cost correction in our algorithm.

2.4 Evaluation

General accuracy of the cleavage prediction algorithm is often measured using F_1-score (harmonic mean of precision and recall) or Matthews correlation coefficient. However, in order to avoid relying on threshold-dependent values of Type I and II classification errors, it is preferable to treat the task as a ranking task and measure the area under the ROC curve (AUC), easily obtained through the following formula [25]:

$$\mathrm{AUC} = \frac{S_0 - n_0(n_0 + 1)/2}{n_0 n_1} \tag{1}$$

where n_0 and n_1 are the numbers of positive and negative instances, respectively, and S_0 is the sum of the ranks of all positive instances in the list of instances ranked by prediction scores (where higher scores/ranks indicate strong prediction of cleavage). AUC scores provide a good way to estimate and compare performances of different prediction algorithms in a threshold-independent way. Looking at the actual ROC curve plot can also provide useful indications: for practical purposes, it may be desirable to train a predictor than can identify a few candidate cleavage sites with a high precision rate, at the possible expense of recall (i.e., picking the ROC curve with the steepest start, for comparable values of AUC).

Overfitting is one of the main pitfalls of supervised-learning, leading to classifiers that generalize poorly to new data. During evaluation, considerable care must be given to run optimization steps within a cross-validation loop (*k*-fold, leave-one-out, etc.). In order to avoid this, our evaluation covered the optimization of all algorithm parameters (such as kernel parameters), through the use of nested cross-validation.

2.5 CalCleaveMKL Overview

CalCleaveMKL, the MKL-based prediction method presented in this chapter has been implemented and can be used online at http://calpain.org. In addition to CalCleaveMKL, the *calpain.org* website contains a curated database of known calpain-related sequences (calpain, calpain substrates, and calpastatin), as well as a statistical overview of the different datasets.

By browsing to the *Predict* subsection of the *Substrates* section, the user is able to apply the main prediction method presented here (based on MKL), along with a few other simpler methods that only use the amino acid sequence as feature: SVM, as well as position-specific scoring matrix (PSSM) (Table 2).

The input form (*see* Fig. 3) takes a single peptide sequence, in FASTA format. That is, the first line must start with a ">" character and provide name and information on the sequence, with following lines containing strings of amino acid letters.

Because the secondary structure prediction step of the MKL method can be time-consuming, the tool works asynchronously, and requires an email address to send a notification once the task is complete.

The result page (*see* Fig. 4) offers a ranked list of scored positions where a cleavage is likely to occur (along with *p*-values), while

Table 2
Overview of main algorithms

Type of algorithm	Input type	Power	Model readability	Computational cost	General performance	Overfitting
PSSM	Vector	Linear	High	Low	Low	High
Decision tree	usu. vector	Nonlinear	High	High	Low	High
ANN	Vector	Nonlinear	None	High	Medium	High
HMM	Variable-length string	usu. linear	Poor	High	Low	High
SVM: dot-product	Vector	Linear	High	Low	Medium	Low
SVM: RBF	Vector	Nonlinear	None	Low	High	Low
SVM: spectrum kernel	Variable-length string	Nonlinear	Poor	Medium	Low	Low
SVM: MKL	Multiple	Nonlinear	Poor	Medium	High	Low

Comparison of main approaches to cleavage prediction, based on a few general criteria. Ratings in each category are given relative to other methods. The tool presented in this chapter offers either SVM MKL or SVM RBF methods

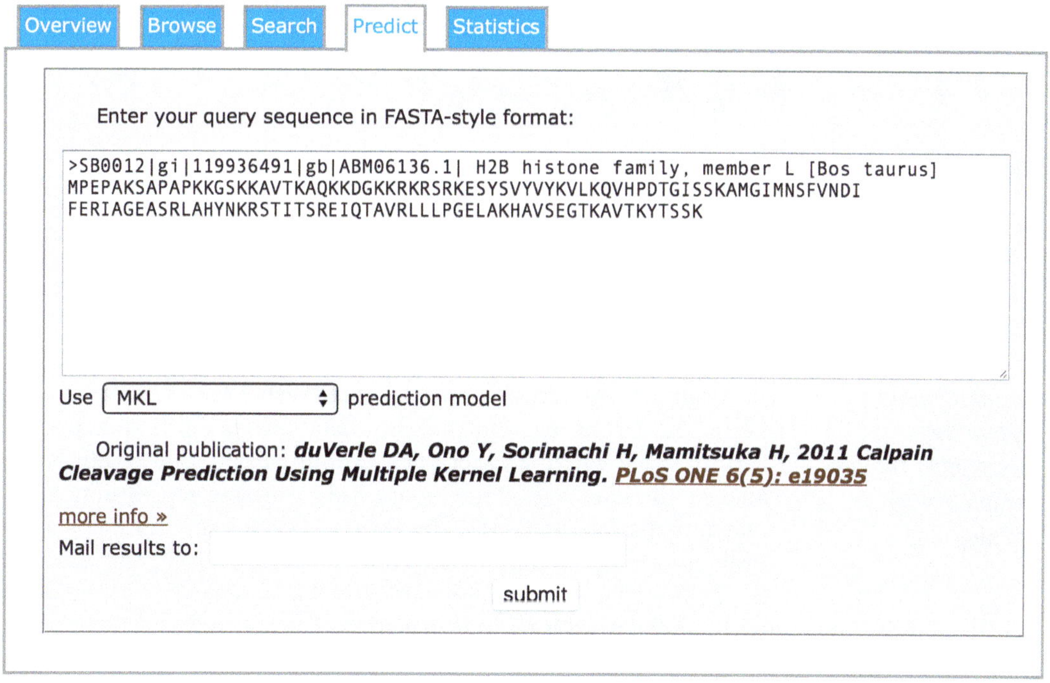

Fig. 3 Calpain prediction tool: Input form

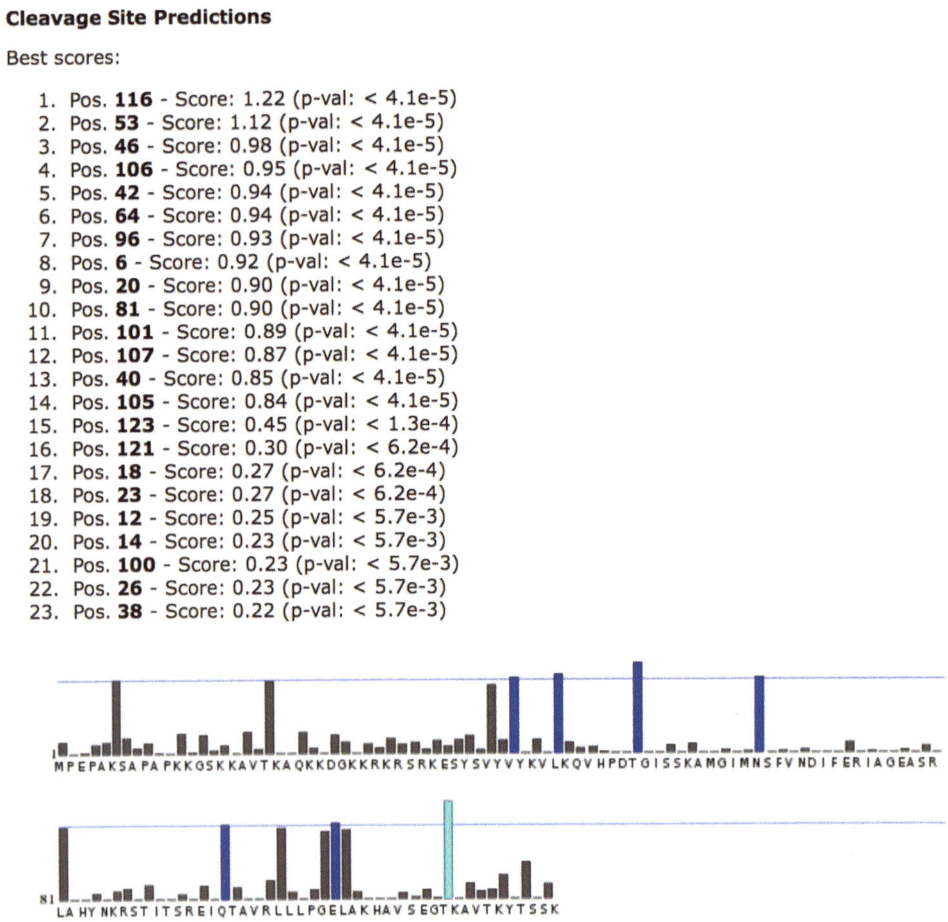

Fig. 4 Calpain prediction tool: Result page

a graphic rendering of the amino acid sequence outlines the scores at each position (with positions in the top 5% of all scores highlighted in blue, and top score in light blue).

2.6 Optimizing Feature Set Contribution Through Multiple Kernel Learning

At the heart of kernel methods, the "kernel-trick" makes use of kernel functions to remap input data into a high dimensional feature space where a variety of methods can be used to efficiently analyze the data (e.g., find a margin-maximizing separating hyperplane, in the case of SVM). The choice of such a kernel function not only affects separability of data in the feature space, but can also help efficiently filtering in or out certain characteristics of the input without the need for additional steps.

A kernel function does not explicitly calculate data coordinates in the feature space, but instead computes the inner products between the images of all pairs of input vectors in that space.

Given a kernel function, $k(\mathbf{x}_i, \mathbf{x}_j)$, and a set of labeled training instances $(\mathbf{x}_i, \mathbf{y}_i)_{i=1,\ldots,N}$ $(\mathbf{y}_i \in \{1, -1\})$, training an SVM means learning the weights (α_i) in the decision function:

$$f(\mathbf{x}) = \text{sign}\left(\sum_{i=1}^{N} \alpha_i \mathbf{y}_i k(\mathbf{x}_i, \mathbf{x}) + b\right) \tag{2}$$

where b is the bias.

Judicious choice of kernel function (see below) gives great flexibility regarding the nature of features that can be used (real values, binary values, strings, etc.), but it can sometimes be desirable to combine features of different structure or dimension within the same classifier. In such case, a standard solution is to find a common encoding that can be satisfyingly applied to each set of features in order to produce a unique input vector for each instance. Going with such an approach, however, means losing potentially useful data structure information in the encoding and being forced to use identical kernel parameters for all data sources. Additionally, it is very difficult to extract useful information in terms of feature contribution to the final classifier.

A more elegant solution resides in the use of "multiple kernel learning." Although there exist a variety of methods [11, 34, 52], they all tend to rely on expressing a combined kernel as a linear sum (Eq. 3) of T sub-kernel functions $(k_1(\mathbf{x}, \mathbf{x}')\ldots k_T(\mathbf{x}, \mathbf{x}'))$, leading to the decision function (Eq. 4) and its associated optimization problem.

$$k(\mathbf{x}, \mathbf{x}') = \sum_{j=1}^{T} \beta_j k_j(\mathbf{x}, \mathbf{x}') \tag{3}$$

$$f(\mathbf{x}) = \text{sign}\left(\sum_{i=1}^{N} \alpha_i \mathbf{y}_i \left(\sum_{j=1}^{T} \beta_j k_j(\mathbf{x}, \mathbf{x}')\right) + b\right) \tag{4}$$

Sonnenburg et al. [52] offer a method to reformulate the problem as a "semi-infinite" linear program that can in turn be solved using standard LP techniques.

2.7 Selecting Feature Sets and Kernel Functions

The use of multiple kernel learning gave us the opportunity to not only select a large number of heterogeneous features, but also assign a specifically adapted kernel function to each set. Through review of biological hypotheses and preliminary results we identified a number of feature types and kernel functions most likely to yield good performance for cleavage prediction. We were then able to run several different configurations in order to evaluate which combination produced the best compromise between performance and model complexity.

In addition to the classic Gaussian radial basis function (RBF) kernel (commonly used on such problem, in conjunction with binary-encoded vectors of amino acid positions; *see*, for example, [4]), we examined two other types of kernel functions that offered interesting alternative perspectives on our data:

1. **String kernel**

   Similar to linear or RBF kernel functions, in that it is position-dependent, a typical string kernel function calculates the number of identical k-mers (of length varying between 1 and the kernel order: d) between two sequences of length L and can be defined as:

$$k(\mathbf{x}, \mathbf{x}') = \sum_{l=1}^{d} \sum_{i=1}^{L-l+1} I(u_{i,l}(\mathbf{x}) = u_{i,l}(\mathbf{x}')) \quad (5)$$

   where $u_{i,l}$ refers to the substring of u of length l starting at position i, and $I(\cdot)$ is the indicator function.
   It offers the advantage of working directly on string data (removing the need for binary encoding of sequences and leading to more compact feature vectors) and can be configured to look at k-mers instead of being restricted to single amino acid position in a sequence. These two aspects make it well-suited to examine position-based sequence features.

2. **Spectrum kernel**
   Spectrum kernels are a family of functions based on position-independent k-mer enumerations. In this instance, we use gapped substring kernels, defined as:

$$k(\mathbf{x}, \mathbf{x}') = \Phi_{d,g}(\mathbf{x}) \cdot \Phi_{d,g}(\mathbf{x}') \quad (6)$$

   where $\Phi_{d,g}(\mathbf{x})$ returns a vector of occurrence counts for all k-mers of length at most d and allowing for at most g gaps within string $\mathbf{x}$.

   Using this type of kernel function lets us focus on the search for feature motifs anywhere in the sub-sequence, regardless of position or window size. It is therefore particularly adapted for structural features, such as secondary structure or solvent accessibility: accommodating their typically flexible nature by allowing for looser positioning around the cleavage site.

2.8 Using Calpain Type Specificity

Humans present 15 genes that encode a calpain-like protease domain, generating diverse kinds of calpain homologues with combinations of several functional domains such as Ca^{2+}-binding domains (C2-domain-type and EF-hand-type) and Zn-finger domains. Additionally, calpain homologues are increasingly being found in other organisms including insect, nematode, trypanosome, plant, fungus, yeast, and even some bacteria. The substrates present in our data (Fig. 5) were mainly shared between μ-calpain

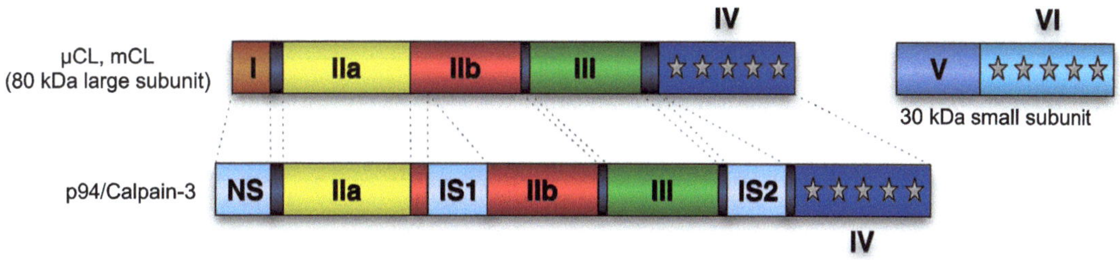

Fig. 5 Schematic structures of major calpain homologues. "Conventional" calpains (μ- and m-calpain) are composed of larger catalytic subunits (calpain-1 and -2) and a smaller regulatory subunit. Some homologues such as skeletal muscle-specific calpain (calpain-3/p94) have slightly diverged properties, including unique insertion sequences (NS, IS1, and IS2) and no requirement for a small subunit. Symbols used are: **I**: N-terminal domain with little homology; **IIa** and **IIb**: protease subdomains containing the active sites Cys and His/Asn, respectively; **III**: C2-like Ca^{2+}-binding domain; **IV** and **VI**: 5-EF-hand Ca^{2+}-binding domain; **V**: Gly-rich hydrophobic domain; **NS**, **IS1**, and **IS2**: p94-specific sequences

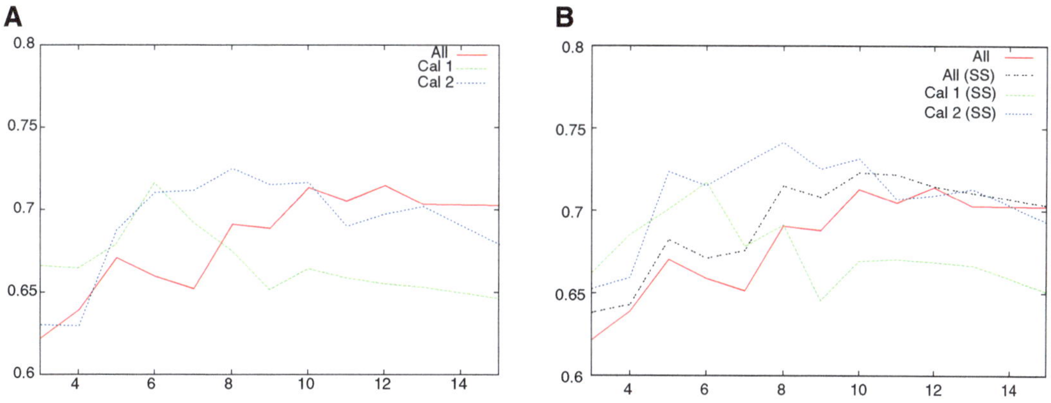

Fig. 6 Linear-kernel SVM performance trained on full set of substrates (All) vs. calpain-1 (Cal 1) and calpain-2 (Cal 2). AUC score as function of symmetrical extension length (number of nucleotides) on each side of putative cleavage site. (**a**) Using only position information. (**b**) Using position and secondary structure (SS) information

(heterodimer of calpain-1 and CAPNS1) and m-calpain (heterodimer of calpain-2 and CAPNS1), two major ubiquitous homologues activated respectively by μM and mM levels of Ca^{2+} concentrations in vitro. In addition to these two types, a muscle-specific calpain known as calpain-3 (also called p94) accounted for two more substrate sequences.

While it is generally considered that members of the calpain family behave similarly in their proteolytic activity [16, 24, 40], preliminary results (Fig. 6) showed that some amount of specificity may exist with regard to substrate sequence and cleavage location. Despite potential issues with reducing the amount of training data even further, we investigated the hypothesis that separating data by calpain type might lead to improved prediction quality.

3 Evaluation

3.1 Training Data and Experimental Setting

The data used in all our experiments (training and evaluation of the models) was obtained from the online calpain database CaMPDB [19], selecting only confirmed substrate sequences ("SB" label). The issue of selection bias in the curated set was addressed by removing redundant sequences (as defined by presenting an alignment with over 95% identity to another sequence in the set) resulting in a set of 90 sequences.

For each sequence, we computed secondary structure and solvent accessibility data using, respectively, PSIPRED [29] and ACCpro [12]. Each amino acid sequence was thus labeled with three classes for secondary structure (α-helix, β-sheet, "other") and two classes for solvent accessibility (above and below a 25% accessibility threshold).

The different kernels were trained and evaluated using the Shogun framework [52] through its Python modular interface. Windows of varying sizes around each cleavage site made the set of positive instances, while negative instances were randomly sampled from every other position in the sequence so as to yield a 10 to 1 ratio between negative and positive instances.

All performance results were measured using area under ROC curve (AUC) with 10×10 cross-validation (average of all AUC values generated from 10 repeats of 10-fold cross-validation).

When used with Gaussian or linear kernels, all feature vectors were extracted from sequence data using a canonical binary encoding (each amino acid position in the primary sequence was matched by 20 binary values in the feature vector). Raw sequence data was used as input for string and spectrum kernels.

Optimal SVM parameters (cost: C and, where applicable, kernel width: τ) were set for each kernel using a grid search (*see* Table 3).

Table 3
SVM parameters

Parameter	Min	Max	Optimal value
C (cost)	0.1	10	1.67
τ (width)	0.001	1000	2.1
d (order of string kernel)	1	6	5
d' (order of spectrum kernel)	1	10	8
g (gaps allowed)	0	3	1

Tested range and optimal values for SVM kernel function parameters. Integer values were tested for the entire range. Non-integer parameters were set using values within their ranges in two-successive grid search of decreasing step value

For each combined kernel the optimal window length parameters (left and right extension around cleavage site position) of each sub-kernel were found by sequentially running a grid search on one set of parameters while freezing the others. This process was iterated until convergence of the top cross-validated AUC score, yielding locally optimal parameters (between 2 and 8, depending on kernel configuration) in reasonable computational time.

We first explored the impact of calpain-type specificity by running limited cross-validation experiments on the full set of substrate sequences, then on two subsets made of substrates cleaved by calpain-1 and calpain-2, respectively (the set of substrates cleaved by calpain-3 was too small to be efficiently analyzed and was therefore ignored in this part).

As a baseline we computed optimal AUC scores using a single Gaussian kernel: first on sequence data alone, then on sequence and secondary structure concatenated together using encoding and window length parameters described in [4] and finally on a variant of [4] using a grid search to find optimal extension length parameters.

Finally, AUCs were computed for three configurations of combined kernel, using the method developed by Sonnenburg et al. [52] to simultaneously optimize sub-kernel weights and matrices.

3.2 Preliminary Results

We analyzed the impact of using different extension sizes on either side of cleavage sites. In particular, we looked for pronounced asymmetrical features. In order to keep the size of input features down and avoid unnecessary noise, it was critical to accurately narrow down sequence regions directly or indirectly involved in substrate recognition and cleavage for each type of feature (sequence, secondary structure and solvent accessibility).

When comparing single-kernel performances across calpain-type (Fig. 6) we can observe that, while AUC performance peaks at 6 amino acids around the cleavage site for the μ-calpain (Cal-1) set, performance on the m-calpain (Cal-2) set increases until at least the 8th amino acid.

This trend is even more visible when considering asymmetrical extension lengths (Fig. 7), where we can clearly see important differences between μ-calpain and m-calpain. Along the Υ-axis, for $X \approx 5$ (that is, with P1′–P5′ fixed and extending toward P1, P2, P3,etc.), μ-calpain has a peak around 6 and quickly decays after that. On the other hand, m-calpain predictions perform well until around the 20th amino acid. This probably indicates that μ-calpain recognizes a relatively short stretch of the N terminal side of substrates (until around P6), whereas m-calpain uses longer portion of the N-terminal side of substrates (P20 and beyond). In other words, μ-calpain probably recognizes substrates mainly by domain II (Fig. 8), while m-calpain also uses domain III for recognition of (at least some) substrates. This may indicate that μ-calpain has more

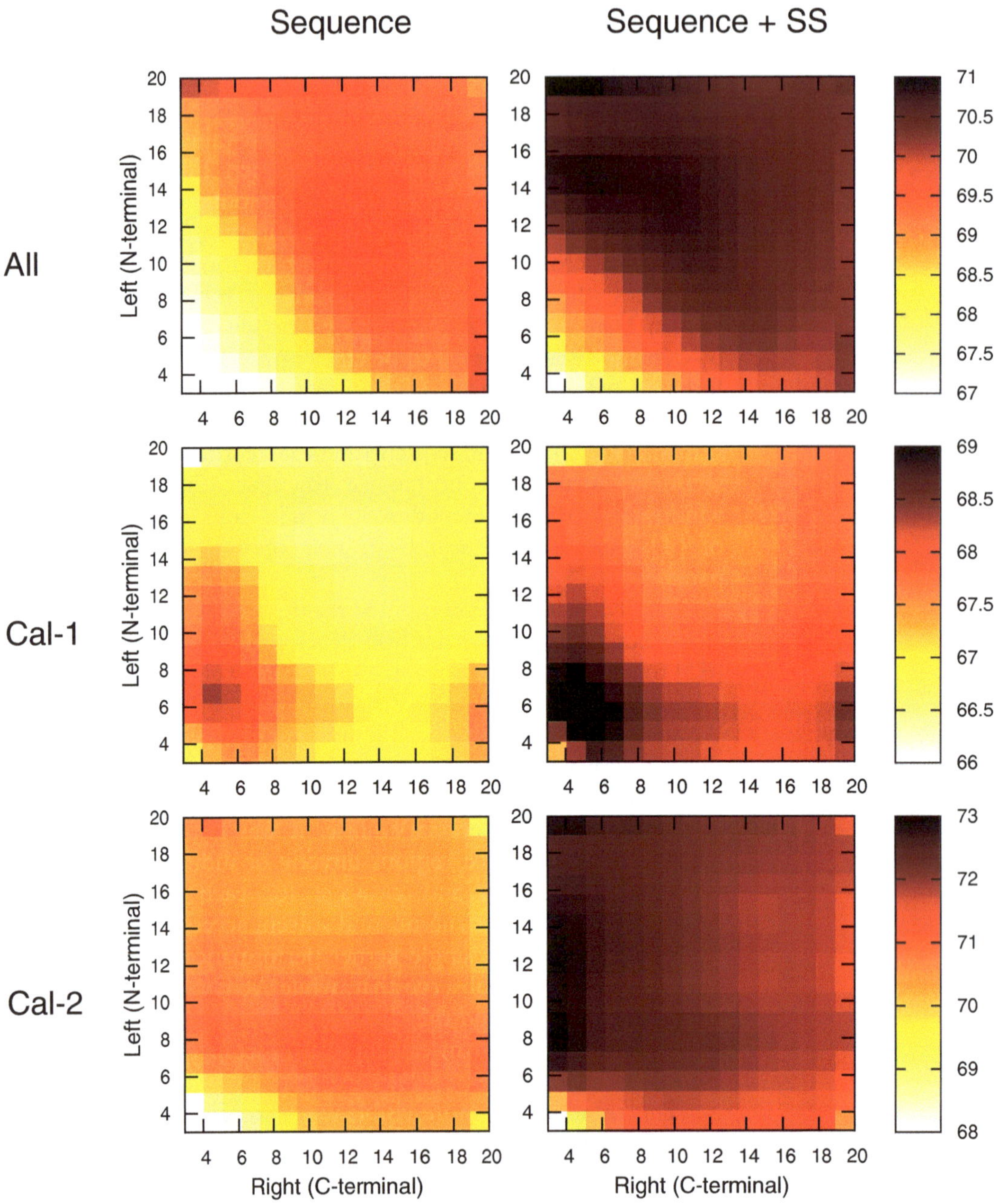

Fig. 7 AUC (with a linear-kernel SVM) as function of cleavage extension length (left and right side of cleavage site) in number of nucleotides. Left column uses sequence only, while right column uses secondary structure information (SS) as well

affinity to short peptides than m-calpain. Considering that μ- and m-calpain have similar catalytic velocity ($V_{\max}$), this would imply that μ-calpain has larger turn-over numbers for short peptides. While in our own experience (unpublished work) μ-calpain has indeed showed higher activity to peptide substrates than

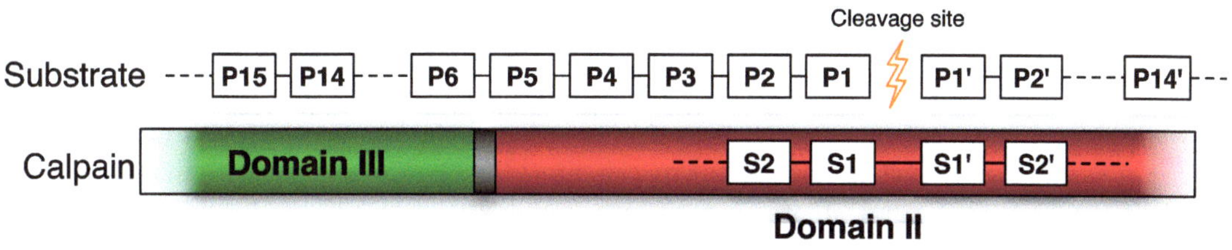

Fig. 8 Schematic representation of contact region between calpain and substrate sequence. Domain II is the protease domain of calpain, while domain III binds Ca^{2+}. Amino acid sequences of domain III are less conserved than those of domain II, which are highly conserved not only between μ- and m-calpains but also among all calpain family members

Table 4
AUC results with single Gaussian kernel methods

	Cal 1	Cal 2	All
Position (I)	77.77 (0.88)	77.09 (1.28)	76.86 (1.05)
Position + SS* (I′)	73.25 (2.00)	74.13 (1.57)	75.39 (1.11)
Position + SS** (I″)	73.25 (2.00)	77.22 (1.19)	75.39 (1.11)

Position: Residue position information, with a Gaussian RBF kernel ($\tau = 2.1$, $C = 1.67$) and canonical binary encoding.
SS: Secondary structure, with a spectrum kernel (k between 2 and 5, allowing up to 1 gap)
*Using same encoding and window length as [4]
**Using same encoding as [4] but with optimal window parameters obtained through grid search Results are shown as: % AUC (% SEM)

m-calpain (over fivefold), current literature presents arguments both partly in favor [49] and against [31] this hypothesis.

One interesting difference was reported from 3D structural studies: when the protease domain (subdomains IIa and IIb, Fig. 5) from either type of calpain was isolated, expressed, and used for proteolytic assay, the domain from μ-calpain showed over 1000-fold more activity than m-calpain. Structurally, this phenomenon could be explained by interferences of the active site with Trp106, due to the lack of interaction between subdomain IIa and domain III resulting in instability of the Gly197-Gly210 loop [36]. In contrast, whole 3D structure of m-calpain, composed of domains I-IV+VI, showed stabilized Gly197-Gly210 loop and no interference with Trp106 [37].

It is possible that some substrates may interact with domain III of m-calpain, resulting in disruption of the interaction between Gly197-Gly210 loop and domain III, which would interfere with the interaction between domain IIa's S2-S3 sites and substrates (with corresponding P2-P3 residues). In this case, m-calpain would use over S4 sites for substrate recognition, which may explain the difference we observed between m- and μ-calpain.

In all cases (Cal-1, Cal-2, and "All"), the addition of secondary structure information (SS) to the linear kernel's input features not only resulted in higher AUC across the board but also attenuated the previously observed impact of features length specificity within each calpain subset (Fig. 7, right column). However, it is interesting to note that this attenuation is less pronounced for μ-calpain (Cal-1) than for m-calpain (Cal-2): this may indicate that m-calpain has stronger substrate specificity at the secondary structure level than μ-calpain. There is no experimental evidence that may support this hypothesis but it is consistent with the previously mentioned specificity of m-calpain over wider substrate areas (since secondary structure only makes sense for oligopeptides of about 10 residues or more).

Finally, there is an imprecise but significant "line" along the X-axis at $Y \approx 6$ for both μ- and m-calpain. This may indicate that, if substrates do not interact well with domain III (i.e., P6-P14), they interact more with domain II via P6′–P14′, implying that calpain uses different ranges of its structure to recognize different substrates: a possible explanation for calpain's ability to recognize such a variety of substrates with a single molecule.

When switching to a nonlinear single kernel (Gaussian RBF), performance increased significantly (Table 4). However, in contrast to the linear model results and despite previous findings [4], the addition of secondary structure information not only failed to bring significant improvement but, in most cases, lowered AUC results (Table 4, I′ and I″). This performance hit could be caused by the extra noise brought by the addition of overly rigid position-specific secondary structure information, compounded by the high dimension and sparsity of the resulting feature space: a type of problem often better handled by simpler linear model over complex kernels (although the added discriminative power of the nonlinear model still results in overall better performance on sequence alone). The unsatisfying compromise of having to choose a single kernel and common encoding for both types of information further hinted at the potential benefit of our multiple kernel approach. Interestingly, AUC results for the calpain-2 subset showed much better resistance (if no significant improvement) to the addition of secondary structure features: a further confirmation that specificity by calpain type might exist, with at least different use of secondary structure information.

3.3 Multiple Kernel Learning Results

We obtained a top AUC score of 83.36% on the full training set ("All"), using MKL with a combined kernel containing position, string, and secondary structure information (Table 5). Despite using no more input data (sequence and secondary structure) than single-kernel methods (Table 4, I″), our method resulted in a considerable AUC increase from the baseline score of 76.86% (pairwise T-test p-value $< 2.2e^{-16}$, between 10×10 cross-validation AUC results, with equal variance assumption). Although the addition of solvent accessibility seemed to improve scores, the

Table 5
AUC results with MKL methods

	Cal 1	Cal 2	All
String + SS (II)	82.39 (0.70)	80.20 (0.67)	81.46 (0.54)
Position + String + SS (III)	84.28 (0.67)	83.09 (0.79)	83.36 (0.54)
Position + String + SS + SA (IV)	85.46 (0.66)	83.23 (0.68)	83.71 (0.59)

In addition to legends from Table 11.4:
String: Sequence string, with a string kernel (position-based substrings of length 1–6). SA: 25% solvent accessibility, with a spectrum kernel (*k* between 1 and 7, allowing up to 1 gap)

Table 6
Pairwise *T*-test comparison

Test	*p*-Value	Conclusion
(I) vs. (II)	$< 2.2e^{-16}$	Significant
(II) vs. (III)	$< 2.2e^{-16}$	Significant
(III) vs. (IV)	0.6029	Inconclusive
(II) vs. (IV)	$< 6.327e^{-9}$	Significant

p-Values for pairwise *T*-test comparisons between results from different combination of kernels, using sets of 10×10 AUC results, under assumption of equal variance

Table 7
Substrate labeling by calpain sub-type

	Calpain-1	Calpain-2	Calpain-3	All types
Substrate sequences	46	49	2	90
Cleavage sites	94	114	4	220

Values from each calpain gene types do not add up to the figures for "All Types," due to some substrates being cleaved by more than one type, while other sequences are missing calpain type labeling

increase was not significant compared to the introduction of secondary structure alone (Table 6).

Due to the limited availability of cleavage data (less than 90 distinct sequences for all calpain types, *see* Table 7), it could be expected that breaking down the general set into smaller calpain gene product subsets (of approximately half the size) would hurt performances; yet, results were not only stable within each subset, but, in some cases, improved. Interestingly, solvent accessibility brought a significantly higher score to the calpain-1 subset (85.46%), indicating that the mechanisms of substrate recognition by μ-calpain might rely more heavily on this property than m-calpain.

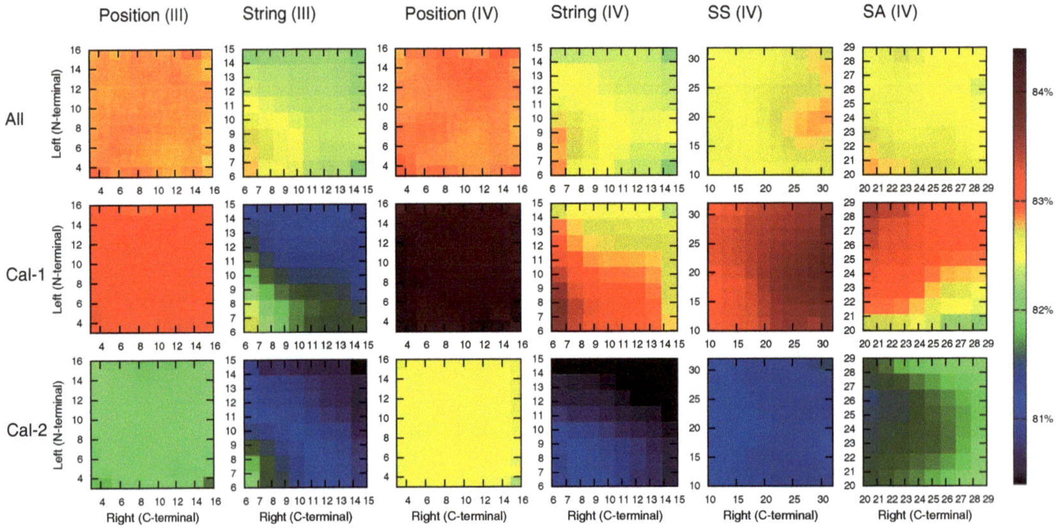

Fig. 9 AUC as function of cleavage extension length. AUC values produced by MKL prediction method, when varying extension length for one feature set at a time (all other parameters at their optimal value). *See* Tables 4 and 5 for notations

Table 8
MKL weights

	Position	String	SS	SA
String + SS (II)	–	1.0	0.09	–
Position + String + SS (III)	0.80	0.59	0.07	–
Position + String + SS + SA (IV)	0.78	0.59	0.06	0.18

Optimal training weights obtained for each combination of kernels (on full calpain set) using MKL training algorithm described in [52]

In most instances, optimal window length parameters showed a strong asymmetry between left- and right-side extension around cleavage site (Fig. 9).

Analyzing the final weights (Table 8) for each sub-kernel in the MKL method (computed on normalized kernel matrices), we were able to confirm what raw AUC results (Table 5) plainly suggested in terms of feature selection: similar orders of magnitude between the weight for position-based features and those for secondary structure and solvent accessibility can be formally interpreted as an indication that the latter still bring discriminative power to the combined classifier [52].

3.4 Validation with Mutant Calpastatin Sequences

Calpastatin is an endogenous inhibitor protein specific to calpain [23, 30]: after activation by Ca^{2+}, calpain is recognized by calpastatin, which binds to its active site while remaining uncleaved, thus

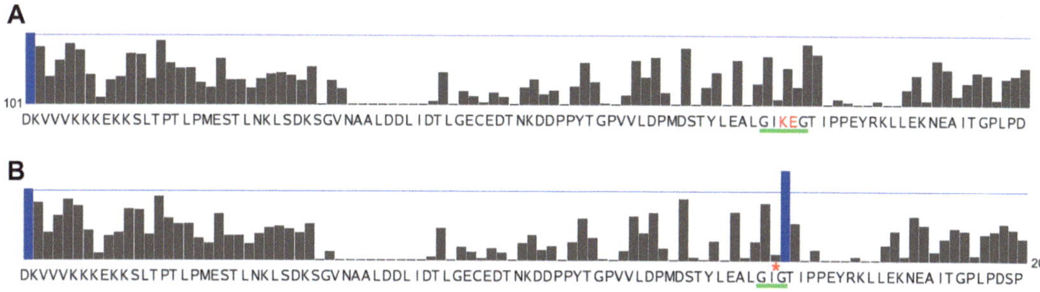

Fig. 10 Cleavage prediction on calpastatin sequences. Normalized MKL prediction scores using position, string, and SS feature sets. (**a**) On wild-type *Rattus norvegicus* calpastatin (gi 13540322). (**b**) On a mutant of calpastatin, obtained by deletion of Lys176 and Glu177 (highlighted in red in sequence (**a**) and marked by a red star in sequence (**b**)). Results were cropped to residues [101–200] in the sequence. Thin blue line marks 5% top scores threshold. Thick green lines highlight "loop-out" area of calpastatin sequences (shortened in the mutant by deletion of Lys176 and Glu177) where cleavage would likely occur

inhibiting proteolytic activity. In their experimental work, Moldoveanu et al. were able to induce proteolysis in several mutant sequences of calpastatin by deleting one or two specific residues (Lys176, Glu177, or both) from a sequence of wild-type rat calpastatin (gi 13540322) [37]. Both sequences of calpastatin (wild type and mutants) being phylogenetically unrelated to all substrate sequences in our training set, and they provided a good opportunity for qualitative validation on the generalization power of our prediction method.

The 3D structure of co-crystallized m-calpain and calpastatin [26, 37] indicated that Leu172-Gly173 and Thr179-Ile180 are at the P2-P1 and P1′–P2′ positions, respectively. Deletion of Lys176 and Glu177 makes this mutant calpastatin a good substrate, strongly suggesting, in the absence of further experimental confirmation, that the cleavage site is at the C-terminus of Ile175 and/or Gly178. Indeed, our program predicted a sharp peak between Gly178 and Thr179 for this mutant calpastatin (Fig. 1). The results presented in Fig. 10 showed not only that our prediction model correctly identifies the binding site in the original calpastatin sequence as a poor candidate for cleavage, but most importantly, detected a sharp signal increase on the same site after the sequence had been altered to allow cleavage by calpain, closely matching what has been experimentally observed by Moldoveanu et al. [37].

4 Summary

Through the use of a novel extension to the classic SVM framework, we were able to significantly improve cleavage prediction performance, as measured by a critical AUC increase: from less than 77% (RBF position-based score for the full calpain set) to over 83% (combined kernel using secondary structure on top of sequence information). The demonstrated inability of single-kernel methods to benefit from the addition of extra features such as secondary structure, presumed to be helpful [48], provides a strong argument in favor of MKL: by allowing seamless integration of heterogeneous features while retaining their respective structure, MKL can yield satisfying performance on even critically small training sets.

Furthermore, we presented results strongly favoring the hypothesis that subtypes of calpain behave differently with regard to substrate recognition and cleavage, dispelling previous assumption that proteolytic action was identical across all types of calpain (treating subtypes separately lead to significant performance increase in the case μ-calpain where AUC was improved by a further 2–85.46%).

Future extensions of this method will likely focus on general cleavage prediction for other types of proteases. Additionally, promising results have been obtained on string prediction, using an approach based on hidden Markov model profiles [53], which we are planning to apply to the prediction of cleavage position.

An online implementation of CalCleaveMKL is available at http://calpain.org.

Acknowledgements

The authors would like to thank Prof. H. Sorimachi of the Department of Advanced Science for Biomolecules at Tokyo Metropolitan Institute of Medical Science, for providing many insights on the biological aspects of calpain proteolysis and providing us with the crystallography view of calpain–calpastatin docking illustrating this chapter.

References

1. Akbani R, Kwek S, Japkowicz N (2004) Applying support vector machines to imbalanced datasets. In: Machine Learning: ECML 2004, pp 39–50
2. Backes C, Kuentzer J, Lenhof H, Comtesse N, Meese E (2005) Grabcas: a bioinformatics tool for score-based prediction of caspase-and granzyme b-cleavage sites in protein sequences. Nucleic Acids Res 33(Suppl 2):W208
3. Banik N, Chou C, Deibler G, Krutzch H, Hogan E (1994) Peptide bond specificity of calpain: proteolysis of human myelin basic protein. J Neurosci Res 37(4):489–496
4. Barkan D, Hostetter D, Mahrus S, Pieper U, Wells J, Craik C, Sali A (2010) Prediction of protease substrates using sequence and structure features. Bioinformatics 26 (14):1714–1722
5. Barrett A, Rawlings N, Woessner J (1998) Handbook of proteolytic enzymes. Academic, New York
6. Bartoli M, Richard I (2005) Calpains in muscle wasting. Int J Biochem Cell Biol 37 (10):2115–2133
7. Bertipaglia L, Carafoli E (2007) Calpains and human disease. Subcell Biochem 45:29–53
8. Cai Y, Chou K (1998) Artificial neural network model for predicting HIV protease cleavage sites in protein. Adv Eng Softw 29(2):119–128

9. Cai Y, Lin S, Chou K (2003) Support vector machines for prediction of protein signal sequences and their cleavage sites. Peptides 24 (1):159–161
10. Carillo S, Pariat M, Steff A, Jariel-Encontre I, Poulat F, Berta P, Piechaczyk M (1996) PEST motifs are not required for rapid calpain-mediated proteolysis of c-fos protein. Biochem J 313(Pt 1):245
11. Chapelle O, Vapnik V, Bousquet O, Mukherjee S (2002) Choosing multiple parameters for support vector machines. Mach Learn 46 (1):131–159
12. Cheng J, Randall A, Sweredoski M, Baldi P (2005) SCRATCH: a protein structure and structural feature prediction server. Nucleic Acids Res 33(Web Server Issue):W72
13. Chou K (1996) Review: prediction of HIV protease cleavage sites in proteins. Anal Biochem 233(1):1–14
14. Craik C, Largman C, Fletcher T, Roczniak S, Barr P, Fletterick R, Rutter W (1985) Redesigning trypsin: alteration of substrate specificity. Science 228(4697):291
15. Croall D, Ersfeld K (2007) The calpains: modular designs and functional diversity. Genome Biol 8(6):218
16. Croall D, Chacko S, Wang Z (1996) Cleavage of caldesmon and calponin by calpain: substrate recognition is not dependent on calmodulin binding domains. Biochim Biophys Acta Protein Struct Mol Enzymol 1298(2):276
17. Cuerrier D, Moldoveanu T, Davies P (2005) Determination of peptide substrate specificity for μ-calpain by a peptide library-based approach. J Biol Chem 280(49):40632
18. Demon D, Van Damme P, Berghe T, Vandekerckhove J, Declercq W, Gevaert K, Vandenabeele P (2009) Caspase substrates: easily caught in deep waters? Trends Biotechnol 27:680–688
19. duVerle D, Takigawa I, Ono Y, Sorimachi H, Mamitsuka H (2010) Campdb: a resource for calpain and modulatory proteolysis. In: Genome informatics. International Conference on Genome Informatics, vol 22, p 202
20. duVerle D, Ono Y, Sorimachi H, Mamitsuka H (2011) Calpain cleavage prediction using multiple kernel learning. PLoS One 6(5):e19035. http://dx.doi.org/10.1371/journal.pone.0019035
21. duVerle DA, Mamitsuka H (2011) A review of statistical methods for prediction of proteolytic cleavage. Brief Bioinform 13(3):337–349
22. Friedrich P, Bozóky Z (2005) Digestive versus regulatory proteases: on calpain action in vivo. Biol Chem 386(7):609
23. Goll D, Thompson V, Taylor R, Zalewska T (1992) Is calpain activity regulated by membranes and autolysis or by calcium and calpastatin? BioEssays 14(8):549–556
24. Goll D, Thompson V, Li H, Wei W, Cong J (2003) The calpain system. Physiol Rev 83:731–801
25. Hand D, Till R (2001) A simple generalisation of the area under the ROC curve for multiple class classification problems. Mach Learn 45 (2):171–186
26. Hanna R, Campbell R, Davies P (2008) Calcium-bound structure of calpain and its mechanism of inhibition by calpastatin. Nature 456(7220):409–412
27. Harris F, Biswas S, Singh J, Dennison S, Phoenix D (2006) Calpains and their multiple roles in diabetes mellitus. Ann N Y Acad Sci 1084:452
28. Horikawa Y, Oda N, Cox N, Li X, Orho-Melander M, Hara M, Hinokio Y, Lindner T, Mashima H, Schwarz P *et al* (2000) Genetic variation in the gene encoding calpain-10 is associated with type 2 diabetes mellitus. Nat Genet 26(2):163–175
29. Jones D (1999) Protein secondary structure prediction based on position-specific scoring matrices1. J Mol Biol 292(2):195–202
30. Kawasaki H, Emori Y, Suzuki K (1993) Calpastatin has two distinct sites for interaction with calpain-effect of calpastatin fragments on the binding of calpain to membranes. Arch Biochem Biophys 305(2):467–472
31. Kelly J, Cuerrier D, Graham L, Campbell R, Davies P (2009) Profiling of calpain activity with a series of FRET-based substrates. Biochim Biophys Acta Proteins Proteomics 1794 (10):1505–1509
32. Kikuchi H, Imajoh-Ohmi S, Kanegasaki S (1993) Novel antibodies specific for proteolyzed forms of protein kinase C: production of anti-peptide antibodies available for in situ analysis of intracellular limited proteolysis. Biochim Biophys Acta Protein Struct Mol Enzymol 1162(1–2):171–176
33. Kimura Y, Saya H, Nakao M (2000) Calpain-dependent proteolysis of NF2 protein: involvement in schwannomas and meningiomas. Neuropathology 20(3):153–160
34. Lanckriet G, De Bie T, Cristianini N, Jordan M, Noble W (2004) A statistical framework for genomic data fusion. Bioinformatics 20(16):2626–2635
35. Liu J, Liu M, Wang K (2008) Calpain in the CNS: from synaptic function to neurotoxicity. Sci. STKE 1(14)

36. Moldoveanu T, Hosfield C, Lim D, Jia Z, Davies P (2003) Calpain silencing by a reversible intrinsic mechanism. Nat Struct Mol Biol 10(5):371–378
37. Moldoveanu T, Gehring K, Green D (2008) Concerted multi-pronged attack by calpastatin to occlude the catalytic cleft of heterodimeric calpains. Nature 456(7220):404–408
38. Molinari M, Anagli J, Carafoli E (1995) PEST sequences do not influence substrate susceptibility to calpain proteolysis. J Biol Chem 270 (5):2032
39. Ono Y, Shimada H, Sorimachi H, Richard I, Saido T, Beckmann J, Ishiura S, Suzuki K (1998) Functional defects of a muscle-specific calpain, p94, caused by mutations associated with limb-girdle muscular dystrophy type 2A. J Biol Chem 273(27):17073
40. Ono Y, Kakinuma K, Torii F, Irie A, Nakagawa K, Labeit S, Abe K, Suzuki K, Sorimachi H (2004) Possible regulation of the conventional calpain system by skeletal muscle-specific calpain, p94/calpain 3. J Biol Chem 279(4):2761
41. Qian N, Sejnowski T (1988) Predicting the secondary structure of globular proteins using neural network models. J Mol Biol 202 (4):865–884
42. Richard I, Broux O, Allamand V, Fougerousse F, Chiannilkulchai N, Bourg N, Brenguier L, Devaud C, Pasturaud P, Roudaut C *et al* (1995) Mutations in the proteolytic enzyme calpain 3 cause limb-girdle muscular dystrophy type 2A. Cell 81(1):27–40
43. Rögnvaldsson T, You L (2004) Why neural networks should not be used for HIV-1 protease cleavage site prediction. Bioinformatics 20 (11):1702–1709
44. Rolius R, Antoniou C, Nazarova L, Kim S, Cobb G, Gala P, Rajaram P, Li Q, Fung L (2010) Inhibition of calpain but not caspase activity by spectrin fragments. Cell Mol Biol Lett 15(3):395–405
45. Saido T, Suzuki H, Yamazaki H, Tanoue K, Suzuki K (1993) In situ capture of mu-calpain activation in platelets. J Biol Chem 268 (10):7422
46. Saido T, Yokota M, Nagao S, Yamaura I, Tani E, Tsuchiya T, Suzuki K, Kawashima S (1993) Spatial resolution of fodrin proteolysis in postischemic brain. J Biol Chem 268 (33):25239
47. Saido T, Sorimachi H, Suzuki K (1994) Calpain: new perspectives in molecular diversity and physiological-pathological involvement. FASEB J 8(11):814
48. Sakai K, Akanuma H, Imahori K, Kawashima S (1987) A unique specificity of a calcium activated neutral protease indicated in histone hydrolysis. J Biochem 101(4):911
49. Sasaki T, Kikuchi T, Yumoto N, Yoshimura N, Murachi T (1984) Comparative specificity and kinetic studies on porcine calpain I and calpain II with naturally occurring peptides and synthetic fluorogenic substrates. J Biol Chem 259 (20):12489
50. Shen H, Chou K (2008) Hivcleave: a web-server for predicting human immunodeficiency virus protease cleavage sites in proteins. Anal Biochem 375(2):388–390
51. Song J, Tan H, Shen H, Mahmood K, Boyd S, Webb G, Akutsu T, Whisstock J (2010) Cascleave: towards more accurate prediction of caspase substrate cleavage sites. Bioinformatics 26(6):752
52. Sonnenburg S, Rätsch G, Schäfer C, Schölkopf B (2006) Large scale multiple kernel learning. J Mach Learn Res 7:1565
53. Sonnhammer EL, Eddy SR, Birney E, Bateman A, Durbin R (1998) Pfam: multiple sequence alignments and HMM-profiles of protein domains. Nucleic Acids Res 26 (1):320–322
54. Sorimachi H, Ishiura S, Suzuki K (1997) Structure and physiological function of calpains. Biochem J 328(Pt 3):721
55. Stabach P, Cianci C, Glantz S, Zhang Z, Morrow J (1997) Site-directed mutagenesis of α II spectrin at codon 1175 modulates its μ-calpain susceptibility. Biochemistry 36(1):57–65
56. Suzuki K, Hata S, Kawabata Y, Sorimachi H (2004) Structure, activation, and biology of calpain. Diabetes 53(Suppl 1):S12
57. Thompson T, Chou K, Zheng C (1995) Neural network prediction of the HIV-1 protease cleavage sites. J Theor Biol 177(4):369–379
58. Tompa P, Buzder-Lantos P, Tantos A, Farkas A, Szilágyi A, Bánóczi Z, Hudecz F, Friedrich P (2004) On the sequential determinants of calpain cleavage. J Biol Chem 279 (20):20775
59. Von Heijne G (1986) A new method for predicting signal sequence cleavage sites. Nucleic Acids Res 14(11):4683
60. Wang K (2000) Calpain and caspase: can you tell the difference? Trends Neurosci 23 (1):20–26
61. Wang N, Chen W, Linsel-Nitschke P, Martinez L, Agerholm-Larsen B, Silver D, Tall A (2003) A PEST sequence in ABCA1 regulates degradation by calpain protease and stabilization of ABCA1 by apoA-I. J Clin Investig 111(1):99–107

62. Wee L, Tan T, Ranganathan S (2006) Svm-based prediction of caspase substrate cleavage sites. BMC Bioinf 7(Suppl 5):S14
63. Wee L, Tan T, Ranganathan S (2007) Casvm: web server for SVM-based prediction of caspase substrates cleavage sites. Bioinformatics 23(23):3241
64. Wells A, Huttenlocher A, Lauffenburger D (2005) Calpain proteases in cell adhesion and motility. Int Rev Cytol 245:1–16
65. Yang Z (2005) Prediction of caspase cleavage sites using Bayesian bio-basis function neural networks. Bioinformatics 21(9):1831
66. Yang Z, Chou K (2004) Bio-support vector machines for computational proteomics. Bioinformatics 20(5):735
67. Zhang Z, Biesiadecki B, Jin J (2006) Selective deletion of the NH2-terminal variable region of cardiac troponin T in ischemia reperfusion by myofibril-associated μ-calpain cleavage. Biochemistry 45(38):11681–11694

Chapter 12

Identification of Calpain-Activated Protein Functions

Maria del Carmen Lafita-Navarro and Maralice Conacci-Sorrell

Abstract

As opposed to proteasome-mediated proteolysis that leads to protein degradation, calpain proteases carry out limited proteolytic cleavages of their substrates. The cleavage of some substrates can produce active fragments that perform functions that are different from those performed by the full-length proteins. Therefore, cleavage by calpains can operate as a posttranslational modification and increase the functional diversity of target proteins. Nevertheless, activation of protein function by calpain cleavage is still an understudied area in molecular biology. Identifying and functionally characterizing by products generated by calpain cleavage could lead to the discovery of biomarkers and the identification of novel drug targets for the treatment of human diseases. This chapter contains a workflow designed to experimentally characterize novel calpain substrates, including identification of potential calpain targets via Western blotting, characterization of calpain cleavage sites, and the study of cellular functions played by such cleaved products. We will employ MYC as an example for these experiments.

Key words Calpains, Cleavage, Calcium, Protein activation, Degradation, MYC, MYC-nick, Posttranslational modification

1 Introduction

Calpains are a family of evolutionary conserved cysteine proteases composed of 15 genes in humans. The conventional calpain family members, calpain-1 (CAPN1) and calpain-2 (CAPN2), previously referred to as μ and m calpains, respectively, are ubiquitously expressed and activated by micro- to millimolar amounts of calcium [1]. Other calpain family members are predominantly tissue specific: for example, *CAPN3* is expressed in skeletal muscle, whereas *CAPN8* and *CAPN9* are expressed in the gut [2]. The protocols described in this chapter will focus on methods designed to study proteolytic targets of the conventional calpains 1 and 2. However, these methods can be optimized for studies involving other calpains or proteases.

Examples of protein activation by conventional calpains include the neuronal receptor mGluR1a, the cell adhesion molecule ß-catenin, and the transcription factor MYC [2]. The cleavage of

Jeannette S. Messer (ed.), *Calpain: Methods and Protocols*, Methods in Molecular Biology, vol. 1915,
https://doi.org/10.1007/978-1-4939-8988-1_12,

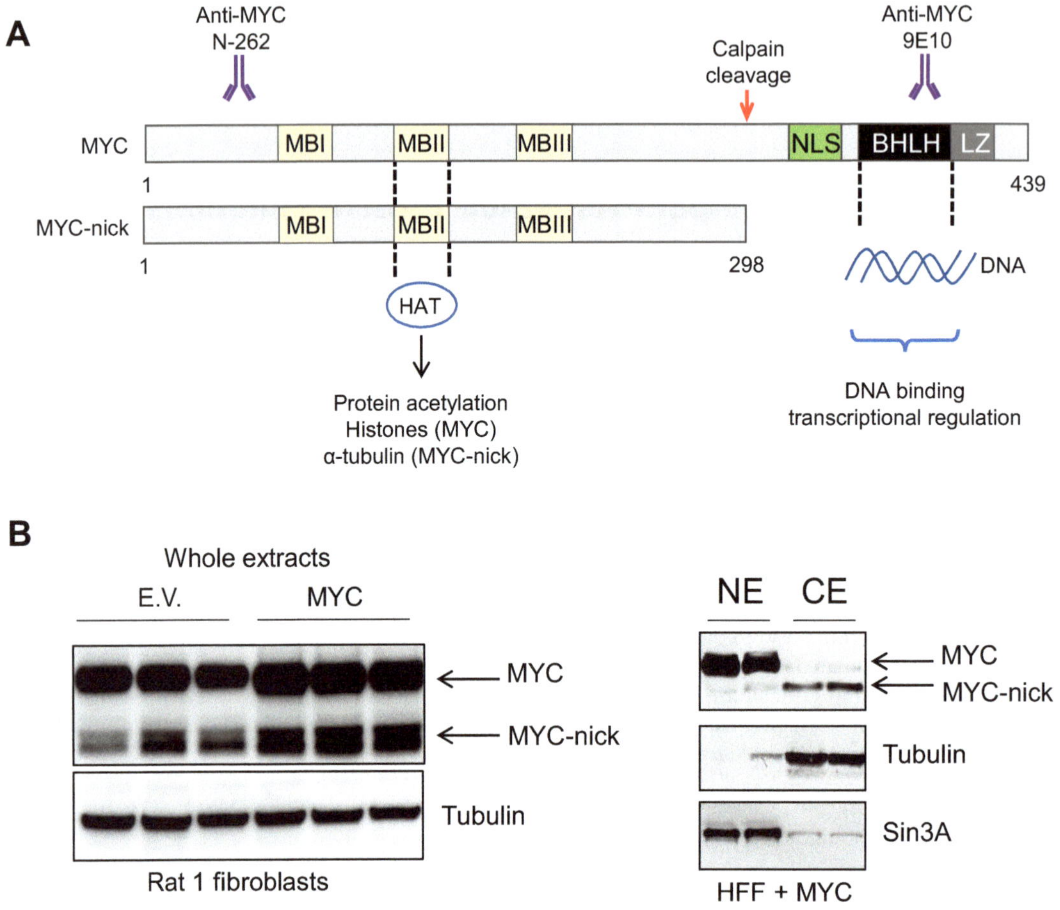

Fig. 1 The cleavage product of MYC, MYC-nick, is localized in the cytoplasm. (**a**) Schematic representation of MYC and MYC-nick proteins. MB: MYC boxes, BHLH: basic helix-loop-helix, LZ: leucine zipper, and HAT: histone acetyltransferases. (**b**) MYC-nick is present in the cytoplasmic fraction, while full-length MYC is present in the nuclear fraction of cells. Whole lysates of TGR1 rat fibroblasts showing MYC and MYC-nick in triplicates (left). Nuclear extracts (NE) or cytoplasmic extracts (CE) of HFF human fibroblasts overexpressing MYC showing full-length MYC and MYC-nick localization in duplicates (right)

mGluR1a by calpains activates its function leading to increased excitotoxicity in primary rat cortical neurons [3]. Targeting of ß-catenin by calpains leads to cleavage of the N-terminal domain and stabilization of the soluble pool of this transcription factor, thereby facilitating nuclear translocation and transcriptional activation [4]. The cleavage of MYC by calpains, discovered by our group, converts this 55 KDa nuclear transcription factor into a cytosolic protein of ~42 KDa named MYC-nick [5–7] (Fig. 1a). While MYC is a pro-proliferative transcription factor, MYC-nick promotes survival and migration. We discovered that MYC-nick recruits acetyltransferases and promotes acetylation of cytoplasmic proteins, which in turn mediates migration and chemotherapy resistance of cancer cells [6, 8, 9].

Due to their profound physiological significance, deregulation of calpains can lead to multiple pathologies. For example, mutations in *CAPN3* cause limb girdle muscular dystrophy type 2A (LGMD2A). These mutations block the proteolytic activity of CAPN3 and trigger death of muscle cells [10]. Moreover, mutations in the CAPN10 gene are correlated with type 2 or non-insulin-dependent diabetes mellitus (NIDDM) [11]. This polymorphism most likely decreases the activity of CAPN10, leading to insulin resistance [10]. In addition, calpains are thought to be involved in neurodegenerative and cardiovascular disorders as well as in cancer [12]. Thus, identifying calpain substrates and defining their biological roles have the potential of providing new insights in the treatment of multiple pathologies.

Because calpains do not have a clear consensus cleavage motif [12–14], it is essential to combine bioinformatics approaches with experimental validation (proposed here) in order to define cleavage sites on potential calpain substrates [15]. Therefore, the goal of this chapter is to provide comprehensive methods for the characterization of potential calpain substrates and the study of their function. We will describe the following approaches: (1) identification and subcellular localization of potential calpain-derived by-products by Western blot, (2) validation of calpain substrates by in vitro cleavage with recombinant or in vitro synthesized protein, (3) generation of cleavage-deficient mutants, and (4) study of the cellular functions performed by calpain-derived by products.

2 Materials

Prepare all solutions using deionized water.

2.1 Nuclear and Cytoplasmic Fractionation of Cultured Cells

1. Lysis Buffer A: 10 mM HEPES pH 7.9, 10 mM KCl, 1.5 mM $MgCl_2$, 0.5% NP40. This is a hypotonic buffer.
2. Lysis Buffer B: 20 mM Hepes-KOH pH 7, 400 mM NaCl, 1.5 mM $MgCl_2$, 0.2 mM EDTA, 15% glycerol. This is a hypertonic buffer.
3. Cold 1× PBS.
4. Sonicator (i.e., Bioruptor® Plus sonication device, Diagenode).
5. Refrigerated bench centrifuge or microcentrifuge.

2.2 Western Blot

1. 5× Laemmli sample buffer: 100 mM Tris–HCl pH 6.8, 8% 2 mercaptocthanol, 4% SDS, 0.1% bromophenol blue, 20% glycerol.
2. Polyacrylamide-SDS (SDS-PAGE) gel with an appropriate acrylamide percentage. We recommend 4–12% gradient gels (*see* **Note 1**).

3. Protein molecular weight marker (i.e., Precision Plus Protein™ Dual Color Standards, Bio-Rad).
4. MES-SDS running buffer: 50 mM MES (2-[N-morpholino] ethanesulfonic acid), 50 mM Tris base, 1 mM EDTA, 0.1% (w/v) SDS, pH 7.3.
5. Transfer buffer: 25 mM Tris–HCl pH 7.6, 192 mM glycine, 20% methanol, 0.03% SDS.
6. Nitrocellulose membrane (i.e., Amersham Protran Supported 0.45 NC, GE Healthcare Life Sciences).
7. TBS-T: 20 mM Tris pH 7.5, 150 mM NaCl, 0.05% Tween-20.
8. Blocking buffer: TBS-T + 5% BSA.
9. Primary and secondary antibodies incubation solution: TBS-T + 1% BSA.
10. ECL detection system (i.e., Pierce™ ECL Western blotting Substrate, ThermoFisher, and ChemiDoc™ MP Imaging System, Bio-Rad).

2.3 In Vitro Transcribed and Translated (IVT) Protein

1. cDNA of interest cloned under the control of SP6 promoter for in vitro translation using wheat germ system. Recommended pCS2+, which contains an SP6 promoter for IVT and a CMV promoter for mammalian expression.
2. TNT® Coupled Wheat Germ Extract System (*see* **Note 2**).
3. S^{35}-labelled methionine (i.e., MP biomedicals).

2.4 Recombinant Protein

Prepare proteins in bacterial expression systems. Alternatively purchase recombinant proteins from a commercial supplier.

2.5 In Vitro Cleavage Assay

1. Knockout cells for the gene of interest.
2. Lysis Buffer A without protease inhibitors. *Do not* add protease inhibitors if lysates are going to be used for cleavage experiments.
3. Buffer G: 30 mM Tris pH 7.0, 100 mM NaCl, 10 mM $CaCl_2$, 2 mM $MgCl_2$, 50 μM ATP, 1 mM DTT, 5% glycerol.
4. Calpain inhibitors: Calpastatin, calpain inhibitor XII, MG132 (*see* **Note 3**).
5. 4–12% gradient gels as recommended for Western blot.
6. Gel drier (Bio-Rad)
7. X-ray film and cassette for radio autography.

2.6 Mammalian Cell Transfection

1. cDNA of interest should be cloned in a mammalian expression vector (i.e., pCS2+).
2. Mammalian transfection reagent (i.e., Lipofectamine 2000).

3 Methods

3.1 Identification and Subcellular Localization of Potential Calpain-Derived By Products by Western Blot

3.1.1 Nuclear-Cytoplasmic Fractionation

1. Seed three million cells in a 10 cm dish and allow them to grow for 36–48 h.
2. Wash cells with cold 1× PBS twice. Harvest cells and transfer to a 1.5 ml tube.
3. Spin cells at 1500 rpm (211 × *g*) for 5 min. Discard supernatant.
4. Add 300 μl *Lysis Buffer A* plus protease inhibitors (*see* **Note 4**) to the cells, and pipet up and down several times on ice. Keep on ice for 30 min, vortexing briefly every 5 min.
5. Spin cells at 1500 rpm (211 × *g*) for 5 min at 4 °C. Collect supernatant, and transfer to another 1.5 ml tube; this contains the cytoplasmic fraction.
6. Add 100 μl *Lysis Buffer B* plus proteases inhibitors (*see* **Note 4**) to the pellet, and pipet up and down several times on ice. Keep it on ice for 15 min, vortexing briefly every 5 min.
7. Sonicate lysates for 5–10 min at 4 °C (i.e., high power setting for 5–10 cycles, 30 s on and 30 s off in the Bioruptor® Plus sonication device, Diagenode).
8. Centrifuge cells at maximum speed for 15 min at 4 °C. Collect supernatant, and transfer to another 1.5 ml tube; this contains the nuclear fraction.
9. Determine protein concentration (i.e., using the Pierce™ BCA protein Assay kit), and use 20 μg protein for Western blot (described below).

3.1.2 Detection of Cleaved Products by Western Blot

1. Dilute protein lysates to a final concentration of 1 μg/μl in the same buffer (A or B). Add Laemmli sample buffer to a final concentration of 1× (5 μl of 5× Laemmli to a final volume of 25 μl).
2. Denature proteins by heating at 95 °C for 5–10 min. Vortex the samples and briefly spin the tubes in a microcentrifuge.
3. Load samples and a protein marker in a polyacrylamide-SDS (SDS-PAGE) gel (*see* **Note 1**).
4. Separate proteins by electrophoresis using MES-SDS running buffer at 160 V constant voltage.
5. After electrophoresis, transfer the proteins to a nitrocellulose membrane using the semidry or wet electrophoresis method as desired (i.e., wet transfer electrophoresis at 400 mA constant amperage for 1 h or semidry transfer electrophoresis using the Trans-Blot® Turbo™ Transfer System, Bio-Rad, at 1.3 A and 25 V for 15 min).
6. Do the following **steps 7–12** in agitation.

7. After protein transfer, block membrane with blocking solution for 1 h at RT to avoid non-specific antibody bindings.
8. Wash with TBS-T for 10 min. Repeat twice.
9. Incubate membrane with primary antibody following manufacturer's instructions (i.e., 1:1000 dilution in TBS-T buffer o/n at 4 °C) (*see* **Note 5**).
10. Wash 3× with TBS-T buffer for 10 min.
11. Incubate membrane with HRP-conjugated secondary antibody of interest following manufacturer's instructions (i.e., 1:10,000 dilution in TBS-T buffer 1 h at RT) (*see* **Note 5**).
12. After incubation, wash membranes with TBS-T for 10 min. Repeat three times.
13. Develop membrane using the ECL system following manufacturer's instructions. In the example used here, full-length MYC is localized in the nuclear fraction as opposed to the MYC-nick fragment that is present in the cytoplasmic fraction (Fig. 1b).

3.2 Validation of Calpain Substrates by In Vitro Cleavage with Recombinant or In Vitro Synthesized Protein

3.2.1 Recombinant Protein

1. Purchase recombinant protein of interest from a commercial supplier.
2. Use 0.1 μg of purified recombinant protein for the cleavage assay.

3.2.2 In Vitro Transcription (IVT) and Translation of Target Protein

1. Synthesize the protein using commercially available in vitro translation systems following the manufacturer's instructions (i.e., Promega TNT wheat germ IVT kit (*see* **Note 3**). Radiolabelled S^{35} methionine can be used to detect the in vitro synthesized protein by film exposure without using specific antibodies. Otherwise Western blot should be used to detect the protein of interest.

Optimized reaction	1 μl DNA (diluted at 1 μl/μl)
	1 μl methionine or S^{35}-labelled methionine
	20 μl SP6 wheat germ
	3 μl water
	Incubate 1 h and 30 min at 30 °C
	IVT proteins can be stored at −80 for up to 4 months

2. To check the levels of synthesized protein, run 3 μl of each sample in an acrylamide gel as used for Western blot.
3. Dry gel overnight using a gel drier.
4. Place the dried gel in a cassette with an X-ray film. Expose it overnight at – 80 °C.
5. Develop film.
6. Use 1–2 μl of IVT protein for the cleavage assay.

3.2.3 *In Vitro Cleavage Assay*

Do all the following steps *on ice* unless otherwise indicated. Volumes and quantities are related to one sample. Scale accordingly for a greater number of samples.

Cleavage Reaction Using IVT Protein

1. Use 10 μg of cytoplasmic extract lysate (at a final concentration 1 μg/μl) of cell of interest (calpain source). Do not add protease inhibitors for the cleavage assay unless otherwise indicated.
2. Mix the 10 μl of cytoplasmic extract lysates with 10 μl of buffer G.
3. Add 1 μl of IVT-synthesized protein to the mix, and incubate for different time points at 30 °C (i.e., 30 min, 1 h, 2 h, 4 h).
4. After incubation, add Laemmli sample buffer to a final concentration of 1×.
5. Run samples in an acrylamide gel, dry the gel, and expose to an X-ray film at −80 °C o/n.
6. Additional samples with calpain inhibitors or calcium chelators (i.e., Bapta or EGTA) should be used as controls to determine whether protein cleavage is mediated by calpains. Some examples of calpain inhibitors are calpeptin, calpain inhibitors VI and XII, and MG132 which targets cysteine proteases. In the case of MYC, the calpain inhibitor XII, calpain inhibitor VI, calpeptin, or the chelators Bapta and EGTA (Fig. 2b in the current chapter and Fig. 3a-b-d-e in Sorrell et al. [6], respectively) demonstrate that the cytoplasmic MYC-nick is generated by the activity of calpains. Inhibitors of other proteases such as lactacystein or epoxomicin do not affect the conversion of MYC into MYC-nick confirming that it is a specific effect of calpains on MYC (Fig. 2h in Sorrell et al. [6]).

Cleavage Reaction Using Purified Recombinant Proteins

1. Use 10 μg of cytoplasmic extract lysates (to a final concentration 1 μg/μl) of the cells that are *knocked out* for the gene of interest (*see* **Note 6**). Do not add protease inhibitors for the cleavage assay unless otherwise indicated.
2. Mix the 10 μl of cytoplasmic extract lysates with 10 μl of buffer G.

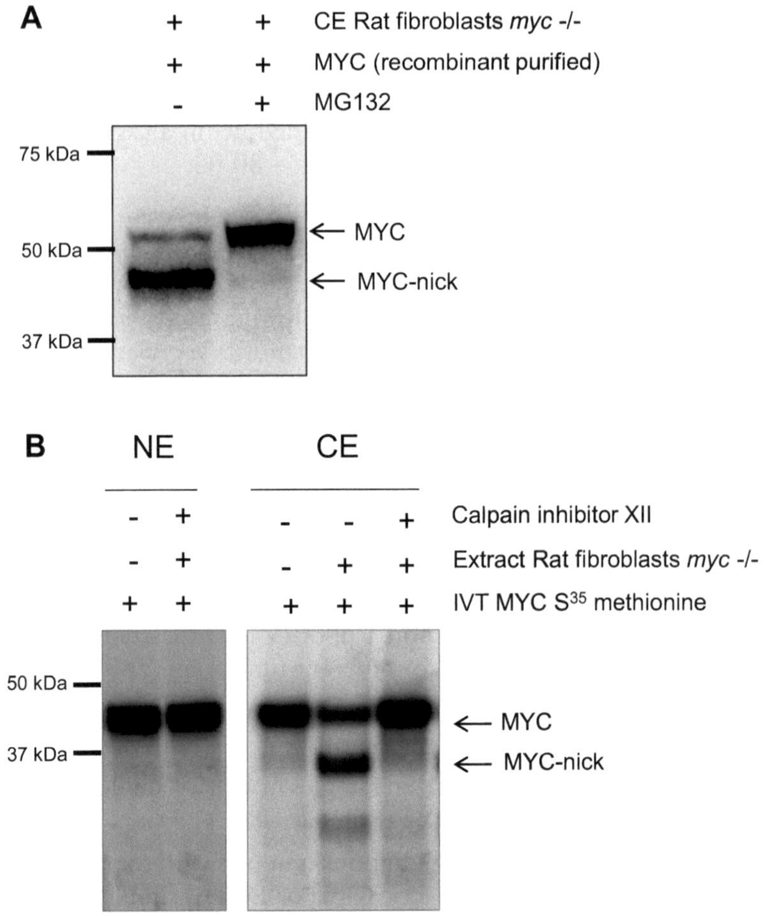

Fig. 2 In vitro cleavage of MYC produces MYC-nick. (**a**) Recombinant full-length MYC is cleaved by calpains generating MYC-nick. The addition of MG132 impairs cleavage. (**b**) S^{35}-labelled in vitro (IVT) synthesized MYC is cleaved by calpains generating MYC-nick. The addition of calpain inhibitor XII prevents the conversion of MYC into MYC-nick

3. Add 0.1 μg of purified recombinant protein to the mix, and incubate for different time points at 30 °C (i.e., 30 min, 1 h, 2 h, 4 h).
4. After incubation, add Laemmli sample buffer to a final concentration of 1×, and boil for 5 min.
5. Run samples in an acrylamide gel and proceed with Western blot. In our example, MYC recombinant protein is cleaved into MYC-nick when incubated with cytoplasmic extracts of rat1 fibroblasts myc−/−. This cleavage is abolished when protease inhibitors are added to the reaction (Fig. 2a).

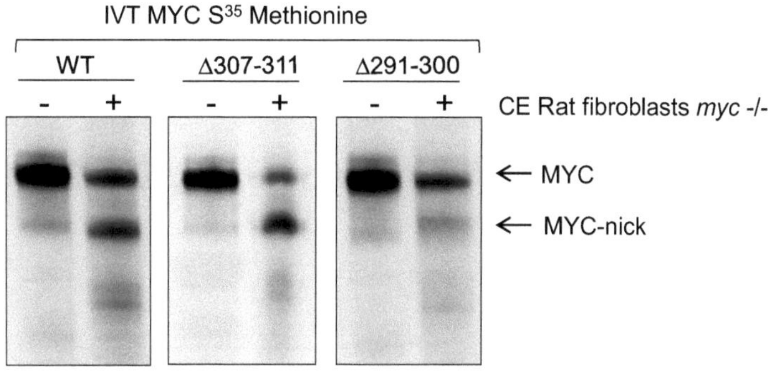

Fig. 3 Cleavage-resistant mutants of MYC cannot produce MYC-nick. S^{35}-labelled in vitro (IVT) synthesized MYC mutants lacking regions from 307 to 311 (Δ307–311) or from 291 to 300 (Δ291–300) amino acids. The Δ291–300 mutant is not cleaved by calpains, indicating that this region is necessary for cleavage

3.3 Generation of Cleavage-Deficient Mutants

1. Clone the cDNA sequences with different point or deletion mutations of interest into a mammalian expression vector (*see* **Note 7**).
2. Express the cDNA mutants in mammalian cells, and detect the fragment by Western blot using specific primary antibodies. In our studies, MYC mutants lacking domains throughout the amino acid sequence were generated (Fig. 3i in Sorrell et al. [6]). Using these mutants, we determined that the region comprising amino acids 298 to 300 in the full-length MYC protein is necessary for calpain cleavage (Fig. 3).

3.4 Determine the Cellular Functions Performed by Calpain-Derived By Products

1. Generate mammalian expression vectors encoding the cleaved fragment (i.e., cDNA sequence corresponding to the MYC-nick amino acid sequence).
2. Transfect the expression vector into mammalian cells (as well as an empty vector as a negative control in another set of cells). Use a cell line where the gene of interest has been knocked out (i.e., rat1 fibroblasts *myc−/−* in our example).
3. Check the expression and localization of the fragment by nuclear-cytoplasmic fractionation and Western blot.
4. Analyze the phenotype of the cells overexpressing the fragment of interest (cell morphology, proliferation, death, viability in different conditions, etc.). Overexpression of MYC-nick in rat1 *myc−/−* fibroblast cells promotes microtubule and actin filament lengthening (Fig. 4). This indicates that the MYC-nick fragment has a function that is independent of the full-length MYC protein [6, 8, 9]. For a comprehensive study, it is key to compare the phenotypes produced by full-length protein, cleavage-resistant mutants, and cleaved fragments.

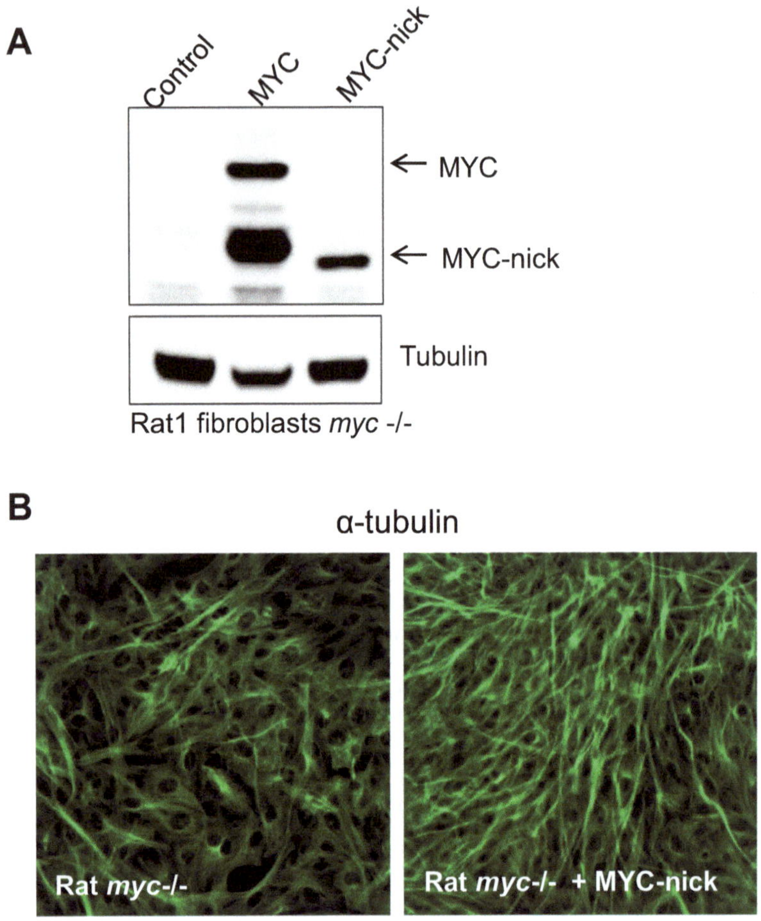

Fig. 4 MYC-nick expression alters the morphology of *myc*−/− fibroblasts. (**a**) Overexpression of MYC-nick in rat1 fibroblasts where the *myc* gene was deleted (*myc*−/−). Full-length MYC is used as a control. (**b**) MYC-nick overexpressing cells display lengthened microtubules

4 Notes

1. Use acrylamide percentage according to the molecular weight of the protein of interest to study. Gradient gels from 4 to 12% work for most proteins.
2. In vitro protein translation kits containing reticulocyte-based methodology have high protease activity [17] that can interfere with the in vitro cleavage assay described here. As such, we recommend only using wheat germ systems.
3. MG132 is known to be a proteasome inhibitor. However, MG132 also targets other cysteine proteases like calpains [16]. Therefore, MG132 can be used to inhibit calpain activity.

4. To prevent inhibition of calpain activity, do not use protease inhibitors if cell extracts are going to be used for the in vitro cleavage assay.
5. Cytoplasmic and nuclear protein markers should be used to ensure that the extracted fractions are clean in cytoplasmic or nuclear proteins, respectively. Examples of cytoplasmic and nuclear loading markers are tubulin and histone 3, respectively.
6. Nuclear extracts lack calpain activity and can be used as a negative control.
7. Antibodies targeted against different regions of the protein of interest can be used to help determine which region of the protein is cleaved. This can facilitate the generation of mutants. The N-262 antibody (Santa Cruz Biotechnology) that targets the amino terminal region of MYC recognizes full-length MYC and MYC-nick. However, the 9E10 antibody (Santa Cruz Biotechnology) that is targeted against the C-terminal region of MYC recognizes full-length MYC but not MYC-nick (Fig. 1). In addition, tagging the C-terminal and N-terminal regions of the protein (e.g., with a flag-tag) can also facilitate the detection of the new products by Western blot.

References

1. Goll DE, Thompson VF, Li H, Wei W, Cong J (2003) The calpain system. Physiol Rev 83 (3):731–801. https://doi.org/10.1152/physrev.00029.2002
2. Ono Y, Sorimachi H (2012) Calpains: an elaborate proteolytic system. Biochim Biophys Acta 1824(1):224–236. https://doi.org/10.1016/j.bbapap.2011.08.005
3. Xu W, Wong TP, Chery N, Gaertner T, Wang YT, Baudry M (2007) Calpain-mediated mGluR1alpha truncation: a key step in excitotoxicity. Neuron 53(3):399–412. https://doi.org/10.1016/j.neuron.2006.12.020
4. Abe K, Takeichi M (2007) NMDA-receptor activation induces calpain-mediated beta-catenin cleavages for triggering gene expression. Neuron 53(3):387–397. https://doi.org/10.1016/j.neuron.2007.01.016
5. Vennstrom B, Sheiness D, Zabielski J, Bishop JM (1982) Isolation and characterization of c-myc, a cellular homolog of the oncogene (v-myc) of avian myelocytomatosis virus strain 29. J Virol 42(3):773–779
6. Conacci-Sorrell M, Ngouenet C, Eisenman RN (2010) Myc-nick: a cytoplasmic cleavage product of Myc that promotes alpha-tubulin acetylation and cell differentiation. Cell 142 (3):480–493. https://doi.org/10.1016/j.cell.2010.06.037
7. Eisenman RN (2000) The max network: coordinated transcriptional regulation of cell growth and proliferation. Harvey Lect 96:1–32
8. Conacci-Sorrell M, Ngouenet C, Anderson S, Brabletz T, Eisenman RN (2014) Stress-induced cleavage of Myc promotes cancer cell survival. Genes Dev 28(7):689–707. https://doi.org/10.1101/gad.231894.113
9. Anderson S, Poudel KR, Roh-Johnson M, Brabletz T, Yu M, Borenstein-Auerbach N, Grady WN, Bai J, Moens CB, Eisenman RN, Conacci-Sorrell M (2016) MYC-nick promotes cell migration by inducing fascin expression and Cdc42 activation. Proc Natl Acad Sci U S A 113(37):E5481–E5490. https://doi.org/10.1073/pnas.1610994113
10. Branca D (2004) Calpain-related diseases. Biochem Biophys Res Commun 322 (4):1098–1104. https://doi.org/10.1016/j.bbrc.2004.07.126
11. Horikawa Y, Oda N, Cox NJ, Li X, Orho-Melander M, Hara M, Hinokio Y, Lindner TH, Mashima H, Schwarz PE, del Bosque-Plata L, Horikawa Y, Oda Y, Yoshiuchi I, Colilla S, Polonsky KS, Wei S, Concannon P, Iwasaki N, Schulze J, Baier LJ, Bogardus C, Groop L, Boerwinkle E, Hanis CL, Bell GI (2000) Genetic variation in the gene encoding calpain-10 is associated with type 2 diabetes

mellitus. Nat Genet 26(2):163–175. https://doi.org/10.1038/79876

12. Ono Y, Saido TC, Sorimachi H (2016) Calpain research for drug discovery: challenges and potential. Nat Rev Drug Discov 15 (12):854–876. https://doi.org/10.1038/nrd.2016.212
13. Tompa P, Buzder-Lantos P, Tantos A, Farkas A, Szilagyi A, Banoczi Z, Hudecz F, Friedrich P (2004) On the sequential determinants of calpain cleavage. J Biol Chem 279 (20):20775–20785. https://doi.org/10.1074/jbc.M313873200
14. Lynch DR, Gleichman AJ (2007) Picking up the pieces: the roles of functional remnants of calpain-mediated proteolysis. Neuron 53 (3):317–319. https://doi.org/10.1016/j.neuron.2007.01.014
15. Collins GA, Goldberg AL (2017) The logic of the 26S proteasome. Cell 169(5):792–806. https://doi.org/10.1016/j.cell.2017.04.023
16. Lee DH, Goldberg AL (1998) Proteasome inhibitors: valuable new tools for cell biologists. Trends Cell Biol 8(10):397–403
17. Mumford RA, Pickett CB, Zimmerman M, Strauss AW (1981) Protease activities present in wheat germ and rabbit reticulocyte lysates. Biochem Biophys Res Commun 103 (2):565–572

Part IV

Interrogation of Calpastatin, The Endogenous Calpain Inhibitor

Chapter 13

Production and Purification of Recombinant Calpastatin

Roberta De Tullio and Monica Averna

Abstract

The production of recombinant calpastatin in *E. coli* has become an efficient tool to obtain discrete amounts of a specific calpastatin species that can be present concomitantly with other calpastatin fragments/forms in the same tissue or cell type in a given condition. Indeed, at present, it is still difficult to distinguish the various calpastatin species for several reasons among which: calpastatins differ only at the N-terminus, can undergo calpain-dependent cleavage generating discrete fragments, and show anomalous electrophoretic mobility. Another benefit of using recombinant calpastatin is that, as the wild-type forms, it is heat resistant and thus can be efficiently isolated taking advantage of a simple quick purification step. Finally, the lack of posttranslational modifications makes recombinant calpastatin species particularly suitable for studying in vitro the biochemical features of specific parts of the inhibitor that following controlled posttranslational modifications change their functional interaction with calpain. In this chapter, we describe, starting from the mRNA sequence, how to produce rat calpastatin Type I in *E. coli*. We use routinely the same method, with minor modifications, for the production of other calpastatin species deriving from different tissues or organisms and calpastatin constructs having only specific domains. The possibility to obtain large amounts of a single calpain inhibitor form is a great advantage for studying the calpain/calpastatin system in vitro.

Key words Recombinant calpastatin, Recombinant proteins, *Escherichia coli*, Calpastatin assay

1 Introduction

Calpastatin, the natural inhibitor of calpain, exists in different forms that derive from alternative splicing events involving the first exon translated [1] as well as exons belonging to the regulatory non-inhibitory domain and the inhibitory domains [2–6]. Calpastatin can be considered a "suicide" inhibitor as it can be cleaved by calpain into different species that, before complete degradation, retain the ability to inhibit calpain [7]. Moreover, the cAMP- or Ca^{2+}-dependent phosphorylation of calpastatin in the regulatory domain [8, 9] can determine the intracellular localization and, consequently, the availability of the inhibitor for interaction with calpain. Finally, calpastatin is characterized by anomalous migration in SDS-PAGE [10, 11], and consequently it is hard to distinguish among all these species if they are present concomitantly in a given

Jeannette S. Messer (ed.), *Calpain: Methods and Protocols*, Methods in Molecular Biology, vol. 1915,
https://doi.org/10.1007/978-1-4939-8988-1_13,

sample. Thus, a different approach for characterizing the biochemical role/function of the different calpastatin forms/species is that of using recombinant constructs that are known to be composed of a single species. These calpastatins can be used in in vitro experiments designed to provide information on the role of that calpastatin form in vivo. Here we describe how to produce recombinant rat calpastatin Type I in *E. coli*. The same method, with minor modifications related to the nature of the calpastatin species produced, is routinely used in our laboratory [6, 9].

2 Materials

2.1 RNA Isolation

1. To prepare total RNA from tissues: RNeasy mini kit (Qiagen).
2. To concentrate RNA by precipitation: 70% ethanol v/v prepared with DEPC water.

2.2 cDNA Synthesis

To synthesize cDNA: ThermoScript® RT-PCR system (Invitrogen).

2.3 PCR and TA Cloning of Rat Calpastatin Type I in pGEM-T Vector

1. To obtain Type I calpastatin transcript by PCR, use the following primers: forward primer, Sn expr EcoRI 5′-AA-GAATTC-**ATG**TCCCGGCCCGGCC-3′, and reverse primer Asn expr XhoI 5′-AA-CTCGAG-CAGATACCAAAATGCATAAACTT-**TA**TGTTGC-3′.
2. Platinum® Taq PCR_x DNA Polymerase kit and Taq DNA Polymerase, recombinant (Invitrogen).
3. 6× agarose loading buffer: 30% glycerol, bromophenol blue.
4. TBE 5× (1 L): 54 g TRIS, 27.5 g boric acid, 20 mL 0.5 M EDTA pH 8.0 (final pH should be 8.3).
5. To extract DNA from agarose gel: QIAquick Gel Extraction Kit (Qiagen).
6. For TA cloning and transformation: pGEM-T vector system, LigaFast™ Rapid DNA Ligation System, bacterial strain JM109 (Promega).
7. 6× ligation loading solution: 50 mM EDTA, 1% SDS, 50% glycerol, bromophenol blue.
8. DNA markers: O'GeneRuler 1 kb DNA Ladder, ready-to-use (Thermo Fisher Scientific, SM1163).

2.4 E. coli Transformation

1. LB medium (1 L): dissolve 10 g tryptone, 5 g yeast extract, 10 g NaCl in dH_2O. Adjust pH to 7.0 with NaOH. Sterilize by autoclaving.
2. Ampicillin solution: dissolve 100 mg ampicillin in a 50% ethanol solution and store it at −20 °C.

3. To prepare LB plates with ampicillin (15 plates, 100 mm): dissolve 7.5 g agarose in 500 mL LB medium and sterilize by autoclaving. Allow the medium to cool to about 50 °C and add ampicillin to a final concentration of 100 μg/mL. Immediately pour 30–35 mL medium into each petri dish under a sterile hood. Let the agarose harden. Seal the agarose plates with Parafilm™ and store them at 4 °C for up to 3–4 months.

2.5 Identification of Bacterial Clones Containing the Calpastatin Transcript in pGEM Vector

1. 2× cracking lysis solution: 100 μL of 2 M NaOH, 400 μL of 50% sucrose (p/v), 50 μL 10% SDS (p/v), dH_2O to 1 mL. This solution should be freshly prepared and used within a week.
2. Cracking loading solution: 1.6 μL of 3.7 M KCl and 0.4 μL bromophenol blue.
3. To purify plasmids from transformed *E. coli* clones: QIAprep Spin Miniprep kit (Qiagen).

2.6 Cloning of Calpastatin in the Expression Vector pGEX-6P-1

1. Restriction endonucleases EcoRI and XhoI.
2. To produce recombinant calpastatin in bacteria: GST Gene Fusion System, pGEX-6P-1 vector, bacterial strain BL21 (GE Healthcare Life Sciences).

2.7 Preparation of GST-Calpastatin Fusion Protein

1. Isopropyl-β-D-thiogalactoside (IPTG) (Sigma-Aldrich, I5502) solution: prepare 100 mM IPTG in dH_2O and store it at −20 °C.
2. Leupeptin (Sigma-Aldrich, L2884) solution: prepare 10 mg/mL leupeptin in dH_2O and store it at −20 °C.
3. 4-(2-Aminoethyl) benzenesulfonyl fluoride hydrochloride (AEBSF) (Fluka, Pefabloc® SC, 76307) solution: prepare 0.1 M AEBSF in dH_2O and store it at −20 °C in aliquots.
4. Aprotinin (Sigma-Aldrich, 10820) solution: prepare 10 mg/mL aprotinin in dH_2O and store it at −20 °C.
5. Lysozyme (Sigma-Aldrich, L6876) solution: Freshly prepare 10 mg/mL lysozyme in dH_2O.
6. Protease Inhibitor Cocktail (100×) (Cell Signalling technology, 5871S).
7. Bacteria lysis buffer 3×: 300 mM Tris/HCl pH 8.0, 30 mM EDTA pH 7.0, 450 mM NaCl, 3% Triton X-100.
8. Electrophoresis buffer: 50 mM Tris, 0.38 M glycine, 0.1% p/v SDS.
9. 6× SDS-PAGE loading buffer: 525 μL Tris/HCl 1 M pH 6.8, 0.15 g SDS, 620 μL glycerol, 250 μL 2-mercaptoethanol, 105 μL dH_2O.

2.8 Purification of Recombinant Calpastatin from the Bacterial Cleared Lysate by Affinity Chromatography

1. To purify calpastatin-GST fusion protein: Glutathione-agarose (Sigma-Aldrich, G 4510).
2. To separate calpastatin from the GST-calpastatin fusion protein: PreScission™ protease (GE Healthcare Life Sciences).
3. Buffer A: 100 mM Tris–HCl pH 8.0, 1 mM EDTA pH 8.0, 150 mM NaCl.
4. Buffer B (PreScission protease digestion buffer): 50 mM Tris–HCl pH 7.0, 1 mM EDTA pH 7.0, 150 mM NaCl.
5. Buffer C: buffer B + 1 mM dithiothreitol (DTT) (BioRad, 1610610).
6. Protein markers: Prestained Protein SHARPMASS™ VI (EuroClone).

2.9 Evaluation of Recombinant Calpastatin Activity

1. To concentrate recombinant calpastatin: stirred ultrafiltration cell (Millipore-Amicon, product, 8050 (50 mL)) with YM-30 Ø43 mm ultrafiltration membrane (Millipore-Amicon, 13722) having a molecular weight cutoff of 30,000.
2. Buffer D: 50 mM NaBorate pH 7.5, 0.1 mM EDTA, 0.5 mM 2-mercaptoethanol.

3 Methods

3.1 RNA Isolation

As RNAs are very susceptible to degradation due to the presence of RNase, all the solutions and reagents used during the isolation of total RNA should be handled with gloves and exclusively with RNase-free disposable material. To avoid degradation, it is better to start in the morning and obtain cDNA quickly, about 3 h later.

1. Use about 25–50 mg of starting material depending on the tissue or cell type (approximately 3–5 × 10^6 cells). This can be either freshly collected and immediately used or stored at −80 °C if it will be used later. Here we describe the isolation of total RNA from rat brain. Excise a sample from the frozen tissue (about 20 mg), and after determining the precise amount, proceed rapidly with its disruption into small pieces using a lancet, directly into a suitably sized vessel.
2. Add the lysis buffer supplied by the kit and follow the instructions of the producer. The kit that we reference is based on the purification of RNA by centrifugation on mini spin columns and requires multiple centrifugation steps. All the purification steps are carried out at room temperature and should be performed as quickly as possible.
3. Total RNA is obtained from the spin column following a three times elution with 20 μL DEPC water each (*see* **Note 1**). An aliquot (5 μL) of the sample is used for determining RNA

concentration at 260/280 nm, and the remaining is immediately concentrated by precipitation in ethanol.

4. To carry out RNA precipitation, set the centrifuge at 2–4 °C and let it cool. In the meantime, add 1/10 volume of the RNA sample (5.5 μL) of NaAcetate 3 M pH 5.2 and 3 volumes (180 μL) of ethanol (100% v/v, RNase-free, stored at −20 °C).
5. Mix thoroughly, and to precipitate the RNA, store the sample at −80 °C for at least 15 min (*see* **Note 2**) before proceeding with the next step.
6. Centrifuge the sample at 2 °C, 13,000 × *g* for 15 min (*see* **Note 3**). Immediately remove carefully all the supernatant using a tip or a Pasteur pipette with a capillary tip, from the side of the tube opposite of the RNA precipitate.
7. Add carefully (without disturbing the invisible pellet) 100 μL of ethanol 70% v/v, and centrifuge at 2 °C, 13,000 × *g* for 5 min. Immediately remove (*see* **Note 4**) the supernatant; use a capillary pipette to remove any drops of liquid that adhere to the walls of the tube.
8. Let the tube stand open at room temperature to dry out any residual supernatant (*see* **Note 5**). Dissolve the RNA pellet (which is often invisible) at a final concentration of 0.5 μg/μL in DEPC-treated water.

3.2 cDNA Synthesis

Even if the sequence of the different forms of calpastatin contains about 50% of GC bases with respect to AT, these are often localized in GC-rich regions at the 5′-terminus. For this reason, cDNA is prepared using a thermostable reverse transcriptase that is active at 60 °C.

1. Combine freshly prepared total RNA (5 μg) in a 0.2 mL PCR tube, RNase-free, with oligo$(dT)_{20}$ primer (0.5 μg). Follow the instructions of the supplier, but before starting the RT reaction, to avoid RNA secondary structure, incubate this mixture at 65 °C for 5 min, and then rapidly cool it on ice. After 1 min add the remaining of the master reaction, containing also the ThermoScript® RT, mix thoroughly, and incubate at 60 °C for 1 h (*see* **Note 6**). The synthesized cDNA is then stored at −20 °C.

3.3 PCR and TA Cloning of Rat Calpastatin Type I in pGEM-T Vector

To obtain by PCR the transcript of calpastatin Type I from rat brain (GenBank: BC091239.1), use the gene-specific primers Sn expr EcoRI and Asn expr XhoI. The primers contain restriction sites for EcoRI and XhoI, respectively (*see* **Note 7**).

1. For PCR (25 μl reaction) use the following parameters: cDNA (0.3 μL) is first heated for 3 min at 98 °C in the presence of the

two primers (1 pmol each) and 1.2 μL 10× $PCR_{\times}$ buffer (add water to 12 μL).

2. Rapidly cool the sample on ice for 1 min. Add 13 μL of a mixture containing 50 mM $MgSO_4$ (1 μL), 10× $PCR_{\times}$ buffer (1.3 μL), 10 mM dNTP (0.5 μL), and 1 unit of Platinum® Taq PCR_x DNA polymerase high fidelity.
3. Heat the sample for 2 min at 95 °C to activate the enzyme, and then perform 35 rounds of amplification consisting of 95 °C for 40 s, 55 °C for 30 s, and 72 °C for 2 min. End the PCR procedure with a final elongation step at 72 °C for 7 min (*see* **Note 8**).
4. To check for the presence of the calpastatin amplicon (it should be about 2000 bp long), add to 20 μL of the PCR reaction 4 μL of 6× agarose loading buffer. Save the rest of the PCR sample at −20 °C or on ice. Separate the PCR products by electrophoresis on 1% agarose gel (50 mL) in TBE buffer 1×, containing 20 μL ethidium bromide 0.5 mg/mL. Load a lane with DNA ladder mix for determining the size of the amplicon(s). If the PCR reaction contains only one amplicon having the correct size, then 2–3 μL of it can be directly used for TA cloning and transformation.
5. Alternatively, if the PCR reaction contains multiple amplification bands, it is advisable to extract the band of interest from the agarose gel and use it for TA cloning (*see* **Note 9**). To maximize gel extraction efficiency, recover the selected band from the DNA purification spin column following a three times elution with 10 mM Tris–HCl pH 8.5 (20 μL each). Immediately, concentrate by precipitation in ethanol the DNA contained in the sample (use the procedure described in Subheading 3.1 for RNA precipitation). Dissolve the "invisible" pellet in 5–8 μL of water. To estimate the amount of the band to be cloned, load 1–2 μL of sample on agarose gel, and compare the intensity of the band with a known amount of standard DNA marker.
6. Carry out TA cloning following the instructions of the supplier using a 1:3 molar ratio of vector/insert DNA (*see* **Note 10**).

Before transformation, it is possible to check if ligation has properly occurred by submitting an aliquot of the ligation reaction to agarose electrophoresis. This step is particularly important when the insert of interest is either in low amounts or is long or is GC-rich, and consequently, due to DNA supercoiling, overhang pairing is more difficult.

You need about 20 ng of DNA to clearly detect a single band by agarose electrophoresis. In a standard TA cloning, the ligation reaction (10 μL volume) contains 100–150 ng of pre-ligated

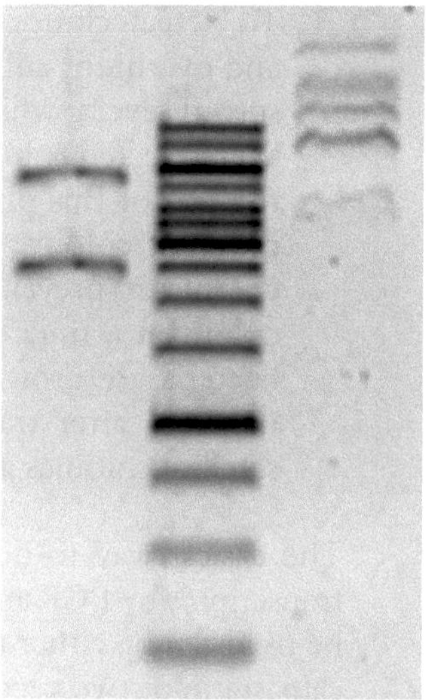

Fig. 1 Ligation of rat Type I calpastatin sequence in pGEX-6-P1 vector. Agarose gel electrophoresis of pGEX-6-P1 vector and Type I calpastatin before and after ligation. *Lane 1*: Type I calpastatin transcript (2300 bp) and pGEX-6-P1 vector (4900 bp) before ligation. *Lane 2*: DNA markers. The bold reference bands correspond to 6000, 3000, and 1000 bp. *Lane 3*: Type I calpastatin transcript and pGEX-6-P1 vector after ligation

DNA; thus we usually load 5–6 μL of the sample as 2 μL will be used for *E. coli* transformation.

Check the efficiency of the ligation by mixing 6 μL of the ligation reaction with 1.5 μL 6× ligation loading solution (*see* **Note 11**), and run it on agarose gel (0.9%) along with the same mixture of the two fragments lacking ligase (not ligated). After ligation several high molecular weight bands should appear. If the two samples look the same, your ligation has not worked (Fig. 1).

3.4 *E. coli* Transformation

Carry out transformation of bacteria by heat shock, following the conditions recommended by the supplier. The features of pGEM vector allow the detection of recombinant clones by color screening which identify blue and white colonies as those containing either "empty" or "full" vectors, respectively. Using the crude PCR product for transformation may lead to the production of a reduced number of white colonies containing the calpastatin transcript due to the preferential incorporation of short PCR products (i.e., primer dimers). On the other hand, the PCR may also contain different calpastatin amplicons resulting from internal splicing events. Therefore, it is advisable to screen multiple white colonies in order to identify clones that contain the calpastatin transcript of interest. The transformation step, if carried out exactly as indicated by the supplier, usually yields a large number of white colonies (*see* **Note 12**).

1. To detect clones containing calpastatin, after transformation and overnight culture (*see* **Note 13**), you should first pick and spread several white colonies on a new LB-ampicillin plate. To keep selected colonies well separate one from another, draw, on the top of this plate many numbered squares, each one for a different colony. Use sterile 10 μL new pipette tips between colonies to prevent contamination, and be sure to pick only one colony at a time. Incubate the plate at 37 °C for 4–5 h until bacteria are grown. It is best to spread white colonies in the morning after transformation and overnight growth, before satellite colonies appear or become large.

3.5 Identification of Bacterial Clones Containing the Calpastatin Transcript in pGEM Vector

The easiest way to detect the colonies containing the calpastatin transcript is by PCR and agarose gel electrophoresis. The primers to be used can be either those reported in Subheading 3.3 specific for calpastatin or two vector specific primers (indicated by the supplier) flanking the cloned insert. In the latter case, PCR will give information on all the clones having Amp resistance and containing recombinant vectors not necessarily bearing the insert of interest. To carry out such PCRs, you can directly use as DNA template the residual material remained in the pipette tip used to spread each white colony as described in Subheading 3.4.

1. Prepare numbered PCR tubes for the colonies to be tested and add 10 μL dH_2O each. To recover genetic material from bacterial cells, rinse the tip used to spread bacteria by pipetting up and down in each PCR tube.
2. To lyse bacteria heat PCR tubes for 5 min at 95 °C and cool them on ice for 3 min. Add 15 μL of a mixture containing the two primers (1 pmol each), 50 mM $MgCl_2$ (1 μL), 10× PCR buffer (2.5 μL), 10 mM dNTP (0.5 μL), and 1 unit of recombinant Taq DNA polymerase. Heat the samples for 1 min 95 °C, and then perform 35 rounds of amplification consisting of 95 °C for 30 s, 55 °C for 30 s, and 72 °C for 2 min. Conclude the PCR procedure by a final elongation step at 72 °C for 7 min.
3. To check for the presence of the correct amplicon (it should be about 2200 bp long excluded the vector bps flanking the cloned insert), add 5 μL of 6× agarose loading buffer to each sample and separate the PCR products by electrophoresis on 1% agarose gel in TBE 1× buffer. Load a lane with DNA ladder mix for determining the size of the amplicon.

If your PCR product contains multiple amplicons having different sizes and you have obtained from the TA cloning many white colonies, or if you want to quickly check transformed colonies for the presence of an insert, it is possible to carry out a first inexpensive screen by performing an agarose electrophoresis directly on bacteria lysates. This method also called "colony cracking" is not as informative as a PCR in terms of insert size, but it will allow you to identify plasmids of different sizes and also spot the "empty" ones (the insert must be at least 400–500 bp long). The following protocol is a modified version of that described in the first edition of *Molecular Cloning* (T. Maniatis, E. F. Frisch and J. Sambrook; Cold Spring Harbor Laboratory Press, 1982).

1. Dispense 50 μL of 10 mM EDTA pH 8.0 in numbered PCR tubes, one for each colony to be tested. Pick a little amount of each colony with a sterile tip from the LB plate containing the white colonies you have already expanded (*see* Subheading 3.4), and rinse the tip pipetting up and down in each tube. Remember to prepare at least one tube containing a blue colony as "empty" control (*see* **Note 14**). You can save 5 μL of this mixture to carry out colony PCR on the "positive" clones resulting from cracking.
2. Add to each sample 50 μL of cracking lysis solution. Close all tubes and mix thoroughly (you can briefly vortex; avoid DNA shearing).
3. Heat the samples at 70 °C for 10 min, add 2 μL of cracking loading solution, mix and leave the tubes on ice for 10 min.
4. Cool down the centrifuge at 2–4 °C, and centrifuge samples at 4 °C, 12,000 × *g* for 10 min. Withdraw carefully, without disturbing the pellet, 40–50 μL of supernatant from each tube, and submit to agarose electrophoresis (agarose gel 0.7% in TBE 1× buffer, 100 V, at least 1-h migration to effectively separate plasmids having different sizes).
5. Following electrophoresis, you will see in each lane (Fig. 2) different bands: an intense higher molecular mass band corresponding to bacterial genomic DNA, almost equal in intensity and migration in all colonies; a lower band

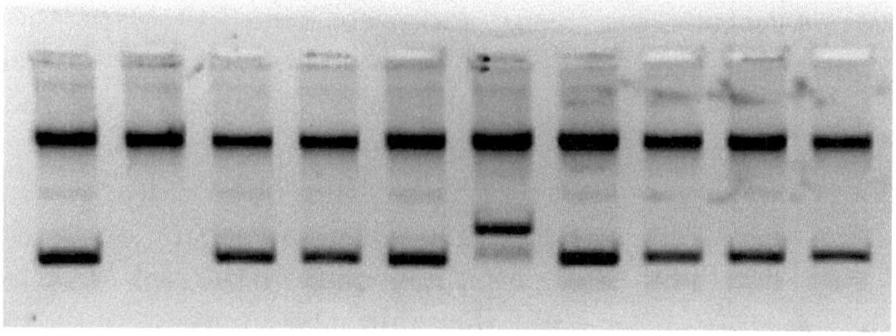

Fig. 2 Screening of bacterial clones transformed with Type I calpastatin in pGEX-6-P1. Transformed bacterial clones grown overnight at 37 °C in Amp-containing plates have been submitted to colony cracking for assessing if they contain the recombinant plasmid. Samples of each colony have been loaded onto an agarose gel. Empty plasmids migrate faster than those containing recombinant DNA. The bacterial colony loaded in Lane 2 does not produce any plasmids

corresponding to plasmid DNA, much more faint, of various sizes depending on the insert length; and a RNA smear, with possibly one or two more definite bands around 500 bp long (not visible in the figure). Once you have identified the colonies possibly bearing the correct calpastatin insert (in this figure only the colony in the sixth lane), perform PCR on the relevant ones, using the specific primers for calpastatin (*see* Subheading 2.3).

Select three colonies containing calpastatin by PCR, and before proceeding with cloning into the expression vector, determine the entire sequence of each colony on the isolated plasmids to establish the splicing form of the cloned calpastatin, the correct insertion of the insert, and the specific restriction sites necessary for cloning into the expression vector pGEX-6P-1.

6. Pick a small amount of three positive clones, and separately transfer each to 5–8 mL LB medium containing 100 μg/mL ampicillin. Grow each clone in a 50 mL sterile disposable culture tube and leave the tube cap slightly open to ensure aeration. Incubate the cultures overnight (12–16 h) at 37 °C under vigorous shaking.
7. The next day mix in a sterile cryogenic vial 825 μL overnight cell culture with 175 μL sterile glycerol (87%), and store cells at −80 °C (glycerol stock).
8. Collect by centrifugation (20 °C 1000 × *g* for 10 min) the remaining bacterial cells, discard all traces of supernatant, and proceed with the purification of the plasmid (*see* **Note 15**). This step can be carried out easily with any of the commercial kits available.
9. Check the sequence of the insert by Sanger sequencing.

3.6 Cloning of Calpastatin in the Expression Vector pGEX-6P-1

Once you have identified the bacterial clone containing the correct calpastatin sequence, you must cut the insert from the pGEM vector and paste it into the expression vector pGEX-6P-1. Since the sequences of both pGEM and calpastatin are cut, neither by EcoRI nor XhoI (*see* **Note 7**), and the restriction recognition sites for these endonucleases have been incorporated at the extremities of the calpastatin sequence, it is possible, following digestion with EcoRI and XhoI and separation by electrophoresis, to release the calpastatin insert from the pGEM vector (*see* **Note 16**). At the same time, the multiple cloning site of the expression vector pGEX-6P-1 contains the consensus sequences for EcoRI and XhoI. Indeed, we have designed the primers for calpastatin to ligate the construct in frame with the glutathione S-transferase (GST) sequence and obtain the expression of the recombinant fusion protein GST-calpastatin.

1. Use for digestion 10 μg of purified pGEM-calpastatin plasmid and "full" pGEX-6P-1. Set up two separate reactions (100 μL total volume) containing 10× Tango buffer (20 μL), EcoRI 10 U/μL (10 μL), XhoI 10 U/μL (5 μL).
2. Incubate reactions at 37 °C for 2 h, and, to check if the reaction is completed, collect 5 μL from each sample, add 1 μL of 6× agarose loading buffer, and carry out electrophoresis (agarose gel 1%). Load a lane with DNA ladder mix to confirm the size of the digestion products. If the reaction is not yet completed, add 2 μL of each restriction enzyme and continue digestion for an additional 1 h (*see* **Note 17**).
3. Stop the reactions by heating the samples at 80 °C for 20 min. Add 20 μL of 6× agarose loading buffer and load each sample in three wells (agarose gel 1%).
4. Carry out electrophoresis until insert and vector are separated (about 1 h migration 60 V). Minimize exposure to UV light as you will excise the calpastatin and pGEX-6P-1 bands from the gel, and following agarose extraction, you will use them for ligation.
5. Cut and collect the two bands in separate tubes. Carry out gel extraction using a DNA gel extraction kit. Recover the two DNA fragments separately from the spin columns following a three times elution with 30 μL 10 mM Tris–HCl pH 8.5 each.
6. Concentrate the DNA by precipitation in ethanol. Add 1/10 volume of the DNA sample (9 μL) of NaAcetate 3 M pH 5.2 and 3 volumes (270 μL) of ethanol (100% v/v, stored at −20 °C). Follow the procedure used in Subheading 3.1 for RNA precipitation.
7. Dissolve pellet in 10–15 μL of 5 mM Tris/HCl pH 8.5 (*see* **Note 18**). To measure the concentration of the calpastatin and pGEX-6P-1 DNA bands, load 1 μL of each sample on agarose gel (1%) together with a lane containing a known amount of DNA ladder mix. Compare the intensity of the purified bands with the intensity of a standard band having similar molecular size.
8. To carry out ligation, the linearized pGEX-6P-1 and calpastatin DNA should be at a vector to insert molar ratio of 1:5. Since pGEX vector is 4900 bp long, we have used 50–100 ng of vector and 100–200 ng of calpastatin DNA construct (approximately 2300 bp long). To calculate the amount of the calpastatin construct to be included in the ligation reaction, use the equation reported in the technical manual of the pGEM-T vector kit. Store the unused EcoRI/XhoI calpastatin construct and the linearized EcoRI/XhoI pGEX-6P-1 vector at −20 °C.

9. Carry out ligation following the instructions of the ligase supplier (Promega), and repeat the steps reported in **Note 10** for maximizing results (*see* **Note 19**). Before transformation it is important to check if ligation has occurred. Submit an aliquot of the ligation reaction (6 μL) to agarose electrophoresis as described at the end of Subheading 3.3. Proceed with transformation following the instructions of the bacteria supplier (*see* **Note 20**).

3.7 Identification of Bacterial Clones Containing the Calpastatin Transcript in pGEX-6P-1 Vector

1. Since you should get only clones bearing full vectors (*see* **Note 19**), spread several colonies in a new LB-ampicillin plate and perform PCR using calpastatin-specific primers (*see* Subheadings 3.4 and 3.5).
2. Add 5 μL of 6× agarose loading buffer to each PCR sample and carry out electrophoresis on a 1% agarose gel. Load a lane with DNA ladder mix for confirming the expected size of the calpastatin amplicon. Choose two or three positive colonies, and purify the plasmid from overnight 10 mL LB-ampicillin cultures (pGEX is a low copy vector). Remember to freeze aliquots of the colonies in glycerol at −80 °C for subsequent expression (*see* Subheading 3.5).
3. Confirm the sequence of the calpastatin insert, and check to see if the sequence of calpastatin is in the correct frame with GST in the vector. The goal is to position the GST upstream of the amino terminus of calpastatin to place a 5′ epitope tag on the protein.

3.8 Preparation of GST-Calpastatin Fusion Protein

Prior to large-scale preparation of calpastatin from a given transformed bacterial clone, it is advisable to screen different pGEX-calpastatin recombinants for the expression of GST-calpastatin fusion protein. To select the most effective clone, set up small-scale cultures, and follow the instructions reported in the GST Gene Fusion System Manual. You can screen crude sonicates for the relative level of expression of GST-calpastatin fusion protein using the GST substrate CDNB (1-chloro-2,4-dinitro benzene).

1. For large-scale preparation of GST-calpastatin fusion protein, scrape off splinters of solid ice from the cryogenic vial previously stored at −80 °C, with a sterile tip, and immediately immerse it in 10 mL LB medium containing ampicillin (100 μg/mL final concentration). Return the frozen culture to storage at −80 °C. Do not allow the contents of the vial to thaw.
2. Use 25 cm^2 sterile cell culture polystyrene flasks for growing the culture and leave the tube cap slightly open to ensure aeration. Incubate the culture overnight (12–16 h) at 37 °C under vigorous shaking.

3. The day after (it is better to start early in the morning), save 1 mL overnight culture for glycerol stock and mix the rest into 100 mL LB without ampicillin (*see* **Note 21**).
4. Let the culture grow at 37 °C with shaking until absorbance at 600 nm reaches 0.5–0.7 (it will take about 3–4 h. Do not go beyond this OD because if bacteria grow too dense, the cells will inhibit the protein expression).
5. Add IPTG to a final concentration of 0.1 mM, and continue shaking for 4 h at a temperature slowly decreasing from 37 to 25–20 °C (*see* **Note 22**). In the meantime, chill about 1 L of dH_2O in an ice bath and set a centrifuge to 2–4 °C (*see* **Note 23**).
6. Following 4 h of induction, transfer the culture to a single 250 mL centrifuge bottle (or to appropriate centrifuge containers), and centrifuge at 4 °C, 7700 × *g* for 10 min. Discard the supernatant, resuspend the bacterial pellet in 250 mL chilled dH_2O, and centrifuge again. Repeat this step twice.
7. Remove the supernatant, drain the pellet, and resuspend it in 2–3 mL of chilled dH_2O (*see* **Note 24**). After measuring the total volume (about 4 mL), transfer the sample to a 10 mL polypropylene tube. Add 2 mL of chilled bacteria lysis buffer 3× and the following protease inhibitors: 60 μL of 10 mg/mL leupeptin, 120 μL of (AEBSF) 0.1 M, 6 μL of aprotinin 10 mg/mL (alternatively use a protease inhibitor cocktail). Finally, add 600 μL lysozyme 10 mg/mL (necessary to disrupt the bacterial wall). Mix thoroughly and leave the sample on ice for 20 min. Add 120 μL of 1 M $MgCl_2$ (60 μL to chelate EDTA + 60 μL to obtain 10 mM $MgCl_2$ final concentration) and 6 μL of 10 mg/mL DNase. Mix thoroughly and leave the sample on ice for 20 min.
8. Sonicate briefly on ice (3 strokes, 15 s each), and centrifuge the lysate at 4 °C, 50,000 × *g* for 10 min. Discard the pellet, transfer the supernatant to a clean tube, and immediately store it at −20 °C. Before freezing, mix an aliquot of the clear lysate (10 μL) with 2 μL of 6× SDS-PAGE loading buffer, heat the sample for 3 min 95 °C, and store it at −20 °C for subsequent analysis by SDS-PAGE.

3.9 Purification of Recombinant Calpastatin from the Bacterial Cleared Lysate by Affinity Chromatography

Before starting the isolation of the recombinant calpastatin-GST fusion protein, swell glutathione-agarose (GSH-agarose) which is usually provided as a lyophilized powder, in dH_2O. Follow the instructions of the provider, and prepare an amount of resin related to its binding capacity (about 5–10 mg glutathione S-transferase per mL of swollen resin).

1. To equilibrate the resin in the binding buffer, put the desired aliquot (500 μL) in a 15 mL polypropylene tube and add

15 mL buffer A. Let the tube rotate end-over-end at 4 °C for 5 min, and then spin the resin at 500 × *g* for 5 min. Carefully remove and discard supernatant without withdrawing agarose beads; use a Pasteur pipette with a capillary tip. Repeat this step three times.

2. Thaw the lysate and carry out a protein assay to determine protein concentration (dilute an aliquot 200–300 times, and use 10, 20, 30, 50 μL sample. Protein concentration will be about 10–20 mg/mL). Work with 50–60 mg crude lysate and 500 μL swollen GSH-agarose (*see* **Note 25**).
3. Dilute the lysate with buffer A to obtain 5 mL final volume, apply the sample to the drained resin equilibrated in buffer A, and mix on an end-over-end rotator for 3–4 h at 4 °C (*see* **Note 26**).
4. Centrifuge the sample for 5 min at 500 × *g*; collect the supernatant, containing non-adsorbed bacterial proteins; and save it on ice, for downstream analysis. This sample could also contain non-adsorbed GST-calpastatin fusion protein that can be reloaded on fresh, regenerated GSH-agarose.
5. Add 2 mL buffer A to the resin, mix end-over-end for 5 min, centrifuge for 5 min at 500 × *g* (washing steps), and collect the supernatant for downstream analysis. Repeat this step twice.
6. Wash the resin 3 times with 5 mL and twice with 10 mL buffer A (*see* **Note 27**). Equilibrate the resin by two washings with 10 mL of buffer B which is the PreScission protease digestion buffer.
7. Carry out a final wash with 5 mL buffer B + 1 mM DTT (buffer C) freshly prepared.
8. Add 5 μL PreScission protease to 1.5 mL buffer C and mix it to the resin containing GST-calpastatin. Let proteolysis proceed overnight at 4 °C on an end-over-end rotator (*see* **Note 28**).
9. The day after, centrifuge the sample at 500 × *g* for 5 min. Carefully remove and save the supernatant on ice. Add to the resin 3 mL buffer C, mix end-over-end for 10 min at 4 °C, centrifuge for 5 min at 500 × *g*, and collect supernatant. Repeat this step twice, saving each supernatant fraction in a separate tube. Wash the resin once more with 1 mL buffer C and use 100 μL supernatant for Bradford protein assay. All these fractions contain purified calpastatin, keep them on ice. Before regeneration of GSH-agarose, save 5 μL of wet resin for SDS-PAGE to confirm both cleavage of the fusion protein and retention of GST and PreScission protease (*see* **Note 29**).

3.10 Evaluation of Recombinant Calpastatin Activity

To measure the inhibitory activity of the recombinant calpastatin, a known amount of purified calpain is assayed in the presence of increasing amounts of calpastatin. The method is based on the detection of $-NH_2$ groups liberated from denaturated globin that is used as calpain substrate. Since the Tris buffer used for recombinant calpastatin purification contains $-NH_2$ groups that interfere with the assay of calpain activity, it is necessary to change the buffer.

1. Pool the fractions that, according to SDS-PAGE analysis, contain calpastatin, and concentrate the sample in a stirred ultrafiltration cell provided with a 30 kD cutoff membrane. Use nitrogen gas for pressurizing the cell since using compressed air can cause pH shifts, due to dissolution of carbon dioxide.
2. When the volume of the sample is about 2–3 mL, fill the ultrafiltration cell (about 50 mL) with 50 mM NaBorate pH 7.5, 0.1 mM EDTA, 0.5 mM 2-mercaptoethanol (buffer D). In order to eliminate buffer C (which contains Tris buffer), repeat this step at least three times (*see* **Note 30**).
3. Transfer the concentrated calpastatin sample (about 2–3 mL volume) to a new tube, and save 20–30 μL for SDS-PAGE (calpastatin before heating).
4. Divide the purified calpastatin into different tubes (200–300 μL aliquots in each tube), heat the samples for 3 min in a boiling water bath, and immediately cool them on ice. Pool together all the heated aliquots, and centrifuge at 4 °C, 50,000 × *g* for 10 min. Transfer the supernatant to a clean tube and discard the pellet that contains denatured proteins (*see* **Note 31**).
5. Use 20–30 μL of purified calpastatin for SDS-PAGE analysis (calpastatin after heating).
   Dilute calpastatin about 200–500 times, and use increasing amounts to assay the inhibitory activity in the presence of a fixed amount of calpain.

4 Notes

1. Elute total RNA in RNase-free 1.5 mL Eppendorf safe-lock tubes. Never use pre-lubricated or DNA low-bind tubes, as the precipitated material is easily resuspended and lost.
2. The precipitated RNA sample can be stored at this temperature for several months before processing [*RNA Methodologies: A Laboratory Guide for Isolation and Characterization*. Robert E. Farrell, Jr.]. RNA can also be precipitated at −20 °C, but in this case, it should be left at this temperature for at least 30 min.
3. Usually the precipitated RNA is scarcely visible as a very thin dust. Before centrifugation, it is advisable to mark the top of

the tube at the side that will correspond to the precipitated material.

4. After the 70% ethanol wash, the pellet does not adhere tightly to the wall of the tube; thus take great care when removing the supernatant.
5. Residual ethanol interferes with RNA resuspension. The tube should appear dry, without any drops, but if the pellet is too dry, then RNA resuspension becomes difficult.
6. The increased temperature of the RT reaction is particularly important to obtain the full-length transcripts of calpastatin. In the past, when the RT reaction was carried out at 37 °C, we have experienced the synthesis of calpastatin transcripts lacking the 5′-terminus.
7. These primers contain restriction sites that are not present in the sequence of Type I calpastatin and allow the possible direct cloning of this calpastatin form into pGEX-6P-1 expression vector. The first ATG codon and the TAA stop codon of the transcript are indicated in bold letters. The two AA bases at the 5′-terminus of each primer increase the cutting efficiency of the restriction enzymes at the extremities of the amplicon. These are important when the PCR product is directly used for cloning in the expression vector.

   If you need to use other pairs of primers having different restriction sites, you should first analyze whether the sequence of interest is cut by those given restriction enzymes. NEBcutter [12] is a software freely available on the web (http://tools.neb.com/NEBcutter) which analyzes DNA sequences for the presence of restriction enzyme sites in a convenient and easy-to-use manner.
8. To perform TA cloning, the DNA polymerase used in the PCR must have a non-template-dependent terminal transferase activity that adds a single deoxyadenosine (A) to the 3′ ends of PCR products. The high-fidelity DNA polymerase used for producing calpastatin Type I amplicon leaves 3′-A overhangs on a portion of the PCR products; however, the cloning efficiency is greatly reduced from that obtained with recombinant Taq DNA polymerase alone, that is not high fidelity. Therefore, to obtain a more efficient TA cloning, we add to the PCR mixture an additional 0.02 units of recombinant Taq DNA polymerase. Since the linearized pGEM vector supplied by Promega has single, overhanging 3′ deoxy-thymidine (T) residues, this “trick” allows the production of a greater number of PCR inserts provided with 3′-A overhangs to ligate efficiently with it.
9. The electrophoresis of the PCR reaction should be carried out soon after PCR in order to minimize time-dependent “loss” of

the 3′-A overhangs. Moreover, if multiple PCR products are present, and you need to extract the calpastatin amplicon from the gel, it is best to limit the UV exposure of the DNA to minimize the risk of damage. Finally, to avoid contamination by different PCR products, be sure to efficiently separate amplicons having similar size.

10. Since the ligase is active at room temperature and the 2× rapid ligation buffer cannot be heated, combine the purified DNA band and the vector prior to ligation, incubate the sample for 10 min at 65 °C, and rapidly cool it on ice to allow DNA partial strand-opening and increase T-A pairing. Then, rapidly introduce 2× ligation buffer, mix by pipetting up and down, add 3 units of ligase, and leave the sample overnight at 4 °C.
11. It is necessary to add SDS to the loading dye since without detergent it is difficult to judge the ligation result because a smear will appear in the agarose. The presence of SDS eliminates DNA–protein interactions and prevents gel shifts. Proteins have to be denatured to release DNA from DNA–protein complexes. Otherwise, the DNA binding proteins may alter the electrophoretic migration and cause DNA to stick to gel wells.
12. Before plating and growing overnight, centrifuge the transformed bacteria at 1000 × *g* for 10 min, discard the supernatant, and resuspend the pellet with 200 μL SOC medium. Plate in different LB plates containing ampicillin, two aliquots of 50 and 150 μL, respectively.
13. Check transformation efficiency and perform positive controls as suggested by the supplier. Use X-Gal stock solutions freshly prepared or stored at −20 °C. If after one-night incubation at 37 °C you get very small blue and white colonies that have grown slowly, and plates should be further incubated to get larger colonies, small "satellite" colonies will appear. They surround larger transformed colonies, and in time, these can grow as large as the original transformed colonies. These are non-transformed, nonresistant cells that grow in the "antibiotic shadow" where ampicillin has been broken down by the large resistant transformed colonies (the Amp resistance protein is secreted by the cells).

    Another problem concerns antibiotic degradation. All antibiotics degrade with time and following repeated freeze–thaw cycles. If antibiotic selection is weak, post-transformation cells would be the result of failed antibiotic selection, and they could grow without retaining any plasmid. This circumstance becomes evident only when the selected colonies are screened for the presence of the insert. Prepare and store ampicillin in a 50% ethanol solution (stock solution 100 mg/mL, final concentration 100 μg/mL), and keep it at −20 °C.

14. Since this method does not involve PCR amplification, it is necessary to pick a quantity of bacteria containing an amount of plasmid sufficient to be seen directly under UV light. The amount of bacteria picked from a single colony before spreading is not sufficient for "cracking" detection. If you have the possibility to prepare an agarose gel with two rows of wells (15 wells for each row), you can screen **30** colonies at the same time. If this is the case, make sure to separate efficiently your bands by performing long migration. You will not be able to clearly tell the size of the insert, but you'll see whether an insert is present or not.
15. Plasmids vary widely in their copy number per cell, depending on their origin of replication that determines whether they are under relaxed or stringent control and depending on the size of the plasmid and its associated insert. Since pGEM is a high-copy-number plasmid, 2–3 mL LB overnight culture is sufficient to purify enough plasmid to determine the sequence of the calpastatin construct. Conversely, you will use about 20–50 μg of plasmid to obtain enough insert to be cloned in pGEX-6P-1 vector.
16. There are several reasons why we have prepared calpastatin constructs from a plasmid and not directly from a PCR product. First, it is necessary to obtain large amounts of the "calpastatin band" as it has to be extracted from the agarose gel, used for sequencing, and digested by EcoRI and XhoI before cloning it into the expression vector. Second, the probability of getting mutations in the calpastatin sequence is much lower if the construct is produced in the plasmid by bacterial growth than following PCR. Third, for cloning, it is necessary to use a construct having the correct cohesive ends that result from EcoRI and XhoI digestion. In fact, unlike TA cloning, this second cloning is unidirectional, and the construct must be in the correct frame with the sequence of the expression vector. If a PCR product is used for direct cloning in pGEX, it is impossible to check if the two restriction enzymes have generated the cohesive ends necessary for cloning.

    For this same reason, if possible, we start for cloning with a purified "full" pGEX-6P-1 plasmid still bearing EcoRI and XhoI restriction sites, and, following digestion, we use the vector and discard the insert.
17. To check the performance of the two enzymes, you can carry out the digestion using the two endonucleases separately. Sometimes, depending on the supplier, the two enzymes are active in different reaction conditions and thus you have to change buffer between the first and the second endonuclease. Read carefully the individual restriction enzyme data sheets before extending reaction times, as those endonucleases that

exhibit star activity should be used under the recommended conditions to inhibit noncanonical cleavage. Also, consider the double digest calculator to set up reaction conditions. Usually, if the two endonucleases come from the same, Company it is possible to refer to a table that indicates the performances of each endonuclease in different buffers and conditions.

Consider that the calpastatin construct is about 2000 bp long. Because the size of the pGEM vector is 3000 bp, you will get, following complete digestion, about 4.5 μg of calpastatin. Since you will cut the band corresponding to calpastatin from the gel, be sure to avoid contamination between calpastatin (2300 bp) and pGEM (3000 bp).

Finally, bear in mind that while a linearized plasmid migrates in electrophoresis as a function of its size (about 5000 bp, calpastatin + pGEM), uncut plasmids will appear to migrate more rapidly than the same linearized plasmid. It is possible for the uncut plasmid to migrate in a position very close to the insert released following double digestion. This means that you could have your calpastatin band contaminated by some uncut pGEM-calpastatin. Consequently, in the subsequent step, cells would be transformed with pGEM-calpastatin instead of pGEX-calpastatin. If this happens, recombinant GST-calpastatin protein will not be produced since pGEM is not an expression vector.

18. You can skip the final ethanol precipitation of the DNA bands extracted from the agarose gel. As suggested by the kit supplier, to obtain increased DNA concentration, you can elute the bands just once in 30 μL buffer. We have sometimes experienced ligation failure when extracted bands were directly used in the ligation reaction. For this reason, it is preferred to elute bands in larger buffer volumes and perform ethanol precipitation, both to concentrate the sample in a small volume and to further purify the insert and vector.
19. This is a unidirectional ligation of cohesive ends; it is fundamental that both insert and vector are cut by endonucleases generating the same palindromic sequences. Moreover, to allow ligation they must be effectively digested at both ends. Practically it is impossible to obtain empty reclosed vectors because the two extremities are not compatible. However, if one of the two endonucleases fails to work, you will find colonies containing reclosed empty vectors, resistant to ampicillin, and producing only GST protein. *See* also **Note 17**.
20. Although you can express the cloned protein in any host *E. coli* strain, since the pGEX vectors carry the lacI^q gene, we have produced the recombinant calpastatin in BL21 *E. coli* cells. Since these are deficient in Lon protease (cytoplasm) and OmpT protease (outer membrane), they are generally preferred for recombinant protein expression. Indeed, following

cloning in JM109, we have experienced the proteolytic degradation of the recombinant calpastatin by bacterial enzymes. Unfortunately, BL21 do not transform well; thus, we suggest alternative strains for cloning and maintenance of the plasmid (TOP10 or DH5α strains).

21. To maximize growth, the LB medium is preheated to 37 °C in a 1 L sterilized glass Erlenmeyer flask, and when the culture is added, the flask is capped with pierced aluminum foil to ensure aeration. Fill the flask to only 20–25% capacity to allow proper shaking of the culture.
22. To trace the production of the GST-calpastatin fusion protein, collect 5 mL of bacteria culture before induction and prepare the clear lysate as described in the text. Submit an aliquot of this sample to SDS-PAGE together with a sample prepared from stimulated cells (*see* **Note 25**). We have carried out many small pilot experiments to establish optimal conditions for obtaining maximal amounts of intact calpastatin. Indeed, the major problem is that calpastatin can be partly degraded even if it is still incorporated into the GST fusion protein. Thus, we use BL21 cells for expression (*see* **Note 20**), but because following IPTG induction they produce GST-calpastatin fusion protein very quickly and in large amounts, the recombinant protein may accumulate in inclusion bodies. For this reason, to slow down the production of the fusion protein and avoid the production of inclusion bodies, during the 4-h induction, we recommend gradually reducing the temperature from 37 °C to 25–20 °C. It is possible to induce protein expression at room temperature, but in this case, induction should be prolonged to 6–8 h.
23. It is better to keep the refrigerated rotor inside the centrifuge. If possible, in order to reduce loss of material, pellet all the bacteria in the same centrifuge container (e.g., use a single 250–500 mL centrifuge bottle). Dissolve the bacterial pellet in a small amount of chilled dH_2O by swirling the side where the pellet is and breaking up the clumps. Then fill the bottle completely with chilled water, mix gently, and centrifuge again.
24. If the final volume of the sample is too large, transfer bacteria to a 30 mL tube and repeat centrifugation. Carry out bacterial lysis immediately after centrifugation in the presence of proteases inhibitors (leupeptin, AEBSF, aprotinin). All of the steps must be conducted on ice using chilled solutions. We add the protease inhibitors and lysozyme directly to the lysing buffer. The mixture is then added to the bacteria dissolved in a small volume of dH_2O. The lysis buffer can be prepared as a 3× solution and stored at −20 °C until use.

25. Crude lysates can be stored at −20 °C in aliquots to be thawed when needed. Sometimes for preparative scale purification, we obtain the lysate from 300–500 mL LB cultures and divide it into 2–3 mL aliquots containing 50–60 mg protein each. Recombinant calpastatin has been prepared successfully from lysates which have been stored at −20 °C for up to 6 months. Since it is not known whether the induced bacteria have produced GST-calpastatin fusion protein, it is advisable, at least for the first time purification, to submit to SDS-PAGE (10%) a small amount (3–5 μL) of clear lysate, before proceeding with purification. An equivalent amount of clear lysate prepared from bacteria collected just prior induction should be loaded for comparison (*see* **Note 22**).
26. The adsorption step can also be carried out overnight at 4 °C immediately after the preparation of the clear bacterial lysate.
27. These washings should contain decreasing amounts of bacterial proteins weakly and non-specifically associated with the matrix. Since they are gradually released from the resin, monitor with Bradford assay 100 μL of each wash by measuring the absorbance at 595 nm until baseline is reached. To confirm that GSH-agarose efficiently binds the fusion protein, it is also advisable to submit aliquots of the washings to SDS-PAGE. Remember to save 5 μL wet resin before digestion. This sample should be used in SDS-PAGE together with an equal amount of resin collected after overnight digestion.
28. It is important to cleave calpastatin from the GST moiety only following extensive washing of the GSH-resin. In this case, the resin should have specifically retained only the fusion protein. After overnight digestion, while the GST protein remains associated with GSH-agarose, free calpastatin is released into the soluble fraction. It is also possible to mix 40 μL PreScission protease with 960 μL of buffer C and incubate the sample for 4 h at 4 °C. We have performed this method but sometimes cleavage was not complete.
29. As also indicated in **Note 25**, it is necessary to check the efficacy of each step before proceeding with the protein purification. Submit an aliquot of the samples saved during purification, to SDS-PAGE (10%). The most important samples that should be analyzed are (1) bacterial lysate (to determine the production of the fusion protein), (2) GST-agarose after washing and before PreScission protease digestion (to check that GST-calpastatin is the only protein bound to the resin), (3) GST-agarose after PreScission digestion and calpastatin elution (this sample is particularly important because the resin should only contain GST, the PreScission protease, and no fusion protein. If fusion protein is still present, you can add 1–2 μL

PreScission and proceed with digestion for a further 2–4 h at room temperature), and (4) aliquots of the samples containing eluted calpastatin (of course these should contain only a single protein band).

30. This technique provides a simple means for rapid microsolute exchange and is typically used as a substitute for dialysis. The same method can also be carried out by using centrifugal filter devices instead of stirred ultrafiltration cells. If this is the case, before each centrifugation step, repeatedly mix the sample with buffer D to reduce the negative effects of concentration polarization (i.e., the buildup of concentrated solutes on the membrane).
31. Calpastatin is a heat-resistant protein; thus the "heating step" is an additional purification stage very important to denature residual bacterial enzymes and proteins still present in the calpastatin sample (compare calpastatin before and after heating by SDS-PAGE). The purified recombinant calpastatin is stable at 4 °C for up to 1 year. Periodically check the integrity of the calpastatin molecule by SDS-PAGE, since calpastatin isolated inhibitory domains retain inhibitory activity following digestion of the full length protein [7]. To preserve calpastatin you can add NaN_3 (1 mM final concentration) to the sample.

Acknowledgment

This work was supported by the grant FRA2015 and FRA2016 from University of Genova to MA and RDT.

References

1. Takano J, Watanabe M, Hitomi K, Maki M (2000) Four types of calpastatin isoforms with distinct amino-terminal sequences are specified by alternative first exons and differentially expressed in mouse tissues. J Biochem 128:83–92
2. Parr T, Sensky PL, Bardsley RG, Buttery PJ (2001) Calpastatin expression in porcine cardiac and skeletal muscle and partial gene structure. Arch Biochem Biophys 395:1–13
3. Lee WJ, Ma H, Takano E, Yang HQ, Hatanaka M, Maki M (1992) Molecular diversity in amino-terminal domains of human calpastatin by exon skipping. J Biol Chem 267:8437–8442
4. De Tullio R, Sparatore B, Salamino F, Melloni E, Pontremoli S (1998) Rat brain contains multiple mRNAs for calpastatin. FEBS Lett 422:113–117
5. Takano J, Kawamura T, Murase M, Hitomi K, Maki M (1999) Structure of mouse calpastatin isoforms: implications of species-common and species-specific alternative splicing. Biochem Biophys Res Commun 260:339–345
6. De Tullio R, Averna M, Stifanese R, Parr T, Bardsley RG, Pontremoli S, Melloni E (2007) Multiple rat brain calpastatin forms are produced by distinct starting points and alternative splicing of the N-terminal exons. Arch Biochem Biophys 465:148–156
7. De Tullio R, Averna M, Salamino F, Pontremoli S, Melloni E (2000) Differential degradation of calpastatin by mu- and m-calpain in Ca^{2+}-enriched human neuroblastoma LAN-5 cells. FEBS Lett 475:17–21
8. Averna M, De Tullio R, Passalacqua M, Salamino F, Pontremoli S, Melloni E (2001) Changes in intracellular calpastatin localization

are mediated by reversible phosphorylation. Biochem J 354:25–30

9. De Tullio R, Cantoni C, Broggio C, Prato C, Stifanese R, Averna M, Antolini R, Pontremoli S, Melloni E (2009) Involvement of exon 6-mediated calpastatin intracellular movements in the modulation of calpain activation. Biochim Biophys Acta 1790:182–187
10. Geesink GH, Nonneman D, Koohmaraie M (1998) An improved purification protocol for heart and skeletal muscle calpastatin reveals two isoforms resulting from alternative splicing. Arch Biochem Biophys 356:19–24
11. Parr T, Sensky MK, Bardsley RG, Buttery PJ (2000) Effects of epinephrine infusion on expression of calpastatin in porcine cardiac and skeletal muscle. Arch Biochem Biophys 374:299–305
12. Vincze T, Posfai J, Roberts RJ (2003) NEBcutter: a program to cleave DNA with restriction enzymes. Nucleic Acids Res 31:3688–3691

Chapter 14

Isolation of Endogenous Calpastatin

Roberta De Tullio and Monica Averna

Abstract

We here describe the purification of calpastatin from human erythrocytes. When calpastatin is purified from tissues, it is necessary to measure its inhibitory activity against calpain in the presence of Ca^{2+} to specifically identify the protein. Thus, the purification steps necessary to obtain the inhibitor protein were originally designed to obtain calpain from the same tissue. For this reason, in addition to calpastatin purification, we also include a method for purifying human erythrocyte calpain and globin. We routinely use these two components for assaying calpastatin inhibition.

Key words Calpastatin isolation, Calpain inhibition, Chromatography, Calpain assay, Human erythrocytes

1 Introduction

Isolation and purification of calpastatin from tissues can be utilized to study the Ca^{2+}-dependent calpain proteolytic system. The calpastatins isolated from a given tissue can include different species and forms that can be separated by specific chromatographic steps such as anion exchange and hydrophobic interaction chromatography. These methods have allowed us to establish that calpastatin can undergo reversible phosphorylation [1] and that the two forms have different specificity and efficiency against the calpains isolated from the same tissue [2]. The phosphorylation of calpastatin has also been demonstrated in vivo [3] and by immunoblot analysis using phosphoserine antibodies [4]. Thus, different methods of investigation are complementary to characterize a biochemical process from different points of view.

The purification of calpastatin takes advantage of the fact that this protein is heat resistant. Heating allows purification of calpastatin while simultaneously inactivating calpain, thus preventing contamination of calpain in the calpastatin preparation. Since to pick up specifically calpastatin among all the other contaminants we measure the inhibitory activity against calpain, the calpastatin

Jeannette S. Messer (ed.), *Calpain: Methods and Protocols*, Methods in Molecular Biology, vol. 1915,
https://doi.org/10.1007/978-1-4939-8988-1_14,

species deprived of inhibitory units present in the tissue cannot be detected with this method. Isolation of endogenous calpastatin is time-consuming and laborious so when a specific form of calpastatin is required, it is more frequently produced recombinantly than isolated from tissues. However, in some cases it may be desirable to isolate calpastatin from a specific tissue since the calpastatins present in a given condition are not only related to the ratio of active/inactive calpains but also to the specific transcription pattern of the inhibitor. With this in mind, in this chapter we describe the purification of human erythrocyte calpastatin by means of specific chromatographic steps. These can be used with minor modifications to isolate naïve calpastatins from other tissues. For the sake of completeness, we also report the purification from human erythrocytes of both calpain, the protease form we use routinely to measure calpastatin inhibition, and denatured globin, the protein used as a substrate in calpain activity assays.

2 Materials

1. Dialyzing tubes: CelluSep T2 (nominal MWCO 6000–8000) (Orange Scientific).
2. Anion exchange resins: DE53 (Whatman®), Source 15Q (GE Healthcare).
3. Gel filtration resin: Sephacryl S-200 HR (GE Healthcare).
4. Hydrophobic interaction resin: Phenyl-Sepharose (GE Healthcare).
5. Stirred ultrafiltration cell (product number 8050, 50 mL) with YM-30 or YM-10 Ø43 mm ultrafiltration membranes having cutoff 30,000 and 10,000, respectively (Millipore Amicon).
6. Fluorescamine (Sigma Chemical Company, Product Number F 9015) solution: Dissolve 30 mg in 100 mL acetone. Store at room temperature protected from light up to 1 week.
7. Trichloroacetic acid solution (TCA): Prepare a 50% (p/v) in dH_2O.
8. Buffer A: 50 mM Na acetate pH 6.7, 0.5 mM 2-mercaptoethanol, 1 mM EDTA, adjusted to pH 6.7.
9. Buffer B: 50 mM Na borate pH 7.5, 0.5 mM 2-mercaptoethanol, 0.1 mM EDTA, adjusted to pH 7.5.
10. Buffer C: 0.5 M Na borate pH 8.5.
11. Globin extraction solution: 5 L chilled acetone containing 10 mL HCl (37%) and 5 mL 2-mercaptoethanol.

3 Methods

3.1 Isolation of Human Erythrocytes from Whole Blood

1. Human erythrocytes derive from either a whole blood donation or a therapeutic phlebotomy. To isolate the erythrocytes, divide the sack contents into 50 mL polypropylene tubes, and centrifuge at 20 °C, 1200 × *g* for 10 min. Withdraw the supernatant (plasma and platelets) and part of the interface between the plasma and the erythrocytes (buffy coat). Be careful not to remove too many erythrocytes; use a vacuum filtration system connected to a Pasteur pipette (*see* **Note 1**).
2. Fill the tubes containing the packed erythrocytes with 0.15 M NaCl, mix briefly and centrifuge the tubes at 4 °C, 1200 × *g* for 10 min. Discard supernatant and the buffy coat still present on top of erythrocytes. Repeat this step three times.
3. Pool the erythrocytes into a few tubes and carry out a final centrifugation at 4 °C, 1200 × *g* for 15 min. Remove the supernatant and proceed with lysis.

3.2 Purification of Calpastatin from Human Erythrocytes

1. Measure the volume of the packed erythrocytes (about 150–200 mL from a therapeutic phlebotomy) and pour the packed cells, drop by drop into 5 volumes of chilled 1 mM EDTA pH 7.0 rapidly rotating on a magnetic stirrer. Let the lysis proceed in gentle rotation for 5 min.
2. Centrifuge the lysate at 4 °C, 14,000 × *g* for 20 min. In the meantime, equilibrate 50–100 g of DE53 resin (wet powder, prepared following the instructions of the supplier) in Buffer A. Leave the resin humid, ready to use.
3. Collect the supernatant in a beaker, and put it on a magnetic stirrer in gentle rotation. Adjust pH to 6.7 with diluted acetic acid and gradually apply the normalized resin to the crude lysate (resin/packed erythrocytes 1 g/5 mL) (*see* **Note 2**). Control that no resin clumps are present, and let absorption proceed in gentle rotation for 15 min.
4. Pour the contents of the beaker into a Büchner funnel provided with a paper filter. Insert the funnel in a Büchner flask associated to a vacuum device and separately collect the flow through (*see* **Note 3**).
5. Wash the resin in the funnel extensively with Buffer A until the absorbance at 280 nm of the flow through is below 0.02 (about 3–4 L of Buffer A).
6. Transfer the wet resin to a clean beaker, dissolve it in Buffer A and pour the suspension over a glass column.
7. Elute the absorbed proteins in 3–4 mL fractions with 500 mL of 0.3 M NaCl dissolved in Buffer A (*see* **Note 4**). Determine the protein content of each fraction at 280 nm.

8. Pool the fractions containing the peak of proteins in a glass graduated cylinder, add 80% $(NH_4)_2SO_4$ (p/v final concentration), mix gently until the salt is completely dissolved. To support efficient protein precipitation, leave the sample at 4 °C for 15 min.
9. Centrifuge the sample at 4 °C, 15,000 × *g* for 20 min. Discard supernatant, dissolve the pellet in the lowest possible volume of Buffer A (5–10 mL), and dialyze the sample against 5 L of Buffer A at 4 °C overnight.
10. Transfer the sample into a Pyrex Erlenmeyer flask (25–50 mL) and heat at 95 °C for 3 min in a boiling water bath (*see* **Note 5**).
11. Centrifuge at 4 °C, 50,000 × *g* for 10 min to remove the precipitated denatured proteins, and load the sample on a DE53 column (1.5 × 15 cm) equilibrated in Buffer A. After washing, elute absorbed proteins with a linear gradient of NaCl from 0 to 0.25 M (250 + 250 mL). Collect fractions of 3 mL and measure the inhibitory activity of calpastatin with 10–20 μL aliquots (*see* Subheading 3.4).
12. Pool the fractions containing calpastatin activity, and concentrate them by ultrafiltration to 1–1.2 mL volume in a stirred ultrafiltration cell provided with a 30 kD cutoff membrane YM-30. Submit the sample to size-exclusion chromatography on a Sephacryl S-200 HR column (1.5 × 120 cm) equilibrated in Buffer B. Collect fractions of 1.3 mL and measure the inhibitory activity of calpastatin on aliquots of 10–20 μL *(see* Subheading 3.4).
13. Concentrate the relevant fractions to 1.5–2 mL by ultrafiltration, and determine protein concentration and the specific inhibitory activity of the purified calpastatin (Fig. 1).

3.3 Purification of Calpain from Human Erythrocytes

To carry out the purification of calpain, follow the method indicated in Subheading 3.2 until **step 6**, and then proceed as indicated below:

1. Elute the absorbed proteins in 3–4 mL fractions with 500 mL of 0.2 M NaCl dissolved in Buffer A.
2. Determine the protein content of each fraction at 280 nm.
3. Pool the fractions containing the peak of proteins, add 50% $(NH_4)_2SO_4$ (p/v final concentration), and mix gently, until the salt is completely dissolved. To support efficient protein precipitation, leave the sample at 4 °C for 15 min.
4. Centrifuge at 4 °C, 15,000 × *g* for 20 min. Discard supernatant, dissolve the pellet in the lowest possible volume of Buffer A (5–10 mL), and dialyze against 5 L of Buffer A at 4 °C overnight.

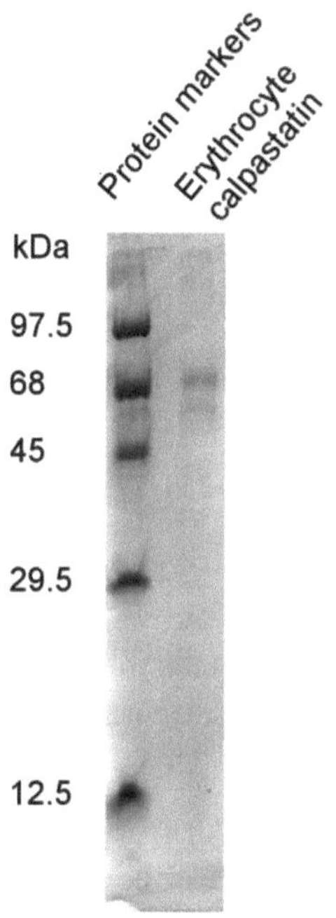

Fig. 1 SDS-PAGE of purified erythrocyte calpastatin. Purified erythrocyte calpastatin (20 μg) loaded on 10% SDS-PAGE and stained with Blue Coomassie

5. Load the sample on a Source 15Q column (1 × 15 cm) equilibrated in Buffer A and after washing, elute absorbed proteins with a linear gradient of NaCl 0–0.3 M (200 + 200 mL). Collect fractions of 2.5–3 mL. Use 50 μL aliquots to measure calpain activity (*see* Subheading 3.4).
6. Pool the fractions containing calpain activity, and concentrate them by ultrafiltration to 5–10 mL volume in a stirred ultrafiltration cell provided with a 30,000 cutoff membrane YM-30. Dialyze the sample against Buffer B.
7. At the end of the dialysis, add NaCl to the calpain sample to have 0.3 M final concentration.
8. Apply the sample to a Phenyl-Sepharose column (1 × 5 cm) equilibrated in Buffer B containing 0.3 M NaCl. After extensive washing, elute absorbed proteins with 100 mL Buffer B without NaCl. Collect 1.5–2 mL fractions (*see* **Note 6**).
9. Collect fractions containing calpain activity, and concentrate the sample to 1–1.5 mL in a stirred ultrafiltration cell.
10. Submit the sample to size-exclusion chromatography on a Sephacryl S-200 HR column (1.5 × 120 cm) equilibrated in Buffer B.
11. Pool the fractions containing calpain activity and concentrate them by ultrafiltration to 1.5–2 mL. Determine protein concentration and calpain-specific activity.

3.4 Assay of Calpastatin Inhibitory Activity in the Presence of Calpain

We routinely assay calpain by measuring the amount of acid-soluble peptide(s) liberated from human denatured globin used as substrate (*see* Subheading 3.5 for preparation of globin). To detect calpastatin, incubate a fixed amount of purified calpain in the absence (total calpain activity) or in the presence of aliquots of the fractions possibly containing the inhibitor. The residual calpain activity detected indicates the amount of calpastatin present in the sample.

1. The assay mixture contains in a final volume of 250 μL, 50 mM Na borate pH 7.5, 1 mM $CaCl_2$, an aliquot (10–20 μL fraction) of calpastatin, and 0.5 mg of denatured globin (add globin separately to the mixture, immediately before starting the assay).
2. Start the reaction with an appropriate amount of calpain and incubate the mixture at 25 °C for 10 min (*see* **Note 7**).
3. Stop the reaction by adding 40 μL of TCA.
4. Centrifuge the samples at 4 °C, 1200 × *g* for 10 min. Transfer carefully, without disturbing the pellet, 200 μL of supernatant in clean borosilicate tubes.
5. Add to each tube 2 mL of Buffer C and, under vigorous rotation, add 200 μL of fluorescamine [5]. Detect the fluorescence of each sample at 384 nm excitation and 475 nm emission within half an hour from addition of fluorescamine (*see* **Note 8**).

3.5 Purification of Globin from Human Erythrocytes

Globin is the protein moiety of hemoglobin and the most represented protein in erythrocytes. It is also the major contaminant during the purification of both calpain and calpastatin from human erythrocytes. Indeed the first chromatography (*see* Subheading 3.2, **steps 1–4**) is aimed to remove hemoglobin from the sample before proceeding with the purification of either calpastatin or calpain.

1. Process human erythrocytes as described in Subheading 3.2, and use for globin purification 100 mL of the flow through obtained from **step 4**. Instead of being discarded (as bith calpain and calpastatin are bound to the resin) is used to prepare globin (*see* **Note 3**).
2. Add drop by drop the flow through to 5 L of globin extraction solution rapidly stirring on a magnetic mixer. Let protein extraction proceed for 10 min under continuous vigorous stirring.
3. At the end, you will see that the solution has become brownish and a white precipitate (globin) is formed on the bottom of the beaker. Discard the supernatant.

4. Collect the precipitate in another beaker, and dissolve it in the lowest possible volume of dH_2O. Repeat again the extraction step (**steps 2** and **3**).
5. Collect the white precipitate in 30 mL polypropylene tubes, and wash it by repeated centrifugation (5 min 80 × *g*) with pure acetone (in order to remove residual HCl) until the pH of the resuspended pellet (use pH-indicator strips) normalizes with that of acetone.
6. Spread the solid pellet in a clean glass container and let acetone dry out overnight.
7. Dissolve the dried pellet in the lowest possible volume of dH_2O, and remove residual HCl by dialyzing the sample against ultrapure water (*see* **Note 9**).
8. Collect all the globin in the same tube, assay protein concentration and store in aliquots at −20 °C until needed.

4 Notes

1. The buffy coat looks like a whitish layer on the erythrocyte pellet that can be removed easily without disturbing the underlying erythrocytes too much.
2. It is important during this absorption step, to maintain the pH at 6.7 by adding diluted acetic acid or NaOH to prevent hemoglobin in the lysate binding to the resin. Hemoglobin is the most represented and contaminating erythrocyte protein. For this reason, during absorption, control pH continuously by submerging the pH-meter electrode in the beaker.
3. This non-adsorbed material contains mainly hemoglobin that can be used in denatured form as substrate for assaying calpain activity. You can store 100–200 mL of the flow through at −20 °C and prepare denatured globin when needed (*see* Subheading 3.5).
4. If you need to purify human erythrocyte calpain, elute the absorbed proteins with 500 mL of Buffer A containing 0.2 M NaCl and proceed as described in Subheading 3.3.
5. Immerse the flask in boiling water, and to allow a rapid increase in the temperature of the calpastatin sample, keep the flask in rapid rotation. Submerge a thermometer in the flask and start a timer when the temperature has reached 95 °C. Keep rotating the flask for 3 min and then rapidly cool the sample by submerging the flask in an ice bath.
6. This chromatography separates calpain that binds to the resin in the presence of NaCl, from residual calpastatin that comes out with non-absorbed proteins.

7. To understand the actual fluorescence deriving from calpain activity, it may be useful to know the initial fluorescence of the sample (zero time) before the reaction occurs. To do so, double the volume of the incubation, and transfer 250 μL of the reaction in another tube containing 50 μL TCA immediately after starting the incubation. Then proceed with **step 4**.
8. The non-fluorescent compound, fluorescamine, reacts rapidly with primary amines in proteins, such as the terminal amino group of peptides, to form highly fluorescent moieties. Thus, an increase in fluorescence means that globin has been digested into small peptides that remain in solution following TCA protein precipitation and centrifugation. If calpastatin is present in the mixture, calpain is inhibited and fluorescence remains low. To know how much calpastatin is present in the fraction, you should carry out a control assay containing only calpain.
9. Use dialysis bags (regenerated cellulose tubular membranes) having cutoff 6000–8000, and change the dialysis water frequently. You must carry out the dialysis until the pH of the sample reaches a neutral value (generally, 2 days of dialysis are required). During dialysis, control the volume of the globin sample periodically because it increases very much and dialysis bags can get very swollen and break down.

Acknowledgments

This work was supported by the grant FRA2015 and FRA2016 from the University of Genova to MA and RDT.

References

1. Pontremoli S, Viotti PL, Michetti M, Salamino F, Sparatore B, Melloni E (1992) Modulation of inhibitory efficiency of rat skeletal muscle calpastatin by phosphorylation. Biochem Biophys Res Commun 187:751–759
2. Salamino F, Averna M, Tedesco I, De Tullio R, Melloni E, Pontremoli S (1997) Modulation of rat brain calpastatin efficiency by post-translational modifications. FEBS Lett 412:433–438
3. Adachi Y, Ishida-Takahashi A, Takahashi C, Takano E, Murachi T, Hatanaka M (1991) Phosphorylation and subcellular distribution of calpastatin in human hematopoietic system cells. J Biol Chem 266:3968–3972
4. Barnoy S, Kosower NS (2007) Calpastatin in rat myoblasts: transient diminution and decreased phosphorylation depend on myogenin-directed myoblast differentiation. Int J Biochem Cell Biol 39:253–261
5. Udenfriend S, Stein S, Böhlen P, Dairman W, Leimgruber W, Weigele M (1972) Fluorescamine: a reagent for assay of amino acids, peptides, proteins, and primary amines in the picomole range. Science 178:871–872

Chapter 15

Immunoblotting for Calpastatin Expression

Monica Averna and Roberta De Tullio

Abstract

Immunoblotting is a procedure routinely used to analyze calpastatin expression. However, immunoblotting alone may not be adequate for this task, since calpastatin isoforms can vary by tissue, can be modified by partial digestion, and can undergo posttranslational modifications. Here we describe a method for more precise evaluation of calpastatin expression by combining immunoblot analysis with an assay for the inhibitory activity of a single calpastatin species isolated by SDS-PAGE and protein elution from the gel.

Key words Immunoblot, Calpastatin, Calpain, Protein elution, Ca^{2+}-dependent proteolysis, Calpastatin phosphorylation

1 Introduction

Full-length calpastatin, the natural inhibitor of calpain, includes the non-inhibitory N-terminal region followed by four repetitive inhibitory domains. The inhibitor exists in four different isoforms that are generated by differential translation regulated by four different promoters [1–5]. Calpastatin also undergoes splicing events involving the L-domain [4, 6–9] and/or the first inhibitory domain [6, 8, 10], generating other calpastatin forms with specific exon composition. Additional calpastatin species can also be generated by calpain-mediated proteolysis. Following an increase in intracellular calcium concentration ($[Ca^{2+}]_i$), active low apparent molecular weight (Mw) (15 kDa) calpastatin fragments, containing only a single inhibitory domain, are produced [11, 12]. Furthermore, calpastatin is characterized by anomalous migration in SDS-PAGE [9, 13], and multiple forms of calpastatin may be present concomitantly in a given sample. For all of these reasons, the identification of the different calpastatin forms expressed in various animal tissues and conditions can sometimes be very difficult. Here we describe a simple but efficient method to identify the different calpastatin species in rat brain by combining immunoblot analysis, using

Jeannette S. Messer (ed.), *Calpain: Methods and Protocols*, Methods in Molecular Biology, vol. 1915,
https://doi.org/10.1007/978-1-4939-8988-1_15,

relevant antibodies, with the direct assay of calpastatin activity following protein separation on SDS-PAGE.

2 Materials

2.1 SDS Polyacrylamide Gel

Prepare all solutions using ultrapure water (resistivity 18 MΩ cm at 25 °C).

1. Resolving gel buffer: 1 M Tris–HCl, pH 8.8. Filter through a 0.22 μm filter and store at 4 °C (*see* **Note 1**).
2. Stacking gel buffer: 1 M Tris–HCl, pH 6.8. Filter through a 0.22 μm filter and store at 4 °C (*see* **Note 1**).
3. 30% acrylamide/bis solution (37.5:1) (Bio-Rad).
4. Ammonium persulfate 10% p/v solution in ultrapure water. Store at 4 °C (*see* **Note 2**).
5. *N*,*N*,*N*′,*N*′-Tetramethylethylenediamine (TEMED). Store at room temperature.
6. 10% sodium dodecyl sulfate (SDS) p/v solution in ultrapure water. Store at room temperature up to 6 months.
7. SDS-PAGE running buffer: 0.05 M Tris, 0.38 M glycine, 0.1% p/v SDS in ultrapure water. Store at room temperature (*see* **Note 3**).
8. SDS-PAGE loading mix: 0.06 mM Tris–HCl pH 6.8, 2% SDS, 10% v/v glycerol, 2% 2-mercaptoethanol, 0.04% p/v bromophenol blue in ultrapure water. Store at room temperature up to 1 month.
9. Bromophenol blue solution: 0.1 g of bromophenol blue in 100 mL of water. Store at −20 °C.

2.2 Immunoblotting

1. Nitrocellulose membrane 0.2 μm.
2. Western blot transfer buffer: 0.025 M Tris, 0.192 M glycine, 20% v/v methanol, 0.05% p/v SDS (*see* **Note 4**) in ultrapure water. Store at 4 °C (*see* **Note 5**).
3. Phosphate-buffered saline (PBS): 0.01 M sodium phosphate, pH 7.5, 0.14 M NaCl in ultrapure water. Store at 4 °C up to 1 month.
4. PBS containing 0.05% v/v Tween 20 (T-PBS).
5. Blocking solution: 5% p/v skim milk powder in T-PBS (*see* **Note 6**).
6. Fiber pads.
7. Blot adsorbent filter paper.

2.3 Isolation of Calpain and Acid-Denatured Globin from Erythrocytes

1. Whole blood.
2. Dialyzing tubes: CelluSep T2 (nominal MWCO 6000–8000) (Orange Scientific).
3. Anion-exchange resins: DE53 (Whatman®), Source 15Q (GE Healthcare).
4. Gel filtration resin: Sephacryl S-200 HR (GE Healthcare).
5. Hydrophobic interaction resin: Phenyl-Sepharose (GE Healthcare).
6. Stirred Ultrafiltration Cell (product number 8050, 50 mL) with YM-30 or YM-10 Ø43 mm ultrafiltration membranes having cutoff of 30,000 and 10,000, respectively (Millipore-Amicon).
7. Fluorescamine (Sigma Chemical Company, Product Number F 9015) solution: dissolve 30 mg in 100 mL acetone. Store at room temperature protected from light up to 1 week.
8. Trichloroacetic acid solution (TCA): prepare a 50% (p/v) in H_2O.
9. Buffer A: 50 mM NaAcetate pH 6.7, 0.5 mM 2-mercaptoethanol, 1 mM EDTA, adjusted to pH 6.7.
10. Buffer B: 50 mM NaBorate pH 7.5, 0.5 mM 2-mercaptoethanol, 0.1 mM EDTA, adjusted to pH 7.5.
11. Buffer C: 0.5 M NaBorate pH 8.5.
12. Globin extraction solution: 5 L chilled acetone containing 10 mL HCl (37%) and 5 mL 2-mercaptoethanol.

2.4 Calpastatin Elution from SDS Polyacrylamide Gel and Analysis of Activity

In addition to the material listed for SDS polyacrylamide gel preparation, use the following materials:

1. Lysis buffer: 0.05 M sodium borate buffer, pH 7.5, containing 1 mM EDTA, 0.5 mM 2-mercaptoethanol. Store at 4 °C up to 1 month.
2. Leupeptin (Sigma-Aldrich, L2884) solution: prepare 10 mg/mL leupeptin in H_2O, and store it at −20 °C.
3. 4-(2-Aminoethyl) benzenesulfonyl fluoride hydrochloride (AEBSF) (Fluka, Pefabloc® SC, 76307) solution: prepare 0.1 M AEBSF in H_2O, and store it at −20 °C in aliquots.
4. Aprotinin (Sigma-Aldrich, 10820) solution: prepare 10 mg/mL aprotinin in H_2O, and store it at −20 °C.
5. Sodium borate buffer: 0.05 M sodium borate, pH 7.5, 0.1 mM EDTA, 0.5 mM 2-mercaptoethanol. Store at 4 °C up to 1 month.

3 Methods

3.1 10% Sodium Dodecyl Sulfate-Polyacrylamide Gel Electrophoresis and Immunoblotting

1. Mix 3.75 mL of resolving buffer with 3.33 mL of acrylamide solution, 2.7 mL of ultrapure water, 100 μL of 10% p/v SDS, and 2.7 μL of TEMED in a 15 mL conical-bottom disposable plastic tube, and maintain this solution at 37 °C for 10 min. Add to the mixture 100 μL of ammonium persulfate, and pour the gel into a 7.3 cm × 10 cm × 1.5 mm gel cassette. Allow space for stacking gel and overlay with 2-butanol (*see* **Note 7**).
2. For the stacking gel, mix 625 μL of stacking gel buffer with 830 μL of acrylamide solution, 3.4 mL of ultrapure water, 50 μL of 10% p/v SDS, and 1.5 μL of TEMED in a 15 mL conical-bottom disposable plastic tube, and maintain this solution at 37 °C for 10 min. Add 50 μL of 10% p/v ammonium persulfate. Eliminate 2-butanol, pour the stacking gel mix over the polymerized gel, and insert a 10-well gel comb without introducing air bubbles.
3. To prepare heated extract from rat brain: collect one brain (1 g); mince the tissue roughly; add 3 volumes of chilled 0.05 M sodium borate buffer, pH 7.5, containing 1 mM EDTA, 0.5 mM 2-mercaptoethanol, 0.1 mg/mL leupeptin, 10 μg/mL aprotinin, and 2 mM Pefabloc® SC, homogenize in a Potter-Elvehjem homogenizer up and down for four times; and lyse the cells by sonication (4 bursts of 10 s each). Carry out all of these steps on ice. Centrifuge at 4 °C, 100,000 × *g* for 10 min, discard the pellet, and collect the clear supernatant. Split the sample in different tubes (200–300 μL aliquots in each tube), heat the samples for 3 min in a boiling water bath, and immediately cool them on ice. Pool all the heated aliquots, and centrifuge at 4 °C, 100,000 × *g* for 10 min. Discard the pellet that contains denatured proteins (*see* **Note 8**), and collect the soluble material (heated extract).
4. Dilute an aliquot (50 μg) of the heated rat brain extract in SDS-PAGE loading mix, heat the sample at 95 °C for 5 min, spin down liquid on the tube walls, and load the sample onto 10% polyacrylamide gel. Apply pre-stained protein ladder mix to a separate lane, for determining the size of the calpastatin species present in the heated extracts. This control sample is also important to confirm the transfer of the proteins from the gel to the nitrocellulose.
5. Perform electrophoresis at 25 mA until the sample has entered the gel, and then increase current to 60 mA until the dye front (bromophenol blue) reaches the bottom of the gel. After separation, remove the gel from the electrophoresis apparatus, and incubate it in Western blot transfer buffer for approximately 10 min in gentle agitation to remove SDS.

6. Cut the nitrocellulose membrane and blot adsorbent filter paper to the dimensions of the gel. Soak nitrocellulose, blot adsorbent filter paper, and fiber pads in Western blot transfer buffer for 10 min (*see* **Note 9**). Place the blot sandwich setup in the correct sequence in module as described by the supplier of the Western blot apparatus, in a tank completely filled with chilled Western blot transfer buffer. Perform the transfer of proteins at 100 V for 2 h (*see* **Note 10**).
7. After the transfer, unclamp the blot sandwich, expose the nitrocellulose membrane, and mark the position of the pre-stained markers, since they may fade away during immunoblotting analysis. Let the nitrocellulose membrane saturate with the blocking solution for 1 h at room temperature with shaking. (Alternatively, you can carry out saturation for 30 min at 37 °C.)
8. Wash the membrane twice in T-PBS, for 10 min each time. Incubate with primary antibodies in T-PBS overnight at 4 °C with shaking. Wash four times in T-PBS, for 10 min each time.
9. Incubate with the appropriate secondary antibody following the manufacturer's instruction, 1 h at room temperature with shaking. Wash four times in T-PBS, for 10 min each time. Wash twice in PBS, for 5 min each time. Develop the immunoreactive bands with an ECL detection system (*see* **Note 11**).

3.2 Isolation of Calpain from Erythrocytes

1. Human erythrocytes derive from either a whole blood donation or a therapeutic phlebotomy. To isolate the erythrocytes, divide the sack contents into 50 mL polypropylene tubes, and centrifuge at 20 °C, 1200 × *g* for 10 min. Withdraw the supernatant (plasma and platelets) and part of the interface between the plasma and the erythrocytes (buffy coat). Be careful not to remove too many erythrocytes; use a vacuum filtration system connected to a Pasteur pipette (*see* **Note 12**).
2. Fill the tubes containing the packed erythrocytes with 0.15 M NaCl, mix briefly, and centrifuge the tubes at 4 °C, 1200 × *g* for 10 min. Discard supernatant and the buffy coat still present on top of erythrocytes. Repeat this step three times.
3. Pool the erythrocytes into a few tubes, and carry out a final centrifugation at 4 °C, 1200 × *g* for 15 min. Remove the supernatant and proceed with lysis.
4. Measure the volume of the packed erythrocytes (about 150–200 mL from a therapeutic phlebotomy), and pour the packed cells, drop by drop, into five volumes of chilled 1 mM EDTA pH 7.0 rapidly rotating on a magnetic stirrer. Let the lysis proceed in gentle rotation for 5 min.
5. Centrifuge the lysate at 4 °C, 14,000 × *g* for 20 min. In the meantime, equilibrate 50–100 g of DE53 resin (wet powder,

prepared following the instructions of the supplier) in Buffer A. Leave the resin humid, ready to use.

6. Collect the supernatant in a beaker and put it on a magnetic stirrer in gentle rotation. Adjust pH to 6.7 with diluted acetic acid, and gradually apply the normalized resin to the crude lysate (resin/packed erythrocytes 1 g/5 mL) (*see* **Note 13**). Control that no resin clumps are present, and let absorption proceed in gentle rotation for 15 min.
7. Pour the contents of the beaker into a Büchner funnel provided with a paper filter. Insert the funnel in a Büchner flask associated to a vacuum device, and separately collect the flow through (*see* **Note 14**).
8. Wash the resin in the funnel extensively with Buffer A until the absorbance at 280 nm of the flow through is below 0.02 (about 3–4 L of Buffer A).
9. Transfer the wet resin to a clean beaker, dissolve it in Buffer A, and pour the suspension over a glass column.
10. Elute the absorbed proteins in 3–4 mL fractions with 500 mL of 0.2 M NaCl dissolved in Buffer A.
11. Determine the protein content of each fraction at 280 nm.
12. Pool the fractions containing the peak of proteins, add 50% $(NH_4)_2SO_4$ (p/v final concentration), and mix gently, until the salt is completely dissolved. To support efficient protein precipitation, leave the sample at 4 °C for 15 min.
13. Centrifuge at 4 °C 15,000 × *g* for 20 min. Discard supernatant, dissolve the pellet in the lowest possible volume of Buffer A (5–10 mL), and dialyze against 5 L of Buffer A at 4 °C overnight.
14. Load the sample on a Source 15Q column (1 × 15 cm) equilibrated in Buffer A, and after washing, elute absorbed proteins with a linear gradient of NaCl 0–0.3 M (200 + 200 mL). Collect fractions of 2.5–3 mL. Use 50 μL aliquots to measure calpain activity. We routinely assay calpain by measuring the amount of acid-soluble peptide(s) liberated from human denatured globin used as substrate (*see* Subheading 3.3 for preparation of globin). The assay calpain mixture contains in a final volume of 250 μL, 50 mM NaBorate pH 7.5, 1 mM $CaCl_2$, and 0.5 mg of denatured globin. The reaction starts with an appropriate amount of calpain and incubation of the mixture at 25 °C for 10 min. Stop the reaction by adding 40 μL of TCA. After centrifugation, add 200 μL of supernatant to 2 mL of Buffer C, and under vigorous rotation, also add 200 μL of fluorescamine. The fluorescence of each sample is detected at 384 nm excitation and 475 nm emission within half an hour from addition of fluorescamine.

15. Pool the fractions containing calpain activity, and concentrate them by ultrafiltration to 5–10 mL volume in a stirred ultrafiltration cell provided with a 30,000 cutoff membrane YM-30. Dialyze the sample against Buffer B.
16. At the end of the dialysis, add NaCl to the calpain sample to have 0.3 M final concentration.
17. Apply the sample to a Phenyl-Sepharose column (1 × 5 cm) equilibrated in Buffer B containing 0.3 M NaCl. After extensive washing, elute absorbed proteins with 100 mL Buffer B without NaCl. Collect fractions containing calpain activity, and concentrate the sample to 1–1.5 mL in a stirred ultrafiltration cell.
18. Submit the sample to size-exclusion chromatography on a Sephacryl S-200 HR column (1.5 × 120 cm) equilibrated in Buffer B.
19. Pool the fractions containing calpain activity, and concentrate them by ultrafiltration to 1.5–2 mL. Determine protein concentration and calpain-specific activity.

3.3 Isolation of Globin from Erythrocytes

Globin is the protein moiety of hemoglobin and the most represented protein in erythrocytes. It is also the major contaminant during the purification of both calpain and calpastatin from human erythrocytes. Indeed the first chromatography step is aimed to remove hemoglobin from the sample before proceeding with the purification of calpain.

1. Process human erythrocytes as described above, and use 100 mL of the flow through obtained from **step** 7 of Subheading 3.2 for globin purification (*see* **Note 14**).
2. Add drop by drop the flow through to 5 L of globin extraction solution rapidly stirring on a magnetic mixer.
3. Let protein extraction proceed for 10 min under continuous vigorous stirring.
4. At the end, you will see that the solution has become brownish, and a white precipitate (globin) is formed on the bottom of the beaker. Discard the supernatant.
5. Collect the precipitate in another beaker, and dissolve it in the lowest possible volume of H_2O. Repeat again the extraction step (**steps 2** and **3**).
6. Collect the white precipitate in 30 mL polypropylene tubes, and wash it by repeated centrifugation (5 min 80 × *g*) with pure acetone (in order to remove residual HCl) until the pH of the resuspended pellet (use pH-indicator strips) normalizes with that of acetone.

7. Spread the solid pellet in a clean glass container and let the acetone dry out overnight.
8. Dissolve the dried pellet in the lowest possible volume of H_2O, and remove residual HCl by dialyzing the sample against ultrapure water (*see* **Note 15**).
9. Collect all the globin in the same tube, assay protein concentration, and store in aliquots at −20 °C until needed.

3.4 Calpastatin Elution from SDS Polyacrylamide Gel and Analysis of Activity

1. Prepare the heated extract from rat brain as described in Subheading 3.1, and submit to 10% SDS-PAGE using aliquots of 100 μg/lane (500 μg total, based on Bradford assay). Apply pre-stained protein ladder mix to a separate lane, for determining the size of the calpastatin species present in the heated extracts.
2. At the end of the electrophoretic run, wash the gel with sodium borate buffer containing 20% v/v methanol to remove SDS (*see* **Note 16**). After 20 min, wash the gel with sodium borate buffer without methanol, and stir gently for additional 20 min.
3. Put the gel on a clean glass plate placed on a sheet of millimeter paper. Trim the lanes accurately, and discard gel in excess. Cut the five gel lanes containing the heated rat brain extract together into 0.3 cm slices. To identify the apparent Mw of the calpastatin species present in the lanes, simultaneously cut the lane containing pre-stained protein ladder, and mark the Mw on the corresponding calpastatin slice.
4. Cut the slices into smaller pieces (*see* **Note 17**) and put them in numbered tubes. Add 0.2 mL of sodium borate buffer, and proceed with protein extraction for 5 h at room temperature in vigorous shaking. Recover the liquid from each fraction, transfer it to a new tube, and store fractions at 4 °C. Repeat protein extraction three times. At the end, you will have about 0.6 mL total.
5. Assay calpastatin inhibitory activity on 200 μL of the recovered solutions (Table 1). To measure calpastatin inhibition, we routinely use human erythrocyte calpain purificated as indicated in Subheading 3.2 because this enzyme is highly sensitive to native and posttranslational modified calpastatins [14]. You can identify which calpastatin species are actually present in rat brain or in other tissues based on the electrophoretic mobility, recognition by anti-calpastatin antibodies, and calpastatin activity when eluted from the gel (Fig. 1).

The data reported in this figure provide information both on the apparent Mw of the active calpastatin forms and, at least partially, on the molecular composition of such inhibitory species. Indeed we can match the results obtained from the immunoblot

Table 1
Assay of calpastatin activity in solution eluted from SDS-PAGE of rat brain-heated extract

Fraction number	Fluorescence Lane a	Fluorescence Lane b
1	187	176
2	175	177
3	**88**	**95**
4	178	165
5	**103**	**98**
6	**99**	**113**
7	125	168
8	136	198
9	**78**	**125**
10	179	196
11	198	174
12	199	189
13	**135**	**138**
14	189	177
15	187	172
16	185	186
17	198	189
18	177	185
19	187	198
20	196	175
21	181	164
22	198	182
23	185	193
24	192	199

In the table are the fluorescence values corresponding to calpain activity assayed in the presence of 200 μL of the recovered solution from each gel slice (*see* Fig. 1). The value of fluorescence obtained from the assay of calpain in the absence of calpastatin is 180 (*see* Subheading 3.2). A decrease in fluorescence reveals the presence of calpastatin in the gel slice (indicated in bold in the table)

analyses, performed with the two different primary antibodies, with those regarding the assay of calpastatin activity. It can be concluded that rat brain contains active calpastatin Type I (no information about Type II) and a cleaved Type I fragment lacking part of the

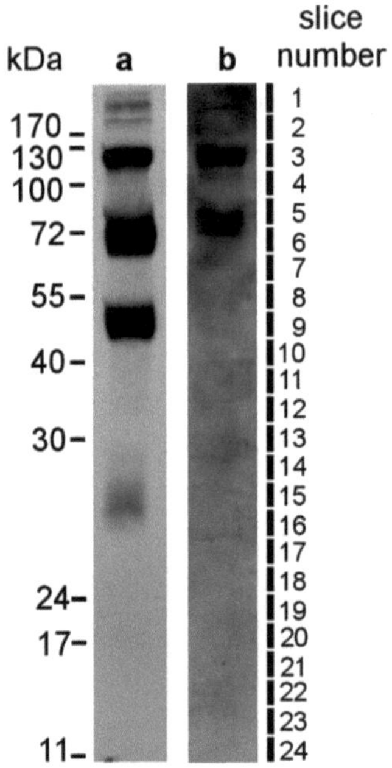

Fig. 1 Immunoblotting of rat brain-heated extract. (Lane **a**) Calpastatin forms present in 50 μg of heated extract and revealed with mAb 35.23 [15] that recognizes a region of calpastatin at the beginning of the exon 8 [10]. (Lane **b**) Calpastatin forms present in 50 μg of heated extract and revealed with anti-exon 1xa antibody [2] that recognizes a region of exon 1xa present in the XL-domain of calpastatin Type I

C-terminus because it is still recognized by anti-1xa antibody. Two other active calpastatin fragments are present: the first one is only recognized by mAb 35.23 and is deprived of the XL-domain, and the second, lacking both the XL-domain and the epitope of 35.23, presumably has lost also the first inhibitory domain [2, 10, 15].

4 Notes

1. In these conditions both buffers can be stored up to 6 months.
2. It is better to prepare small amounts of this solution weekly.
3. The running buffer can be reused three to four times.
4. The addition of 0.05% p/v SDS to Western blot transfer buffer improves the transfer of high molecular weight proteins from the gel to the nitrocellulose membrane, without altering the transfer of the lower Mw proteins.
5. The Western blot transfer buffer can be reused twice.
6. We find that it is best to prepare this fresh each time.

7. Since the unpolymerized acrylamide solution is toxic and may cause cancer, use the fume hood before the polymerization of the gel.
8. Differently from other proteins, calpastatin does not denature at high temperature. Thus, after heating you are able to completely recover native calpastatin.
9. Always wear gloves when handling the gel or the nitrocellulose membrane to minimize protein background.
10. Remember, during transfer of proteins, to immerse the electrophoretic tank in a container filled with water and ice to maintain the Western blot transfer buffer cold enough to avoid heating of the nitrocellulose membrane. The current intensity should not exceed 400 mA, if it does, decrease the voltage. Alternatively, you can carry out the Western blot overnight at 15 V without cooling.
11. We recommend using Amersham™ ECL Select™ Western Blotting Detection Reagent, catalog number: RPN2235 from GE Healthcare.
12. The buffy coat looks like a whitish layer on the erythrocyte pellet that can be removed easily without disturbing the underlying erythrocytes too much.
13. It is important during this absorption step to maintain the pH at 6.7 by adding diluted acetic acid or NaOH to prevent hemoglobin in the lysate binding to the resin. Hemoglobin is the most represented and contaminating erythrocyte protein. For this reason, during absorption, control pH continuously by submerging the pH-meter electrode in the beaker.
14. This non-adsorbed material contains mainly hemoglobin that can be used in denatured form as substrate for assaying calpain activity. You can store 100–200 mL of the flow through at −20 °C, and prepare denatured globin when needed (*see* Subheading 3.3).
15. Use dialysis bags (regenerated cellulose tubular membranes) having cutoff of 6000–8000, and change the dialysis water frequently. You must carry out the dialysis until the pH of the sample reaches a neutral value (generally, 2 days of dialysis are required). During dialysis, control the volume of the globin sample periodically because it increases very much and dialysis bags can get very swollen and break down.
16. Control the size of the gel before this step because it is possible that during the washing, the size of the gel can be a little modified. In this case we have to take into account this size modification during the comparison of calpastatin forms identified by immunoblotting with those identified after elution from SDS-polyacrylamide gel.

17. Cut the gel into small well-defined pieces in order to avoid the crumbling of the polyacrylamide gel in the elution buffer. This could alter the calpastatin assay by raising the background signal.

Acknowledgment

This work was supported by grants FRA2015 and FRA2016 from the University of Genoa to MA and RDT.

References

1. Takano J, Watanabe M, Hitomi K, Maki M (2000) Four types of calpastatin isoforms with distinct amino-terminal sequences are specified by alternative first exons and differentially expressed in mouse tissues. J Biochem 128:83–92
2. Parr T, Jewell KK, Sensky PL, Brameld JM, Bardsley RG, Buttery PJ (2004) Expression of calpastatin isoforms in muscle and functionality of multiple calpastatin promoters. Arch Biochem Biophys 427:8–15
3. Li S, Liang ZG, Wang GY, Yavetz B, Kim ED, Goldberg E (2000) Molecular cloning and characterization of functional domains of a human testis-specific isoform of calpastatin. Biol Reprod 63:172–178
4. Parr T, Sensky PL, Bardsley RG, Buttery PJ (2001) Calpastatin expression in porcine cardiac and skeletal muscle and partial gene structure. Arch Biochem Biophys 395:1–13
5. Raynaud P, Vignoles-Jayat C, Laforêt MP, Leveziel H, Amarger V (2005) Four promoters direct expression of the calpastatin gene. Arch Biochem Biophys 437:69–77
6. Lee WJ, Ma H, Takano E, Yang HQ, Hatanaka M, Maki M (1992) Molecular diversity in amino-terminal domains of human calpastatin by exon skipping. J Biol Chem 267:8437–8442
7. De Tullio R, Sparatore B, Salamino F, Melloni E, Pontremoli S (1998) Rat brain contains multiple mRNAs for calpastatin. FEBS Lett 422:113–117
8. Takano J, Kawamura T, Murase M, Hitomi K, Maki M (1999) Structure of mouse calpastatin isoforms: implications of species-common and species-specific alternative splicing. Biochem Biophys Res Commun 260:339–345
9. Geesink GH, Nonneman D, Koohmaraie M (1998) An improved purification protocol for heart and skeletal muscle Calpastatin reveals two isoforms resulting from alternative splicing. Arch Biochem Biophys 356:19–24
10. De Tullio R, Averna M, Stifanese R, Parr T, Bardsley RG, Pontremoli S, Melloni E (2007) Multiple rat brain calpastatin forms are produced by distinct starting points and alternative splicing of the N-terminal exons. Arch Biochem Biophys 465:148–156
11. Averna M, Stifanese R, De Tullio R, Passalacqua M, Defranchi E, Salamino F, Melloni E, Pontremoli S (2007) Regulation of calpain activity in rat brain with altered Ca^{2+} homeostasis. J Biol Chem 282:2656–2665
12. Averna M, De Tullio R, Salamino F, Minafra R, Pontremoli S, Melloni E (2001) Age-dependent degradation of calpastatin in kidney of hypertensive rats. J Biol Chem 276:38426–38432
13. Takano E, Maki M, Mori H, Hatanaka M, Marti T, Titani K, Kannagi R, Ooi T, Murachi T (1988) Pig heart calpastatin: identification of repetitive domain structures and anomalous behavior in polyacrylamide gel electrophoresis. Biochemistry 27:1964–1972
14. Pontremoli S, Viotti PL, Michetti M, Salamino F, Sparatore B, Melloni E (1992) Modulation of inhibitory efficiency of rat skeletal muscle calpastatin by phosphorylation. Biochem Biophys Res Commun 187:751–759
15. De Tullio R, Averna M, Salamino F, Pontremoli S, Melloni E (2000) Differential degradation of calpastatin by mu- and m-calpain in Ca^{2+}-enriched human neuroblastoma LAN-5 cells. FEBS Lett 475:17–21

Part V

Manipulation of Calpain Expression

Chapter 16

Experimental Manipulation of Calpain Activity In Vitro

Seiichi Kitagawa

Abstract

The calpain activity in cells can be experimentally manipulated in vitro by calpain inhibitors, and various types of calpain inhibitors such as peptide aldehydes and α-mercapto-acrylic acid derivatives are widely used as a valuable tool to elucidate the physiological and pathological roles of calpain. Here I describe the experimental procedures with calpain inhibitors, with human neutrophils being primarily used in this experiment. It should be noted that potent calpain inhibitors not only inhibit the calpain activity but also stimulate cell functions via direct activation of human formyl peptide receptors and/or other G protein-coupled receptors depending on the inhibitors used.

Key words Calpain, Calpain inhibitors, Human formyl peptide receptors, G protein-coupled receptors, Migration, Apoptosis

1 Introduction

Calpain, the calcium-dependent cysteine protease, has been implicated as a regulator of cell migration and apoptosis [1–3], and calpain activation has been shown to be involved in various disorders, including neurodegenerative diseases and myocardial dysfunction associated with myocardial infarction, indicating that calpain is a potential therapeutic target for these disorders [4, 5]. Potent calpain inhibitors are widely used for in vitro and in vivo experiments to elucidate the role of calpain activity [1–6]. The calpain inhibitors commonly used include peptide aldehydes (such as *N*-acetyl-Leu-Leu-Nle-CHO [ALLN] and *N*-acetyl-Leu-Leu-Met-CHO [ALLM]) and α-mercapto-acrylic acid derivatives (such as PD150606 [3-(4-iodophenyl)-2-mercapto-(*Z*)-2-propenoic acid] and PD151746 [3-(5-fluoro-3-indolyl)-2-mercapto-(*Z*)-2-propenoic acid]). The peptide aldehydes such as ALLN and ALLM exhibit competitive inhibition with respect to substrate and interact with the active sites of calpain, whereas α-mercapto-acrylic acid derivatives exhibit uncompetitive inhibition and interact with the calcium-binding domain of the large and small subunits of calpain,

Jeannette S. Messer (ed.), *Calpain: Methods and Protocols*, Methods in Molecular Biology, vol. 1915,
https://doi.org/10.1007/978-1-4939-8988-1_16,

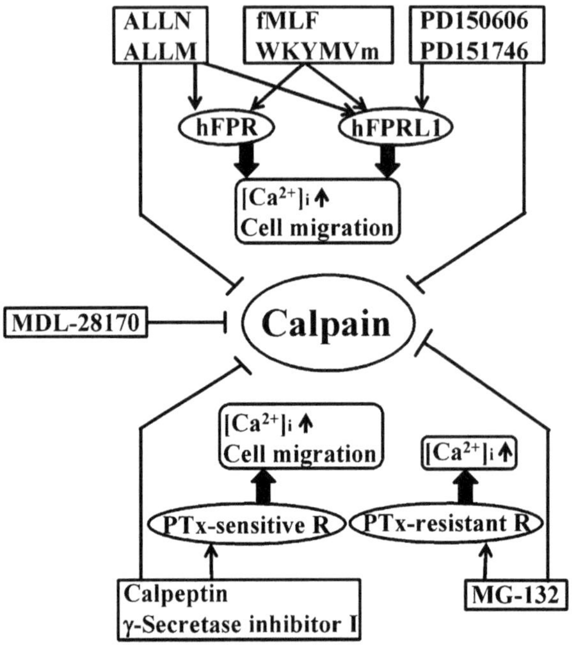

Fig. 1 Possible targets of calpain inhibitors. Potent calpain inhibitors activate human formyl peptide receptors (hFPR and hFPRL1) and/or other GPCRs in addition to inhibition of the calpain activity, which may be responsible for the characteristic effects of calpain inhibitors on cell functions. PTx-sensitive R, pertussis toxin-sensitive receptor, WKYMVm, Trp-Lys-Tyr-Met-Val-D-Met (W peptide)

but not the active sites [7, 8]. Other peptide aldehydes, including MDL-28170, γ-secretase inhibitor I, calpeptin, and MG-132, are also known to be capable of inhibiting the calpain activity; the precise mechanisms by which these compounds inhibit the calpain activity remain elusive. Some compounds such as ALLN, ALLM, and MG-132 are also known to be capable of inhibiting the proteasome activity [9]. Our recent studies show that potent calpain inhibitors not only inhibit the calpain activity but also stimulate cell functions via direct activation of human formyl peptide receptors and/or other G protein-coupled receptors (GPCRs) depending on the inhibitors used [10, 11], indicating that careful interpretation is required depending on the inhibitors and the cells used (*see* **Note 1** and Fig. 1).

Here, I describe a method for the experimental manipulation of calpain activity in vitro using human peripheral blood neutrophils as the source of calpain and various calpain inhibitors. Potent calpain inhibitors suppress ionomycin-induced cleavage of talin and induce delayed spontaneous apoptosis and directed migration of neutrophils, which is accompanied by activation of specific signaling molecules. The calpain activity in intact cells can be easily manipulated or suppressed by potent calpain inhibitors. However, the findings obtained should be carefully interpreted in consideration of the characteristics of the inhibitors.

2 Materials

2.1 Calpain Inhibitors

Potent calpain inhibitors include ALLN (*N*-acetyl-Leu-Leu-Nle-aldehyde, calpain inhibitor I), ALLM (*N*-acetyl-Leu-Leu-Met-aldehyde, calpain inhibitor II), MDL-28170 (*N*-benzyloxycarbonyl-Val-Phe-aldehyde, calpain inhibitor III), MG-132 (*N*-benzyloxycarbonyl-Leu-Leu-Leu-aldehyde, calpain inhibitor IV), γ-secretase inhibitor I (*N*-benzyloxycarbonyl-Leu-Leu-Nle-aldehyde), calpeptin (*N*-benzyloxycarbonyl-Leu-Nle-aldehyde), PD150606 [3-(4-iodophenyl)-2-mercapto-(*Z*)-2-propenoic acid], and PD151746 [3-(5-fluoro-3-indolyl)-2-mercapto-(*Z*)-2-propenoic acid]) (Table 1) (*see* **Note 2**). PD145305 (2-mercapto-3-phenylpropanonic acid), which is an inactive analog of PD150606 and unable to inhibit the calpain activity, can be used as a control.

Table 1
Potent calpain inhibitors

Inhibitors[a]		Synonyms	Chemical structures
α-Mercapto-acrylic acid derivatives	PD150606		3-(4-Iodophenyl)-2-mercapto-(*Z*)-2-propenoic acid ($C_9H_7IO_2S$)
	PD151746		3-(5-Fluoro-3-indolyl)-2-mercapto-(*Z*)-2-propenoic acid ($C_{11}H_8FNO_2$)
	PD145305		2-Mercapto-3-phenylpropanonic acid ($C_9H_{10}O_2S$)
Peptide aldehydes	ALLN	Calpain inhibitor I, MG-101	*N*-Ac-Leu-Leu-Nle-CHO
	ALLM	Calpain inhibitor II	*N*-Ac-Leu-Leu-Met-CHO
	MDL-28170	Calpain inhibitor III	*Z*-Val-Phe-CHO
	MG-132	Calpain inhibitor IV	*Z*-Leu-Leu-Leu-CHO
	Calpeptin		*Z*-Leu-Nle-CHO
	γ-Secretase inhibitor I		*Z*-Leu-Leu-Nle-CHO

N-Ac, *N*-acetyl; *Z*, *N*-benzyloxycarbonyl; CHO, aldehyde

[a]The peptide aldehydes such as ALLN and ALLM exhibit competitive inhibition with respect to substrate and interact with the active sites of calpain, whereas α-mercapto-acrylic acid derivatives exhibit uncompetitive inhibition and interact with the calcium-binding domain of the large and small subunits of calpain, but not the active sites. Refer to the data shown in Fig. 2 for the effective concentrations; for example, those for PD150606 and ALLN are 1–50 μM and 1.3–130 μM, respectively

1. ALLN, ALLM, MDL-28170, γ-secretase inhibitor I, PD150606, PD151746, and PD145305 (Calbiochem, San Diego, CA).
2. Calpeptin and MG-132 (BIOMOL International, Plymouth Meeting, PA).
3. Dissolve each of these compounds in dimethyl sulfoxide (DMSO) (>400-fold concentration of the final working solution), and store the aliquots at −20 °C (*see* **Note 3**).

2.2 Other Reagents

1. Ionomycin (a calcium ionophore) and fMLF (*N*-formyl-Met-Leu-Phe; a chemotactic peptide) (Calbiochem, San Diego, CA). Dissolve ionomycin and fMLF in DMSO, and store the aliquots at −20 °C (*see* **Note 3**).
2. Dextran (molecular mass 170,000–200,000) and *N*-2-hydroxyethyl-piperazine-*N′*-2-ethane-sulfonic acid (HEPES) (Nakarai Chemicals, Kyoto, Japan).
3. Ficoll-Conray solution (Daiichi-Kagaku Co., Tokyo, Japan).
4. Mouse monoclonal antibody against talin (Sigma, St Louis, MO).
5. Whole blood (40–60 ml). Peripheral blood in a syringe containing heparin (5–10 units/ml blood).
6. Phosphate-buffered saline (PBS) and Hank's balanced salt solution (HBSS) (Nissui Pharmaceutical Co, Tokyo, Japan).
7. Trichloroacetic acid.
8. Syringes and 18G needles.
9. Annexin V-FITC and propidium iodide.
10. Antibodies against Mcl-1 and X-linked inhibitor of apoptosis (XIAP).

3 Methods

3.1 Neutrophil Preparation

1. Freshly prepare 1% dextran solution in 0.9% NaCl (*see* **Note 4**).
2. Add 3/4 volume of 1% dextran solution into the syringe (20–30 ml syringe) containing the blood. Mix the solution well, and stand the syringe with the needle (18G) upward for 40–60 min at room temperature.
3. After sedimentation of red blood cells, harvest the upper leukocyte-rich fraction using the bended needle. (Bend a needle carefully and push the inner piston.)
4. Put the leukocyte-rich fraction (6–7 ml) onto the Ficoll-Conray solution (3 ml) in the centrifuge tube, and centrifuge it at 500 × *g* at room temperature to separate neutrophils from monocytes, lymphocytes, and platelets.

5. Aspirate and discard all fractions except the neutrophil fraction located at the bottom.
6. Harvest the neutrophil fraction, suspend cells in PBS (approximately 20 ml), and centrifuge them at 500 × *g* for 5 min at 4 °C.
7. Suspend cells in 1.5 ml of PBS. Add the ice-cold distilled water (10 ml) to the cell suspension, and shake the tube gently for 20–30 s to lyse contaminating erythrocytes (hypotonic lysis of erythrocytes). Add the ice-cold 1.8% NaCl solution (10 ml) to restore the osmolality.
8. Centrifuge the tubes at 500 × *g* for 5 min at 4 °C. Repeat this procedure, if necessary, to lyse the residual contaminating erythrocytes (*see* **Note 5**).
9. Suspend neutrophil fractions in Hank's balanced salt solution (HBSS) containing 10 mM *N*-2-hydroxyethyl-piperazine-*N*′-2-ethane-sulfonic acid (HEPES, pH 7.4), and keep the cell suspensions on ice. Neutrophil fractions contain more than 95% neutrophils, with contaminating cells being almost exclusively eosinophils (*see* **Notes 6** and 7).

3.2 Calpain Inhibition by Potent Calpain Inhibitors in Intact Cells

Treatment of neutrophils with ionomycin (a calcium ionophore) (1 μM, dissolved in DMSO) for 30 min at 37 °C results in the cleavage of talin (a substrate of calpain), indicating that the calpain activity is increased in ionomycin-treated neutrophils. By using this system, one can confirm that the calpain activity is suppressed by

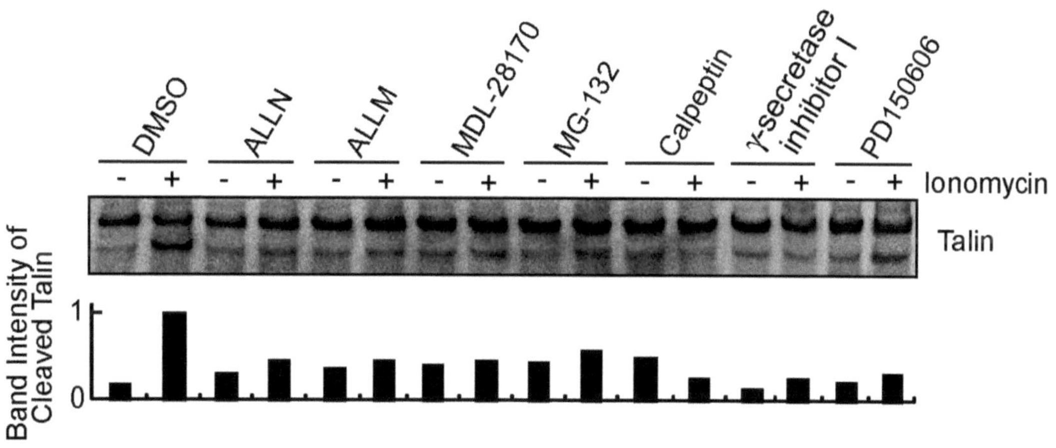

Fig. 2 Confirmation of calpain inhibition by potent calpain inhibitors in intact cells. Neutrophils are pretreated with ALLN (130 μM), ALLM (125 μM), MDL-28170 (261 μM), MG-132 (112 μM), calpeptin (115 μM), γ-secretase inhibitor I (112 μM), PD150606 (50 μM), or DMSO (a vehicle) for 30 min at 37 °C and thereafter stimulated with ionomycin (1 μM) for 30 min at 37 °C. Immunoblotting is performed using antibody against talin. The cell lysates, equivalent to 2.5×10^5 cells, are loaded onto each lane. The densitometric data are shown in the lower panel (Reproduced from [10])

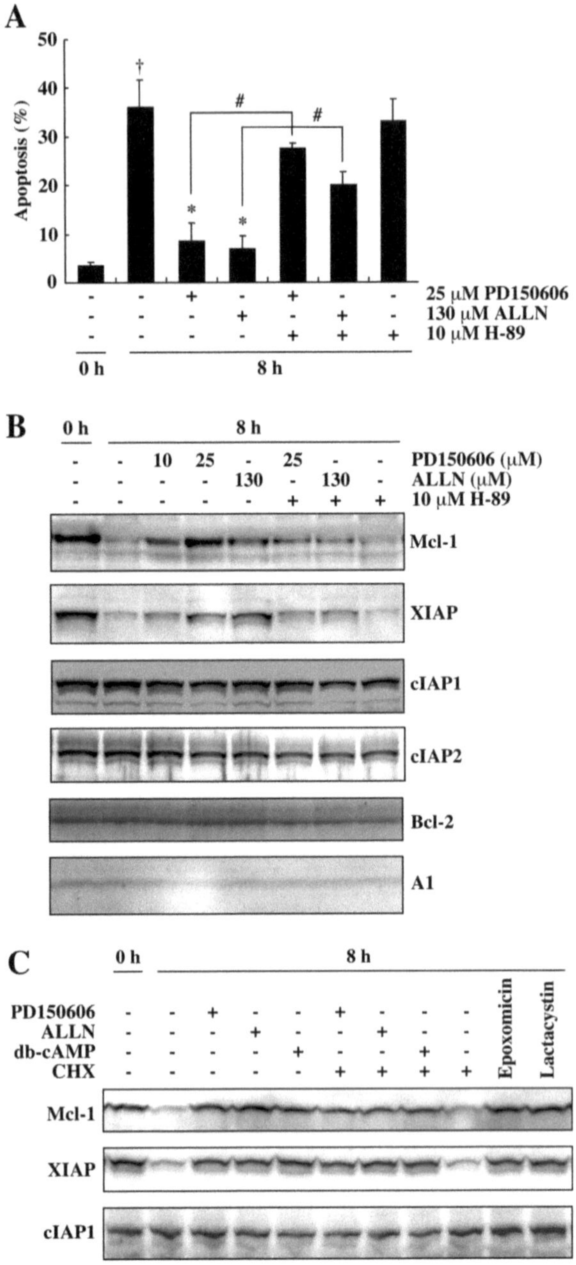

Fig. 3 Calpain inhibitors delay human neutrophil apoptosis via protein kinase A-mediated stabilization of Mcl-1 and XIAP. Neutrophils are incubated in the presence or absence of PD150606 (25 μM), ALLN (130 μM), dibutyryl cyclic AMP (db-cAMP, 1 mM), epoxomicin (1 μM), or lactacystin (5 μM) for 8 h at 37 °C. When required, cells are pretreated with H-89 (10 μM) or cycloheximide (CHX, 10 μg/ml) for 30 min. (**a**) Apoptotic cells are analyzed by flow cytometry. Annexin V-positive, propidium iodide-negative cells are considered as apoptotic cells. †Significantly greater as compared with 0 h controls ($P < 0.01$). *Spontaneous apoptosis is significantly prevented by PD150606 or ALLN ($P < 0.01$). #PD150606- or ALLN-mediated delay of neutrophil apoptosis is significantly prevented by H-89 (protein kinase A inhibitor) ($P < 0.05$). (**b** and **c**) The immunoblotting is performed using antibodies against each protein. The levels of Mcl-1 and XIAP are markedly decreased during spontaneous neutrophil

potent calpain inhibitors in the experimental systems used; i.e., ionomycin-induced cleavage of talin could be suppressed by potent calpain inhibitors (*see* **Note 8**).

1. Add calpain inhibitors (e.g., ALLN [130 μM], ALLM [125 μM], MDL-28170 [261 μM], MG-132 [112 μM], calpeptin [115 μM], γ-secretase inhibitor 1 [112 μM], or PD150606 [50 μM]) or DMSO (a vehicle) to the cell suspensions (1×10^6 cells/ml). The reaction mixture is 1 ml. Add each calpain inhibitor at once, and the inhibitor is present throughout the incubation period.
2. Incubate the reaction mixture for 30 min at 37 °C.
3. Add ionomycin (1 μM), and incubate the reaction mixture for 30 min at 37 °C. As protease activity is very high in human neutrophils, the reactions should be terminated by the addition of trichloroacetic acid (TCA, final concentration 10%), which can effectively prevent the proteolysis during the preparation of cell lysates (*see* **Note 9**).
4. Detect intact talin and its cleaved product by the standard immunoblotting method using antibody against talin. Confirm that ionomycin-induced cleavage of talin is inhibited by potent calpain inhibitors (Fig. 2).

3.3 Delayed Neutrophil Apoptosis by Calpain Inhibitors

Human neutrophils undergo spontaneous apoptosis, which is delayed by calpain inhibitors.

1. Suspend neutrophils in HBSS (1×10^6 cells/ml). Add calpain inhibitors (e.g., PD150606 [10 or 25 μM] or ALLN [130 μM]) or DMSO to the cell suspensions (*see* **Note 10**).
2. Incubate the reaction mixture for 8 h at 37 °C.
3. Stain cells with annexin V-FITC and propidium iodide, and determine apoptotic cells by flow cytometry with FACSCalibur (Becton Dickinson) [3]. Annexin V-positive, propidium iodide-negative cells are considered as apoptotic cells (*see* **Note 11**).
4. Terminate the reaction by the addition of trichloroacetic acid (TCA), and determine the levels of the anti-apoptotic molecules such as Mcl-1 and X-linked inhibitor of apoptosis (XIAP) by the standard immunoblotting method (Fig. 3). Use 5×10^6 cells/ml for this experiment (*see* **Note 12**).

Fig. 3 (continued) apoptosis, and the reduction in the levels of Mcl-1 and XIAP is prevented by calpain inhibitors (PD150606 and ALLN), proteasome inhibitors (epoxomicin and lactacystin), or dibutyryl cyclic AMP. The stabilizing effect of PD150606 and ALLN on Mcl-1 and XIAP is suppressed by H-89. cIAP1, cellular inhibitor of apoptosis 1 (Reproduced from [3])

4 Notes

1. Possible targets of potent calpain inhibitors are shown in Fig. 1 [12]. It should be noted that potent calpain inhibitors not only inhibit the calpain activity but also activate human formyl peptide receptor (hFPR), hFPR-like 1 (hFPRL1), and/or other GPCRs, leading to cell activation. ALLN and ALLM may activate both hFPR and hFPRL1 (both being PTX-sensitive GPCRs), whereas α-mercapto-acrylic acid derivatives (PD150606 and PD151746) may activate hFPRL1, but not hFPR. Calpeptin and γ-secretase inhibitor I may activate other PTX-sensitive GPCRs than hFPR and hFPRL1. MG-132 may activate PTX-resistant GPCRs and fails to induce neutrophil migration. MDL-28170 induces phosphorylation of ERK, p38, and Akt, but not an increase in $[Ca^{2+}]_i$ and migration, in human neutrophils, suggesting that MDL-28170-mediated activation of these molecules may be associated with inhibition of the calpain activity. Chemotactic peptides (fMLF and W peptide [Trp-Lys-Tyr-Met-Val-D-Met]) activate both hFPR and hFPRL1 [12]. Our recent studies suggest that certain calpain inhibitors activate hFPR and/or hFPRL1 via direct binding to the putative fMLF binding site, which is apparently independent of the ability of these inhibitors to inhibit the calpain activity [10–12].

   It should be noted that the concentrations of the calpain inhibitors used in various studies are apparently sufficient to activate hFPR, hFPRL1, and/or other GPCRs. For example, calpeptin (100 μM) and PD150606 (50 μM) are used as calpain inhibitors to prevent XIAP degradation and to delay human neutrophil apoptosis [13]. ALLN (100 μM) and MG-132 (50 μM) are used as proteasome inhibitors to prevent Mcl-1 degradation and to delay human neutrophil apoptosis [14]. In this regard, it should be noted that ALLN and ALLM are originally reported to be potent inhibitors of calpains and cathepsins [7], and thereafter these peptide aldehydes, including ALLN, ALLM, and MG-132, have been shown to be capable of inhibiting the proteasome [9].
2. A synthetic calpastatin peptide (molecular mass 3177.7, a 27-residue peptide, Calbiochem) can be also used. However, this peptide may be less cell-permeable as compared with ALLN (molecular mass 383.5) because of its higher molecular mass, and therefore treatment of cells for longer periods (30–60 min with 50 μM calpastatin peptide) is required for its effects [2].
3. Some inhibitors are hardly soluble in the medium. Add each inhibitor (dissolved in DMSO) directly into the cell

suspensions to obtain the desired concentrations. Usually I add 1–2.5 μl of the stock solutions into 1 ml of the reaction mixture. Final DMSO concentration should be less than 0.25%. Higher concentrations of DMSO may impair cell functions. Pretreatment of cells with the inhibitors for 5 min is sufficient to exert the inhibitory effects. This procedure is also applicable to ionomycin.

4. Neutrophils may be activated when a stock solution of dextran is used.
5. Hypotonic lysis of contaminating erythrocytes is preferred for the functional analysis of neutrophils.
6. Calpain activity can be similarly manipulated in various types of cells. Here, I show the experiments with human neutrophils. Human monocytes and the cell lines such as HL-60 and HEK-293 cells can be also used. It should be noted that the cell responses differ according to the cell types and the inhibitors used.
7. All solutions are prepared using double-distilled water, in which lipopolysaccharide contamination is excluded. When cells are cultivated, all solutions are sterilized.
8. The concentration of intracellular Ca^{2+} ($[Ca^{2+}]_i$) in resting cells is approximately 100 nM, and calpain is rapidly activated by an increase in $[Ca^{2+}]_i$; for example, calpain 1 (μ-calpain) and calpain 2 (m-calpain) are activated by 1 μM Ca^{2+} and 1 mM Ca^{2+}, respectively.
9. The addition of TCA is not necessary when cell types other than human neutrophils are used.
10. Spontaneous neutrophil apoptosis is delayed by calpain inhibitors and dibutyryl cyclic AMP and accelerated by cycloheximide, suggesting that constitutive synthesis of anti-apoptotic proteins is required for neutrophil survival.
11. Other methods can be used for determination of apoptosis.
12. The levels of Mcl-1 and XIAP are selectively decreased during spontaneous neutrophil apoptosis, and the reduction in the levels of Mcl-1 and XIAP as well as spontaneous neutrophil apoptosis is prevented by calpain inhibitors [3, 14, 15]. The stabilizing effects of calpain inhibitors on Mcl-1 and XIAP may be ascribed to cyclic AMP-independent activation of protein kinase A. It should be noted that the reduction in the levels of Mcl-1 and XIAP as well as spontaneous neutrophil apoptosis is prevented not only by calpain inhibitors but also by proteasome inhibitors (such as epoxomicin and lactacystin) (Fig. 3) [3].

References

1. Lokuta MA, Nuzzi PA, Huttenlocher A (2003) Calpain regulates neutrophil chemotaxis. Proc Natl Acad Sci U S A 100:4006–4011
2. Katsube M, Kato T, Kitagawa M, Noma H, Fujita H, Kitagawa S (2008) Calpain-mediated regulation of the distinct signaling pathways and cell migration in human neutrophils. J Leukoc Biol 84:255–263
3. Ozaki Y, Kato T, Kitagawa M, Fujita H, Kitagawa S (2008) Calpain inhibition delays neutrophil apoptosis via cyclic AMP-independent activation of protein kinase A and protein kinase A-mediated stabilization of Mcl-1 and X-linked inhibitor of apoptosis (XIAP). Arch Biochem Biophys 477:227–231
4. Camins A, Verdaguer E, Folch J, Pallas M (2006) Involvement of calpain activation in neurodegenerative processes. CNS Drug Rev 12:135–148
5. Mani SK, Balasubramanian S, Zavadzkas JA, Jeffords LB, Rivers WT, Zile MR, Mukherjee R, Spinale FG, Kuppuswamy D (2009) Calpain inhibition preserves myocardial structure and function following myocardial infarction. Am J Physiol Heart Circ Physiol 297:H1744–H1751
6. Noma H, Kato T, Fujita H, Kitagawa M, Yamano T, Kitagawa S (2009) Calpain inhibition induces activation of the distinct signalling pathways and cell migration in human monocytes. Immunology 128:e487–e496
7. Sasaki T, Kishi M, Saito M, Tanaka T, Higuchi N, Kominam E, Katunuma N, Murachi T (1990) Inhibitory effect of di- and tripeptidyl aldehydes on calpains and cathepsins. J Enzym Inhib 3:195–201
8. Wang KK, Nath R, Posner A, Raser KJ, Buroker-Kilgore M, Hajimohammadreza I, Probert AW Jr, Marcoux FW, Ye Q, Takano E, Hatanaka M, Maki M, Caner H, Collins JL, Fergus A, Lee KS, Lunney EA, Hays SJ, Yuen P (1996) An alpha-mercaptoacrylic acid derivative is a selective nonpeptide cell-permeable calpain inhibitor and is neuroprotective. Proc Natl Acad Sci U S A 93:6687–6692
9. Rock KL, Gramm C, Rothstein L, Clark K, Stein R, Dick L, Hwang D, Goldberg AL (1994) Inhibitors of the proteasome block the degradation of most cell proteins and the generation of peptides presented on MHC class I molecules. Cell 78:761–771
10. Fujita H, Kato T, Watanabe N, Takahashi T, Kitagawa S (2011) Calpain inhibitors stimulate phagocyte functions via activation of human formyl peptide receptors. Arch Biochem Biophys 513:51–60
11. Fujita H, Kato T, Watanabe N, Takahashi T, Kitagawa S (2011) Stimulation of human formyl peptide receptors by calpain inhibitors: homology modeling of receptors and ligand docking simulation. Arch Biochem Biophys 516:121–127
12. Kitagawa S, Kato T, Kitagawa M, Aomatsu M, Fujita H (2013) Biological effects of calpain inhibitors on human phagocyte functions. Chapter V. In: Lashinski EM (ed) Enzymes and enzyme activity. Nova Science Publishers, New York, pp 121–144
13. Kobayashi S, Yamashita K, Takeoka T, Ohtsuki T, Suzuki Y, Takahashi R, Yamamoto K, Kaufmann SH, Uchiyama T, Sasada M, Takahashi A (2002) Calpain-mediated X-linked inhibitor of apoptosis degradation in neutrophil apoptosis and its impairment in chronic neutrophilic leukemia. J Biol Chem 277:33968–33977
14. Derouet M, Thomas L, Cross A, Moots RJ, Edwards SW (2004) Granulocyte macrophage colony-stimulating factor signaling and proteasome inhibition delay neutrophil apoptosis by increasing the stability of Mcl-1. J Biol Chem 279:26915–26921
15. Kato T, Kutsuna H, Oshitani N, Kitagawa S (2006) Cyclic AMP delays neutrophil apoptosis via stabilization of Mcl-1. FEBS Lett 580:4582–4458

Chapter 17

In Vivo Calpain Knockdown Using Delivery of siRNA

Xiaoqian Liu and Guochang Hu

Abstract

Calpain is an intracellular Ca^{2+}-dependent non-lysosomal cysteine protease expressed ubiquitously in mammals. In endothelial cells, dysregulation of calpain has been shown to be involved in a wide variety of pathological conditions such as angiogenesis, vascular inflammation, and diabetes. Cell- or tissue-targeted in vivo delivery of small interfering RNA (siRNA) is a powerful research tool in the analysis of protein function and has been proposed as an attractive therapeutic modality that is applicable against a large number of human diseases including cancer. In this chapter we describe a method to knockdown calpain 1 in mouse pulmonary vascular endothelium using delivery of siRNA/cationic liposome complex. This technique results in a greater than 80% reduction in calpain 1 protein levels 48 h after a single i.v. injection of calpain 1 siRNA (0.5 mg siRNA/kg)/cationic liposome complex. We also describe confocal imaging to verify the loss of calpain 1 expression in pulmonary microvessel endothelial cells and application of this technique in the mouse model of ventilator-induced lung injury.

Key words siRNA, Calpain, Knockdown, Cationic liposome, Endothelial cell, Lung

1 Introduction

Calpain is an intracellular Ca^{2+}-dependent non-lysosomal cysteine protease. In humans, there are currently 15 known calpain isoform genes, which are defined by the presence of a protease domain. The best characterized, ubiquitous isoforms are calpain 1 (or μ-calpain) and calpain 2 (or m-calpain) which require micromolar (1 to 20 μM) and millimolar (0.25 to 0.75 mM) concentrations of intracellular Ca^{2+} for activation, respectively. Calpains participate in a wide variety of physiological and pathological processes. Dysregulation of calpain expression and activity is related to myriad diseases including neurological disorders, cardiovascular diseases, and cancers [1]. Knocking down calpain isoforms in vivo and in vitro has identified important roles for calpains in the prevention and treatment of hyperhomocysteinemia, hyperglycemia [2], diabetes-induced renal injury [3], lung inflammation [4], atherosclerosis [5], and brain injury [6, 7].

Jeannette S. Messer (ed.), *Calpain: Methods and Protocols*, Methods in Molecular Biology, vol. 1915,
https://doi.org/10.1007/978-1-4939-8988-1_17,

Gene silencing mediated by siRNAs, the 21- to 23-nt double-stranded RNA, is a sequence-specific RNA degradation mechanism. In contrast to the tangible and obvious effectiveness of RNA interference in vitro, the success of siRNA-mediated gene silencing in vivo has been limited due to multiscale barriers, including rapid blood clearance of molecules <10 nm in size, siRNA degradation in the bloodstream, nonspecific accumulation in tissues, poor cellular uptake, and inefficient intracellular release [8]. Delivering naked or liposome-complexed siRNA in vivo basically falls into two approaches: local and systemic administrations. Advantages of the local administration strategy for siRNA delivery in vivo include simple siRNA formulation, site-specific delivery for localized effects, and lower doses for efficacy with a lower risk of off-target effects. Systemic delivery by intravenous injection is the most common route of siRNA delivery and more widely applicable to clinical practice [9].

Liposomes are currently considered to be the most successful and effective drug carrier system. The efficacy of liposomes depends on the physicochemical properties of their components and their size, surface charge, and lipid organization [10]. Cationic liposomes have emerged as attractive siRNA carriers for systemic transfection in vivo due to their favorable characteristics such as biodegradability, minimal toxicity, nonimmunogenicity, relative ease of large-scale production, simplicity of use, and high in vitro transfection efficiency. Numerous laboratory studies [11–14] and clinical trials [15–17] have demonstrated the potential of cationic liposomes in gene therapy. Since cholesterol is one of the most promising siRNA conjugates that have shown efficient RNA interference in vivo [13, 18, 19], cholesterol has been utilized as an effective "helper" lipid in dimethyl dioctadecyl ammonium bromide (DDAB) for preparation of the cationic liposome in our laboratory [4, 20, 21]. Cationic liposomes, by virtue of their surface positive charge, bind selectively to endothelial cells via electrostatic interaction and have substantial endothelial uptake [18, 22, 23]. Due to the large pulmonary vascular surface area and delivery of the entire cardiac output to pulmonary circuit, intravenous administration of siRNA/cationic liposome complex can be used to selectively deliver siRNA to pulmonary microvessel endothelial cells [24–34].

In this chapter, we describe a method for preparation of the cationic liposome by mixing cholesterol and DDAB and then delivery of the calpain 1 siRNA/cationic liposome complex into mice through the tail vein to selectively deplete calpain 1 in the pulmonary vasculature. Efficiency of siRNA-mediated depletion of calpain 1 is confirmed by Western blot analysis of pulmonary endothelial and non-endothelial cell lysates and confocal immunofluorescence for direct visualization of protein expression in lung vascular endothelial cells. A mouse model of ventilator-induced lung injury is also described, to examine the effects of in vivo calpain knockdown on lung inflammation.

2 Materials

2.1 Preparation of siRNA/Cationic Liposome Complex

1. Calpain 1 siRNAs: a pool of three target-specific 19–25 nt siRNAs.
2. Control siRNA: ON-TARGETplus Non-targeting control siRNA (Dharmacon).
3. Dimethyl dioctadecyl ammonium bromide (DDAB) stock solution: 400 mg DDAB in 20 mL chloroform.
4. Cholesterol stock solution: 400 mg cholesterol in 20 mL chloroform.
5. 5% (w/v) glucose: 5 mg glucose in 100 mL deionized H_2O; sterilize the solution in an autoclave.
6. Sodium dodecyl sulfate.
7. Rotavapor.
8. Argon gas.
9. Sonicator.
10. Water bath.
11. Round-bottom flask.
12. 0.22 μm syringe filter.
13. Cell scrapers.

2.2 Experimental Animals

1. Mice: house in microisolator cages under specific pathogen-free conditions, feed with autoclaved food, and use in experiments at 8 weeks of age (20–30 g).
2. Separate mice into two groups: calpain 1 siRNA/cationic liposome group and control siRNA/cationic liposome group.

2.3 Intravenous Administration of siRNA/Cationic Liposome Complex

1. Mouse injection restrainer.
2. 1 mL syringe.
3. Sterile 30-gauge needles.
4. Heating device.
5. Gauze sponges.

2.4 Isolation of Lung Endothelial Cells

1. 1× DPBS: 9.55 g of Dulbecco's phosphate-buffered saline (DPBS) powder, 200 mg KCl, 200 mg K_2HPO_4, 8 g NaCl, 1.15 g NaH_2PO_4. Adjust pH to 7.4 and bring to 1 L with deionized H_2O.
2. Collagenase/dispase: 1 mg/mL in DPBS.

3. Isolation buffer: 0.1% BSA, 2 mM EDTA in DPBS, pH 7.4.
4. High glucose (4.5 g/L) Dulbecco's Modified Eagle Medium (DMEM).
5. Dynabeads magnetic beads: Sheep anti-rat IgG Dynabeads.
6. Anti-PECAM-1 antibody.
7. 40 μm cell strainer.
8. 10 mL syringe.
9. MACS Separator.
10. MACS Column.

2.5 Western Blot

1. Modified RIPA lysis buffer: 50 mM Tris–HCl, pH 7.4, 150 mM NaCl, 1% NP-40, 0.5% sodium deoxycholate, 0.1% SDS, 1 mM sodium orthovanadate (Na_3VO_4), 1 mM phenylmethylsulfonyl fluoride (PMSF), and protease inhibitor cocktail (10 μL/mL lysis buffer).
2. 50% (w/v) bis-acrylamide (29:1) solution. Store at room temperature. Keep the solution in a dark glass bottle.
3. Separating gel buffer: 1 M Tris–HCl, pH 8.8. Store at 4 °C.
4. 10% (w/v) sodium dodecyl sulfate (SDS). Store at room temperature.
5. Stacking gel buffer: 0.375 M Tris–HCl, pH 6.8. Store at 4 °C.
6. 5% (w/v) ammonium persulfate (APS).
7. 8% separating gel solution: total 12.54 mL of solution. 2 mL 50% bis-acrylamide, 4.7 mL Tris–HCl pH 8.8, 125 μL 10% SDS, 5.4 mL deionized H_2O, 312.5 μL 5% APS, 3.125 μL *N*,*N*,*N'*,*N'*-tetramethylethane-1,2-diamine (TEMED).
8. 4× Laemmli sample buffer.
9. 4% stacking gel solution. Mix 6.315 mL of solution: 500 μL 50% bis-acrylamide, 2.1 mL Tris–HCl pH 6.8, 62.5 μL 10% SDS, 3.15 mL deionized H_2O, 500 μL 5% APS, 2.5 μL TEMED.
10. 10× running buffer (1 L): 30.3 g (0.25 M) Tris base, 144.0 g (1.92 M) glycine, and 10.0 g (1%, w/v) SDS. Bring to 1 L with deionized H_2O. Store at room temperature.
11. 5× transfer buffer (1.6 L): 29.3 g glycine, 58.2 g Tris base. Bring to 1.6 L with deionized H_2O.
12. Tris-buffered saline with Tween (TBS-T) (1 L): 10 mM Tris–HCl, pH 8.0, 0.15 M NaCl, and 1% (v/v) Tween 20. Bring to 1 L with deionized H_2O.
13. Blocking buffer: 5% (w/v) nonfat dry milk in TBS-T.
14. Primary antibody: rabbit anti-VE-cadherin, mouse anti-calpain 1, and rabbit anti-GAPDH.

15. Secondary antibody: donkey anti-rabbit and donkey anti-mouse IgG conjugated to horseradish peroxidase.
16. Odyssey Fc Imager (LI-COR Biosciences).
17. BCA protein assay kit.
18. Nitrocellulose membrane.
19. SuperSignal West Pico Chemiluminescent Substrate (Thermo Fisher Scientific).
20. ImageJ software (NIH).

2.6 Confocal Immunofluorescence Microscopy

1. 4% (v/v) paraformaldehyde: dilute 16% (v/v) paraformaldehyde to 4% (v/v) with 1× DPBS.
2. Wash solution: 0.1% (v/v) Triton X-100 in DPBS.
3. Blocking buffer: 5% (v/v) donkey serum and 1% (w/v) BSA in wash solution.
4. Antibody dilution buffer: 1% (v/v) donkey serum and 1% (w/v) BSA in wash solution.
5. Primary antibody: rabbit anti-PECAM-1 and mouse anti-calpain 1 antibodies.
6. Secondary antibody: donkey anti-rabbit Alexa Fluor 594 and donkey anti-mouse Alexa Fluor 488.
7. Cryostat with metal grids.
8. Inert support medium: optimal cutting temperature (OCT) compound.
9. Confocal microscopy.
10. 12 mm microscope cover glass.
11. Mounting medium: antifade mounting medium with DAPI.

2.7 Mouse Model of Ventilator-Induced Lung Injury

1. Animal: mice should be housed in microisolator cages, fed with autoclaved food, and used in experiments at 8–12 weeks of age (25–30 g).
2. Anesthetics: ketamine (75 mg/kg), 1% isoflurane.
3. Ventilator: small animal ventilator for rodents.
4. Grouping: mice transfected with control (CON) and calpain 1 (Calpn 1) siRNA should be divided into three groups: control group, CON siRNA-treated mice ventilated with normal tidal volume; mechanical ventilation (MV) group, CON siRNA-treated mice ventilated with high tidal volume; and MV plus Calpn 1 group, Calpn 1 siRNA-treated mice ventilated with high tidal volume.

2.8 Polymorphonuclear Neutrophil (PMN) Counts in Bronchoalveolar Lavage (BAL) Fluid

1. 1 mL sterile syringe.
2. Cytocentrifuge.
3. Diff-Quick dye.
4. Hemocytometer.
5. Light microscope.
6. Slides.

3 Methods

3.1 Preparation of Cationic Liposome

1. Add 315 μL of DDAB stock solution (i.e., 6.3 mg DDAB), 200 μL of cholesterol stock solution (i.e., 4 mg cholesterol), and 9.5 mL of chloroform to a sterilized round-bottom flask, and gently swirl until they are completely mixed.
2. Connect the round-bottom flask to the Rotavapor, and evaporate the chloroform under vacuum at 40 °C, 100 rpm for approximately 20 min (*see* Fig. 1). All of the chloroform should be removed by rotary evaporation yielding a thin lipid film on the sides of a round-bottom flask (*see* **Note 1**). For small volumes (< 1 mL), the chloroform may be evaporated using a dry nitrogen or argon stream in a fume hood.
3. Add 2 mL of 5% glucose to resuspend the lipid. Scratch the flask bottom thoroughly using the cell scrapers to dissolve all of the lipid.
4. Add a pinch of SDS powder to the water of the bath sonicator, sonicate the lipid for at least 20 min using the flask holding stand and just barely touch the bottom of the round-bottom flask to the water in the vat (*see* **Note 2**).
5. Transfer the sonicated liposome from the flask to an Eppendorf tube, extrude, and sterilize the liposome with a 0.22 μm syringe filter (*see* **Note 3**).

3.2 Preparation of the Mixture of siRNA/ Cationic Liposome

1. Prepare the siRNA/liposome complex by addition of 0.25 or 0.5 mg/kg calpain 1 siRNA (stock concentration, 10 μM) into 200 μL (i.e., 20 g mice, 10 μL/g) [3, 18–19] of liposome suspensions. A scrambled siRNA should be employed as a control (*see* **Note 4**).
2. Mix immediately and gently by shaking, and incubate for 15–20 min at room temperature (*see* **Note 5**).

3.3 Delivering siRNA/ Cationic Liposome Complex into Mice by Intravenous Injection

1. Hold the mice in a standard mouse restrainer.
2. Stimulate dilation of the tail veins in mice by placing the tail in warm (30–35 °C) water.

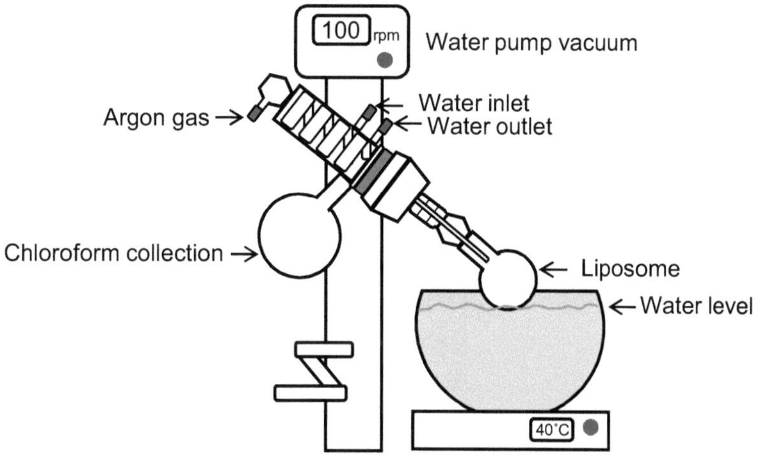

Fig. 1 Scheme of a rotary evaporator. The liposome is placed in the evaporation flask (right) which is rotated (100 rpm) by a motor unit. The flask is lowered into the water bath (40 °C). Chloroform is removed under vacuum and trapped by a condenser

3. Grasp the tail at the distal end. Place the index and middle fingers of the non-dominant hand around the tail above where the needle will be inserted. Hold the lower part of the tail between the thumb and ring fingers below the injection site.
4. Slightly oppose pressure with both sets of fingers to straighten and stabilize the tail. Release pressure immediately before injection.
5. Insert the needle into the vein toward the direction of the head with the dominant hand. Hold the needle and syringe parallel to the tail with bevel side up.
6. Gently pull blood into the syringe to confirm proper placement in the vessel before injecting. Alternately, verify proper placement by advancing the needle smoothly into the vein without any resistance.
7. Slowly inject the total siRNA/liposome mixture (*see* **Note 6**).
8. Remove the needle and apply gentle compression until bleeding has stopped.
9. Return mice to their cage, and monitor for 5–10 min to ensure that bleeding has not resumed.

3.4 Determination of Efficiency of Calpain 1 Knockdown

3.4.1 Western Blot Analysis

1. Anesthetize mice by injection of ketamine/xylazine (100 mg/kg/10 mg/kg, i.p.) at 48 h post-transfection.
2. Sacrifice mice, harvest the lung after perfusing with DPBS (0.6 mM EDTA in DPBS, pH 7.4), followed by perfusing the lung intratracheally with collagenase/dispase solution, and then place the lung into ice-cold DMEM high-glucose (4.5 g/L) medium.
3. Collect the lung tissues in collagenase/dispase solution after being minced, and then incubate at 37 °C for 35 min with shaking.
4. After digestion, filter cell suspension with a 40 μm cell strainer, and centrifuge at 400 × *g* for 10 min.
5. Resuspend the cells in an isolation buffer, and then incubate with the Dynabeads magnetic beads pretreated with PECAM-1 antibody for 20 min at 4 °C with gentle rotation (*see* **Note 7**).
6. Isolate lung endothelial cells, and separate them from the non-endothelial cell fraction by a MACS Column, which is placed in the magnetic field of a MACS Separator.
7. Wash the isolated cells with ice-cold DPBS.
8. Add the cold RIPA buffer to each tube.
9. Homogenize the cells with sonicator, and then lyse the cells on ice for 30 min.
10. Centrifuge at 16,000 × *g*, 4 °C for 15 min.
11. Transfer supernatants to new Eppendorf tubes.
12. Measure protein concentration with a BCA assay kit.
13. Mix 25 μg protein from each sample with 4× Laemmli sample buffer in Eppendorf tubes. Boil the mixture at 95–100 °C for 5 min.
14. Load prepared samples into wells of 8% SDS-polyacrylamide gel electrophoresis (SDS-PAGE) with 4% stacking gel, and make sure not to overflow.
15. Transfer the protein blots that have been separated by SDS-PAGE onto nitrocellulose membrane at 170 mA for 90 min.
16. Place the membranes into the blocking buffer for 1 h at room temperature on a rocking platform.
17. Incubate the membranes with primary antibodies overnight at 4 °C.

    The dilutions of primary antibodies are as follows:
    Anti-rabbit VE-cadherin antibody: 1:1000
    Anti-mouse calpain 1 antibody: 1: 300
    Anti-rabbit GAPDH antibody: 1:2000.

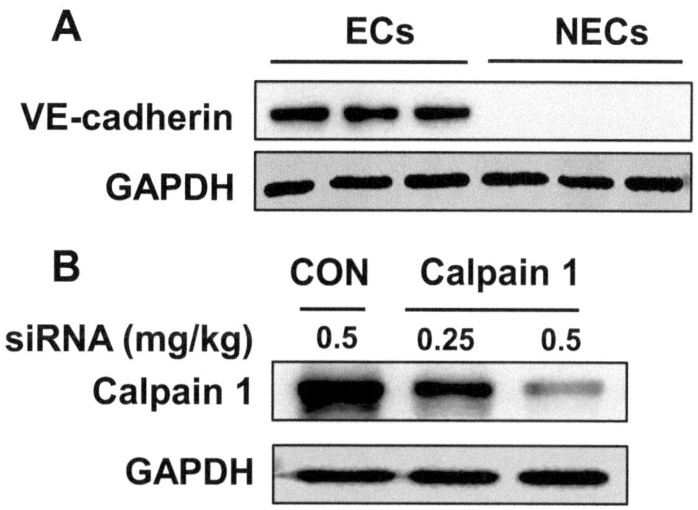

Fig. 2 Expression of calpain 1 protein in calpain-depleted mice. Mice were transfected with control (CON) and calpain 1 siRNA. At 48 h post-transfection, pulmonary endothelial cells (ECs) were isolated as described in Subheading 3. CON siRNA-treated mice were used as the control group. (**a**) The purity of isolated ECs was confirmed by the expression of VE-cadherin. NECs = non-endothelial cells. (**b**), Effective knockdown of calpain 1 expression was confirmed by Western blotting of pulmonary ECs. Results are representative of three independent experiments

18. Remove the primary antibody, and wash the membranes three times with TBS-T on a rocking platform.
19. Incubate the membranes with secondary antibodies (1:2000) at room temperature for 1 h.
20. Discard the secondary antibodies and wash the membrane three times with TBS-T on a rocking platform.
21. Immerse the membranes with SuperSignal West Pico Chemiluminescent Substrate for approximately 1 min.
22. Detect protein bands with Odyssey Fc Imager.
23. Quantify the intensity of immunoblot bands by ImageJ software (NIH) (*see* Fig. 2).

3.4.2 Confocal Immunofluorescence

1. Freeze a fresh, unfixed lung tissue sample, up to 2.0 cm in diameter, in OCT in a suitable tissue mold. Freeze the OCT containing the tissue onto the specialized metal grids that fit onto the cryostat.
2. Cut lung sections 5 μm thick in the cryostat at −20 °C.
3. Immediately transfer the lung section to microscope slide by touching the slide to the tissue at room temperature. Thaw the slides mounted with cryosections (5 μm thick) of mouse lung

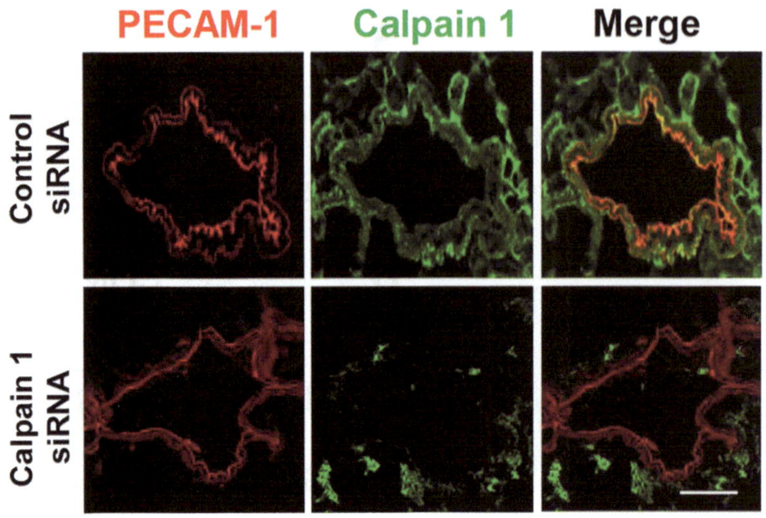

Fig. 3 Confocal images showing downregulation of calpain 1 in pulmonary vasculature with a specific calpain 1 siRNA. All images were acquired from lung tissues treated with control and calpain 1 siRNA (0.5 mg/kg) for 48 h. After fixation and permeabilization, the tissues were stained with anti-PECAM-1 (red) and anti-calpain 1 (green) antibodies followed by Alexa 594-conjugated secondary antibody and Alexa 488-conjugated secondary antibody, respectively. Endothelial calpain 1 expression was ascertained in PECAM-1 positive structure. Scale bars = 20 μm. Shown are representative results of three independent experiments

tissue for 20–30 min at room temperature, if the slides are stored in −80 °C freezer (*see* **Note 8**).

4. Wash the slides in DPBS to remove the OCT compound.
5. Fix slides with 4% paraformaldehyde for 8 min at room temperature.
6. Discard paraformaldehyde, and wash the slides three times with DPBS for 6 min each at room temperature on the rocking platform.
7. Place the slides in a blocking buffer for 60 min at room temperature.
8. Incubate the slides with primary antibodies overnight at 4 °C.

   The dilutions of primary antibodies are as follows:
   Anti-rabbit PECAM-1 antibody: 1:200
   Anti-mouse calpain 1 antibody: 1:40.
9. Remove the primary antibody and wash the slides three times for 6 min each with washing solution.
10. Preincubate the slides with a blocking buffer again for 30 min at room temperature.

11. Incubate the slides with the secondary antibody donkey anti-rabbit Alexa Fluor 594 (1:750) and donkey anti-mouse Alexa Fluor 488 (1:750) for 1 h at room temperature. And from this step, keep the slides in the dark.
12. Wash the slides five times for 6 min each with washing solution.
13. Mount the slides to glass microscope slides properly with mounting medium and avoid trapping air bubbles on the slide.
14. Dry the slides at room temperature for several hours in the dark. Store the slides at −20 °C in the dark for up to several months, if necessary.
15. Acquire images by a laser-scanning confocal microscope (*see* Fig. 3).

3.5 Ventilator-Induced Lung Injury

1. Intravenously inject control siRNA or calpain 1 siRNA into mice.
2. Anesthetize mice with ketamine (75 mg/kg) at 48 h after injection.
3. Perform tracheostomy and insert the intubation cannula into the mouse trachea.
4. Ventilate mice for 2 h with 1% isoflurane in room air to maintain anesthesia, using either high (28 mL/kg) or normal tidal volume (7 mL/kg) [4].

3.5.1 Determination of PMN Counts in BAL Fluid

1. Perform BAL by intratracheal injection of 1 mL of PBS followed by gentle aspiration. Repeat the procedure for three times.
2. Centrifuge the pooled BAL fluid at 400 × *g* for 5 min.
3. Suspend cell pellets in PBS at a final concentration of 1×10^5 cells/mL.
4. Cytospin 200 μL volume of resuspended cells onto slides at 300 rpm for 5 min.
5. Stain the slides with Diff-Quick dye.
6. Examine the slides at a magnification of ×20 and ×40 by light microscopy.
7. Calculate the percentage of PMNs after counting 300 cells in randomly selected fields.
8. Manually measure the total cell count in BAL fluid using a hemocytometer (*see* Fig. 4).

4 Notes

1. When the round-bottom flask is rotated, the water bath temperature should be kept at 40 °C strictly. The optimal lipid film

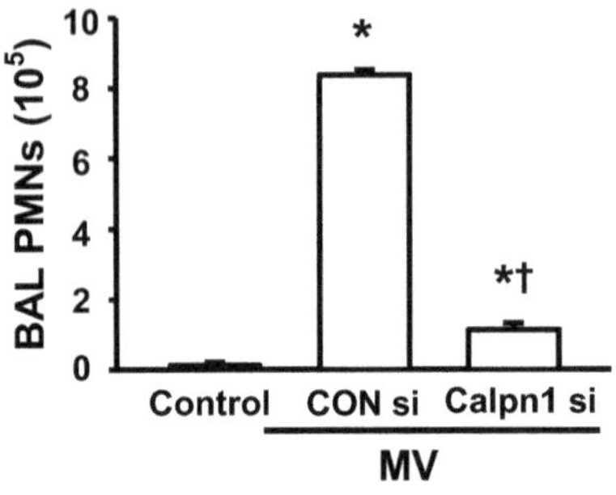

Fig. 4 Decreased lung inflammation by mechanical ventilation in calpain-depleted mice. Mice were transfected with control (CON) and calpain 1 (Calpn 1) siRNA (si). After 48 h, mice ($n = 6$/group) were mechanically ventilated with high tidal volume (28 mL/kg, MV) for 2 h. Control siRNA-treated mice ventilated with normal tidal volume were used as the control groups. PMN counts in BAL fluid were calculated. *$P < 0.05$ vs. control groups. †$P < 0.05$ vs. CON siRNA +MV groups (*see* ref. 4)

is thin and well-distributed. The bottom of the round-bottom flask should just barely touch the water.

2. The ultrasound should not be stopped until the lipid changes from large precipitates to small particles. The lipid suspension should begin to clarify to yield a slightly hazy transparent solution.
3. To obtain liposomes with a narrow size distribution, liposome is extruded by 0.22 μm syringe filter. The diameter of the liposomes synthesized here is <220 nm. To ensure size homogeneity, multiple extrusions (10 times) are needed.
4. An optimal efficient concentration of siRNA and the appropriate ratio of siRNA/cationic liposome are important for higher gene silencing efficiency of the siRNA.
5. The siRNA and liposome should be mixed immediately by pipetting or vortexing to avoid unequal mixing of the lipid and siRNA components that can cause formation of precipitates. The siRNA/cationic liposome complex should be freshly prepared just prior to injection using DNAase, RNAase, and Pyrogen free tubes and tips. Keep sterile operation to the greatest extent.
6. A small volume of siRNA/liposome complex is injected gently in order to determine whether the needle is in the lumen of the vein. The entire volume of siRNA/liposome complex is then injected slowly. There should be little resistance if the mixture of siRNA/liposome is being injected into the tail vein. If there is significant resistance to injection, the needle may be not in the lumen of the vein. Keep sterile operation to the greatest extent.

7. Dynabeads (100 μL/mouse) must be pre-coated with anti-PECAM-1 antibody (6 μg/mouse) by co-incubating for at least 30 min at 4 °C with gentle tilting and rotation. To improve the purity of lung endothelial cells, the Dynabeads to cell suspension ratio must be optimized.
8. To prevent the lung tissues from curling-up or detaching from the surface of the slide, frozen slides are warmed to room temperature for 20–30 min.

Acknowledgments

This work was supported by the NIH Grant HL104092 to GH.

References

1. Ono Y, Saido TC, Sorimachi H (2016) Calpain research for drug discovery: challenges and potential. Nat Rev Drug Discov 15:854–876
2. Cheng Z, Jiang X, Pansuria M, Fang P, Mai J, Mallilankaraman K, Gandhirajan RK, Eguchi S, Scalia R, Madesh M, Yang X, Wang H (2015) Hyperhomocysteinemia and hyperglycemia induce and potentiate endothelial dysfunction via μ-calpain activation. Diabetes 64:947–959
3. Covington MD, Schnellmann RG (2012) Chronic high glucose downregulates mitochondrial calpain 10 and contributes to renal cell death and diabetes-induced renal injury. Kidney Int 81:391–400
4. Liu D, Yan Z, Minshall RD, Schwartz DE, Chen Y, Hu G (2012) Activation of calpain s mediates early lung neutrophilic inflammation in ventilator-induced lung injury. Am J Physiol Lung Cell Mol Physiol 302: L370–L379
5. Miyazaki T, Taketomi Y, Takimoto M, Lei XF, Arita S, Kim Kaneyama JR, Arata S, Ohata H, Ota H, Murakami M, Miyazaki A (2011) m-Calpain induction in vascular endothelial cells on human and mouse atheromas and its roles in VE-cadherin disorganization and atherosclerosis. Circulation 124:2522–2532
6. Bevers MB, Ingleton LP, Che D, Cole JT, Li L, Da T, Kopil CM, Cohen AS, Neumar RW (2010) RNAi targeting micro-calpain increases neuron survival and preserves hippocampal function after global brain ischemia. Exp Neurol 224:170–177
7. Sha D, Jin Y, Wu H, Wei J, Lin CH, Lee YH, Buddhala C, Kuchay S, Chishti AH, Wu JY (2008) Role of mu-calpain in proteolytic cleavage of brain L-glutamic acid decarboxylase. Brain Res 1207:9–18
8. Shim MS, Kwon YJ (2010) Efficient and targeted delivery of siRNA in vivo. FEBS J 277:4814–4827
9. Akhtar S, Benter IF (2007) Nonviral delivery of synthetic siRNAs in vivo. J Clin Invest 117:3623–3632
10. Bozzuto G, Molinari A (2015) Liposomes as nanomedical devices. Int J Nanomedicine 10:975–999
11. Zimmermann TS, Lee AC, Akinc A, Bramlage B, Bumcrot D, Fedoruk MN, Harborth J, Heyes JA, Jeffs LB, John M et al (2006) RNAi-mediated gene silencing in non-human primates. Nature 441:111–114
12. Mikhaylova M, Stasinopoulos I, Kato Y, Artemov D, Bhujwalla ZM (2009) Imaging of cationic multifunctional liposome-mediated delivery of COX-2 siRNA. Cancer Gene Ther 16:217–226
13. Soutschek J, Akinc A, Bramlage B, Charisse K, Constien R, Donoghue M, Elbashir S, Geick A, Hadwiger P, Harborth J et al (2004) Therapeutic silencing of an endogenous gene by systemic administration of modified siRNAs. Nature 432:173–178
14. Wolfrum C, Shi S, Jayaprakash KN, Jayaraman M, Wang G, Pandey RK, Rajeev KG, Nakayama T, Charrise K, Ndungo EM et al (2007) Mechanisms and optimization of in vivo delivery of lipophilic siRNAs. Nat Biotechnol 25:1149–1157
15. Caplen NJ, Alton EW, Middleton PG, Dorin JR, Stevenson BJ, Gao X et al (1995) Liposome-mediated CFTR gene transfer to the nasal epithelium of patients with cystic fibrosis. Nat Med 1:39–46

16. McLachlan G, Ho LP, Davidson-Smith H, Samways J, Davidson H, Stevenson BJ et al (1996) Laboratory and clinical studies in support of cystic fibrosis gene therapy using pCMV-CFTR-DOTAP. Gene Ther 3:1113–1123
17. Gill DR, Southern KW, Mofford KA, Seddon T, Huang L, Sorgi F et al (1997) A placebo-controlled study of liposome-mediated gene transfer to the nasal epithelium of patients with cystic fibrosis. Gene Ther 4:199–209
18. Bardita C, Predescu D, Predescu S (2013) Long-term silencing of Intersectin-1s in mouse lungs by repeated delivery of a specific siRNA via cationic liposomes. Evaluation of knockdown effects by electron microscopy. J Vis Exp 76:50316
19. DN P, Neamu R, Bardita C, Wang M, Predescu SA (2012) Impaired caveolae function and upregulation of alternative endocytic pathways induced by experimental modulation of intersectin-1s expression in mouse lung endothelium. Biochem Res Int 2012:672705
20. Hu G, Vogel SM, Schwartz DE, Malik AB, Minshall RD (2008) Intercellular adhesion molecule-1 dependent neutrophil adhesion to endothelial cells induces caveolae-mediated pulmonary vascular hyperpermeability. Circ Res 102:e120–e131
21. Wang YL, Malik AB, Sun Y, Hu S, Reynolds AB, Minshall RD, Hu G (2011) Innate immune function of the adherens junction protein p120-catenin in endothelial response to endotoxin. J Immunol 186:3180–3187
22. Bai L, Andersson HA, McConnell KI, Chan DL, Hernandez M, Gonzalez J, Liu X, La Francesca S, Sakamoto JH, Serda RE (2015) Silencing of tumor necrosis factor Receptor-1 in human lung microvascular endothelial cells using particle platforms for siRNA delivery. Curr Drug Targets 16:1531–1539
23. Kowalski PS, Zwiers PJ, Morselt HW, Kuldo JM, Leus NG, Ruiters MH, Molema G, Kamps JA (2013) Anti-VCAM-1 SAINT-O-Somes enable endothelial-specific delivery of siRNA and downregulation of inflammatory genes in activated endothelium in vivo. J Control Release 176:64–75
24. Zhou MY, Lo SK, Bergenfeldt M, Tiruppathi C, Jaffe A, Xu N, Malik AB (1998) In vivo expression of neutrophil inhibitory factor via gene transfer prevents lipopolysaccharide-induced lung neutrophil infiltration and injury by a beta2 integrin-dependent mechanism. J Clin Invest 101:2427–2437
25. Zhu N, Liggitt D, Liu Y, Debs R (1993) Systemic gene expression after intravenous DNA delivery into adult mice. Science 261:209–211
26. Thierry AR, Lunardi-Iskandar Y, Bryant JL, Rabinovich P, Gallo RC, Mahan LC (1995) Systemic gene therapy: biodistribution and long-term expression of a transgene in mice. Proc Natl Acad Sci U S A 92:9742–9746
27. Li S, Huang L (1997) In vivo gene transfer via intravenous administration of cationic lipid-protamine-DNA (LPD) complexes. Gene Ther 4:891–900
28. Liu F, Qi H, Huang L, Liu D (1997) Factors controlling the efficiency of cationic lipid-mediated transfection in vivo via intravenous administration. Gene Ther 4:517–523
29. Liu Y, Mounkes LC, Liggitt HD, Brown CS, Solodin I, Heath TD, Debs RJ (1997) Factors influencing the efficiency of cationic liposome-mediated intravenous gene delivery. Nat Biotechnol 15:167–173
30. Song YK, Liu F, Chu S, Liu D (1997) Characterization of cationic liposome- mediated gene transfer in vivo by intravenous administration. Hum Gene Ther 8:1585–1594
31. Templeton NS, Lasic DD, Frederik PM, Strey HH, Roberts DD, Pavlakis GN (1997) Improved DNA: liposome complexes for increased systemic delivery and gene expression. Nat Biotechnol 15:647–652
32. Barron LG, Meyer KB, Szoka FC Jr (1998) Effects of complement depletion on the pharmacokinetics and gene delivery mediated by cationic lipid-DNA complexes. Hum Gene Ther 9:315–323
33. McLean JW, Fox EA, Baluk P, Bolton PB, Haskell A, Pearlman R, Thurston G, Umemoto EY, McDonald DM (1997) Organ-specific endothelial cell uptake of cationic liposome-DNA complexes in mice. Am J Phys 273: H387–H404
34. Liu D, Knapp EJ, Song YK (1999) Mechanisms of cationic liposome mediated transfection of the lung endothelium. In: Huang L, Wagner E, Hung MC (eds) Nonviral vectors for gene therapy. Academic Press, San Diego, CA, pp 313–335

Chapter 18

In Vivo Expression of Mutant Calpains in the Eye Using Lentivirus

Katherine J. Wert and Vinit B. Mahajan

Abstract

Exome sequencing has identified many candidate genes and mutations for human diseases, but the functional validation of these candidates is a time-consuming and costly process. Here, we describe a method which uses lentiviruses to overexpress calpain mutations that may play a role in dominant diseases such as autosomal dominant neovascular inflammatory vitreoretinopathy (ADNIV). The use of lentivirus to deliver the mutant calpain allows for a cost-effective, rapid, and efficient approach to test whether or not a candidate gene mutation from exome sequencing acts as the disease-causing allele for a human disorder. This method also provides for a comparison of different candidate mutations from a single gene identified by exome sequencing, as well as elucidating the mechanisms underlying these complex human disorders. Furthermore, this chapter focuses on two different methods to deliver mutant calpain to the cells of the eye, using either a subretinal or an intravitreal injection of the lentivirus into the mouse eye.

Key words Calpain, *CAPN5*, Lentiviral vectors, Molecular cloning, Subretinal injection, Intravitreal injection, Mouse eye

1 Introduction

Advances in next-generation exome sequencing have allowed for the identification of candidate disease alleles for complex human disorders. However, although many human genes have been identified as having potential roles in disease, the ability to functionally validate these candidate genes and mutations has progressed at a slower pace. Therefore, using lentiviruses to transiently express candidate mutations can provide a more rapid and cost-efficient method to determine whether or not a gene mutation is causative for disease, as well as allowing for the comparison of different disease alleles in transgenic mouse models.

Mutations in CAPN5, encoded by the calpain-5 gene, have been recently determined to lead to autosomal dominant neovascular inflammatory vitreoretinopathy (ADNIV, OMIM 193235, [1–6]. ADNIV is a dominant, heritable, autoimmune disorder,

Jeannette S. Messer (ed.), *Calpain: Methods and Protocols*, Methods in Molecular Biology, vol. 1915,
https://doi.org/10.1007/978-1-4939-8988-1_18,

where the clinical presentation is found only within the eye even though the gene is widely expressed throughout the human body. ADNIV is also unusual since it shows the pathological features of a variety of eye diseases that do not normally occur together, such as retinitis pigmentosa, diabetic retinopathy, autoimmune uveitis, and proliferative vitreoretinopathy [7–9]. Thus, the study of candidate disease alleles and mutations for *CAPN5* can allow for further understanding of many human eye disorders, along with elucidating the disease mechanisms underlying ADNIV. Additionally, if the expression of a single Mendelian mutation is enough to cause the ADNIV phenotype in a mouse model, it is likely that ADNIV patients with this disease allele may be good candidates for retinal gene therapy.

Here, we describe a method in which calpain mutations in the mouse eye are overexpressed using lentivirus. This method can be used to create a preclinical model for the study of ADNIV using mutant *CAPN5* or to study overexpression and/or mutant forms of other calpains and their potential roles in eye disorders. This chapter describes the molecular cloning techniques needed to construct lentiviruses containing mutant calpain genes, as well as two different injection techniques, subretinal and intravitreal injections, to deliver the mutant calpain to the various cells of the eye.

2 Materials

2.1 Construction of Lentiviral Vectors

1. Lentiviral transfer plasmid.
2. Mutant calpain DNA.
3. High-fidelity taq PCR kit.
4. 50× TAE: 242 g Tris-base, 57.1 mL glacial acetic acid, 100 mL of 0.5 M EDTA, add H_2O to final volume of 1 L (*see* **Note 1**).
5. 1× TAE: dilute 20 mL of 50× TAE into 980 mL of sterile H_2O.
6. Ethidium bromide.
7. Gel electrophoresis equipment.
8. Gel chambers and gel combs.
9. 1 kb DNA ladder.
10. 6× DNA gel loading dye.
11. UV light source.
12. Gel DNA purification kit.
13. Restriction enzymes and buffers.
14. Calf alkaline phosphatase.
15. Water bath.

16. T4 DNA ligase and buffer.
17. Chemically competent *E. coli*.
18. Luria broth (LB): 10 g NaCl, 10 g tryptone, 5 g yeast extract, and 950 mL H_2O. Adjust pH to 7.0 using NaOH and bring to final volume of 1 L with H_2O. Autoclave on liquid cycle for 20 min at 15 psi and store at room temperature.
19. LB agar plates: Add 15 g agar to 1 L of LB. Autoclave on liquid cycle for 20 min at 15 psi. Add appropriate antibiotic once cooled to 60 °C and mix. Pour gently into 10 cm bacterial culture plates, and allow to cool and solidify at room temperature. Once solid, store at 4 °C.
20. Appropriate antibiotic: e.g., Sigma-Aldrich, ampicillin at 100 mg/mL stock and 100 μg/mL working concentration, diluted in sterile H_2O and filtered before usage.
21. Endotoxin-free Midi-prep Plasmid Kit.
22. Nanodrop machine.

2.2 Lentivirus Production

1. HEK293T cells.
2. 10 cm and 6-well sterile cell culture plates.
3. DMEM media: DMEM basal media, 15% fetal bovine serum, 2 mM L-glutamine, 1% nonessential amino acids, 1% penicillin/streptomycin, filtered through 0.22 μm filter and stored at 4 °C for up to 14 days.
4. Lipofectamine transfection reagent.
5. Lentiviral packaging plasmid.
6. Lentiviral VSV-G envelope expressing plasmid.
7. Opti-MEM media.
8. 0.45 μm syringe filters.
9. 10 mL disposable syringes.
10. Ultracentrifuge tubes and ultracentrifuge.
11. Sterile phosphate buffered saline (PBS).
12. Polybrene: 10 μg/mL.
13. Fluorescence stereomicroscope.

2.3 Lentivirus Delivery to the Mouse Eye

1. 0.8–1.10 × 100 mm glass capillary tube.
2. Micropipette puller.
3. Sigmacote: silicone.
4. Safety-Lok 25 3/4 gauge × 12″ blood collection set.
5. 1 mL Sub-Q 25 5/8 gauge slip-tip syringe.
6. Curved dressing forceps with serrations.
7. 15-degree microsurgery knife.

8. Bupivacaine.
9. Institutionally approved anesthetics and pain medications.

3 Methods

3.1 Construction of Lentiviral Vectors

1. Purchase a lentiviral transfer plasmid from an appropriate source, such as Addgene. This transfer plasmid should contain a 5′ long terminal repeat (LTR), a packaging signal, and a 3′ LTR. In addition, it should include the promoter of interest for driving mutant calpain expression, whether it be ubiquitous expression of mutant calpain after lentiviral delivery, inducible mutant calpain expression such as using tetracycline-controlled transcriptional activation, or a cell-specific constitutive expression via a cell-specific promoter sequence. In addition, a fluorescence reporter is recommended (Fig. 1). If needed, a promoter sequence can be cloned into the lentiviral transfer plasmid using a similar cloning procedure as described below.
2. Amplify mutant calpain DNA using a polymerase chain reaction (PCR) with primers containing the appropriate restriction enzyme sites on the 5′ and 3′ ends for ligation into the lentiviral transfer plasmid after the promoter sequence (*see* **Note 2**). Primers should be designed to include the restriction sites and a hybridization sequence (*see* **Note 3**). The restriction site is the sequence needed for the specific restriction enzyme (based on the transfer plasmid restriction cut sites specific to the correct location to insert the calpain DNA, i.e., directly after the promoter sequence and not found in other locations in the plasmid backbone) to cut at either end of the amplified mutant calpain DNA (*see* **Note 4**). The hybridization sequence is the 18–21 bp that matches the ends of the mutant calpain DNA to be amplified by the primer (*see* **Note 5**). When designing the primers, it is ideal to have a melting temperature that is between 50 and 60 °C, and primer pairs should have melting temperatures within 5 °C of each other. Additionally, primers perform best when kept at a 40–60% G/C content, beginning and ending with 1–2 G/C pairs. Remember to reverse complement the reverse primer, so that the primers bind facing each other and amplify the gene of interest. Furthermore, since it is important to amplify the correct calpain sequence with only the mutation of interest, high-fidelity taq polymerase should be used for the PCR. A standard PCR protocol is as follows, and this can be modified based on the enzyme, buffer, and melting temperature of the primers:

   PCR Mix:
   2 μL template DNA (10–500 ng).
   5 μL 10× buffer with $MgCl_2$.

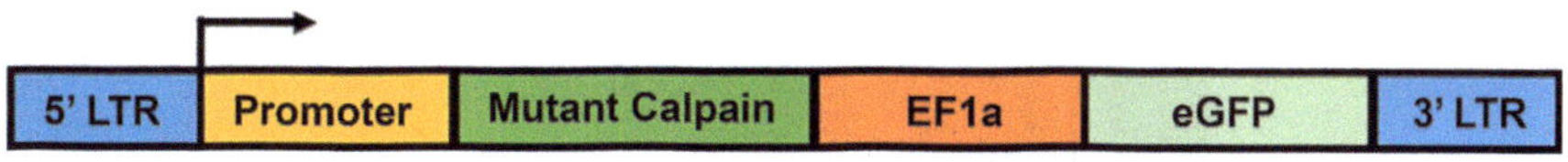

Fig. 1 Schematic of lentiviral transfer plasmid. A lentiviral transfer plasmid containing 5′ and 3′ long terminal repeats (LTRs), as well as the promoter sequence driving the expression of mutant calpain and the eukaryotic elongation factor 1 alpha (EF1a) promoter driving constitutive expression of a green fluorescent protein (eGFP)

1 μL dNTP.
2.5 μL forward primer (10 μM stock).
2.5 μL reverse primer (10 μM stock).
0.2 μL Taq DNA polymerase (5 units/μL).
32.8 μL sterile H_2O for a 50 μL total volume.
The basic PCR protocol rules are to keep the annealing temperature at approximately 5 °C below the primer melting temperatures. The DNA extension step is 1–2 min per kilobase of the expected mutant calpain gene. An example of a PCR run is as follows:
Step 1: Denaturation for 2 min at 94 °C.
Step 2: Denaturation for 30 s at 94 °C.
Step 3: Anneal for 30 s at 55 °C.
Step 4: Extension for 2 min at 74 °C.
Repeat **steps 2–4** for 25–30 cycles.
Step 5: Extension for 5 min at 74 °C.
Step 6: Cool to 4 °C until ready to use.

3. Dissolve 1 g of agarose into 100 mL of 1× TAE buffer using a microwave (*see* **Note 6**). When agarose is completely dissolved, let the solution cool for approximately 5 min (until the flask can be touched by hand), and add 0.5 μg/mL ethidium bromide (EtBr, *see* **Note 7**). Pour the solution into a gel tray with a wide gel comb, and allow to cool at room temperature until solid. Be cautious to avoid air bubbles and pour evenly into the gel tray. Once solid, place the solidified gel into a gel box (an electrophoresis system), and fill the gel box with 1× TAE until the top of the gel is lightly covered. Remove the gel comb, and make sure that air bubbles are removed from the loading lanes.

4. Add 6× loading dye to the PCR samples. Load the PCR product in one gel lane. In a nearby lane, load the appropriate DNA ladder for identifying the correctly amplified calpain cDNA based on its size in kilobases (*see* **Note 8**). Run the gel at approximately 100 V, running from the negative to positive electrode, until the dye has reached near the bottom of the gel (run through approximately 80% of the gel). Turn off the voltage, remove the gel from the gel box using nitrile gloves, and image the gel over a UV light source while wearing the appropriate face and eye protection. If a band is detectable at the correct size, cut out the band from the gel using a scalpel

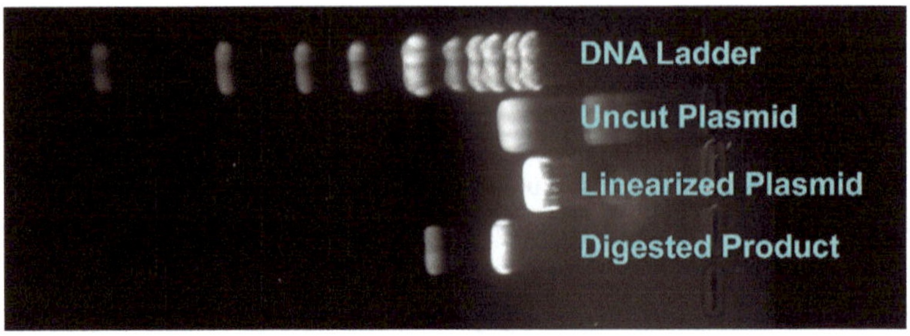

Fig. 2 Gel image with DNA plasmids. An example of a gel imaged over a UV light source showing the 1 kb DNA ladder (lane 1), an uncut linear DNA plasmid (lane 2), a linearized DNA plasmid after a single restriction enzyme cut (lane 3), and a cut DNA plasmid (lane 4, top band) after restriction enzyme digest of an insert sequence (bottom band)

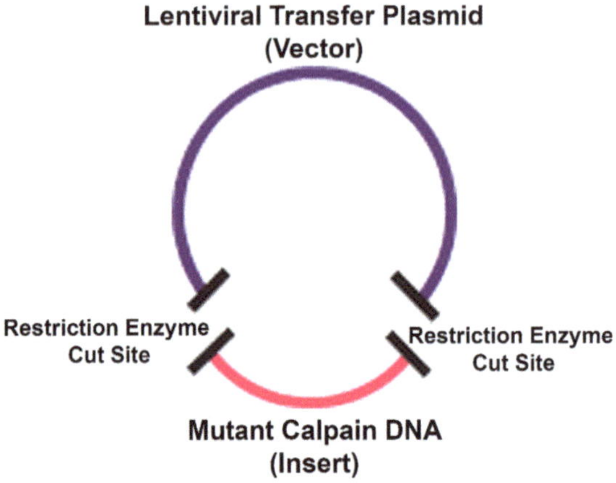

Fig. 3 Schematic of vector and insert digestion and ligation. A vector backbone (the lentiviral transfer plasmid) will be cut by restriction enzymes (sites marked by black bands). The insert DNA sequence (mutant calpain DNA) will be digested with the same restriction enzyme cut sites (black bands). The insert DNA sequence can then be ligated with a T4 DNA ligase to the vector backbone to create a circular plasmid containing both the vector and the insert

blade (Fig. 2). Purify the DNA from the gel using a gel purification kit, as can be purchased from companies such as Qiagen.

5. The lentiviral transfer plasmid should be digested by the appropriate restriction enzymes, to linearize it for ligation of the cDNA (Fig. 3). It is recommended to digest the insert sequence purified in Subheading 3.1, **step 4** at the same time. Heat a water bath to 37 °C. then, add the following in a 1.5 mL Eppendorf tube:

   1–3 μg DNA (one tube for the lentiviral transfer plasmid and another for the mutant calpain DNA).
   1 μL of the restriction enzyme(s).

3 μL of the appropriate 10× buffer for the restriction enzyme (s).
3 μL bovine serum albumin (BSA).
Sterile H_2O to a 30 μL final volume.
Digest in the 37 °C water bath for at least 4 h (*see* **Note 9**), adding calf alkaline phosphatase (CIP) to the lentiviral transfer plasmid after 3 h.

6. Gel purify the digested transfer plasmid and insert following **steps 3** and **4** (*see* **Note 10**).
7. Ligate the digested mutant calpain cDNA into the linearized transfer plasmid. It is recommended to start with a 1:3 (vector-insert) ratio; however this ratio can be modified as needed for ligation efficiency.
   Vector-insert ligation:
   25 ng vector DNA.
   75 ng insert DNA.
   10× or 5× ligase buffer (*see* **Note 11**).
   1 μL T4 DNA ligase.
   H_2O to a 10 μL final volume.

   Ligate overnight at 16 °C.
8. Warm two antibiotic-resistant LB agar plates at room temperature (*see* Subheading 2 for making LB agar plates).
9. Chill 2 vials of chemically competent *E. coli* cells on ice for approximately 15 min. Add 5 μL of the ligated transfer plasmid containing the mutant calpain cDNA to one vial of competent cells and 5 μL of the linearized transfer plasmid alone to another vial of competent cells (vector only control). Very gently mix, but be careful not to damage competent cells. Leave on ice for 30 min. Warm a water bath to 42 °C. After 30 min, heat shock the competent cell vials by placing in the 42 °C water bath for 45 s. Place them back on ice for 2 min, then add 250 μL of LB broth to each tube, and warm while shaking at 37 °C for 1 h. After 1 h of recovery at 37 °C, spread 100 μL of each of the competent cell solutions onto an individual drug-resistant LB agar plate, and place it overnight at 37 °C (*see* **Note 12**). The remaining competent cell solutions can be placed at 4 °C overnight in case the experiment will need to be repeated the following day. Do not keep them longer than 24 h; repeat with new competent cells if a repeat experiment is needed after 24 h.
10. Compare the number of colonies grown overnight on the LB agar plates. Colony numbers should be higher on the plate containing the transfer plasmid with the ligated calpain cDNA, in comparison to the plate containing vector alone. A pipet tip can be used to scrape a small amount of each colony from the plate, and place into 3 mL of

LB. Approximately 10 colonies should be picked with a separate, clean pipet tip and placed into individual, sterile tubes with LB broth (*see* **Note 13**). Allow the bacteria to grow while shaking for 12–18 h at 37 °C. After growth, purify the DNA using an endotoxin-free midi-prep kit, which can be purchased from suppliers such as Qiagen (*see* **Note 14**).

11. Nanodrop the samples to determine DNA concentration, and confirm correct orientation and presence of insert by sequencing. It is also important to ensure that no other mutations occurred during the cloning procedure. At this point, the plasmid is ready for lentivirus production.

3.2 Lentivirus Production

1. Each of the following steps should be performed in sterile tissue culture hoods and incubators, in a biosafety level 2+ (BL2+) room for safety when working with virus. Follow the appropriate institutional guidelines.
2. The day before transfection, plate 2.5×10^6 HEK293T cells in a 10 cm dish (approximately 40% confluency). After approximately 24 h, the cells should reach 70% confluency and be ready for transfection (*see* **Note 15**).
3. Add 5 μg of the mutant calpain lentiviral transfer plasmid DNA (from Subheading 3.1), 7.5 μg of psPAX2 packaging vector (available through Addgene), and 2.5 μg of pMD2.G VSV-G envelope expressing plasmid (available through Addgene) in a 15 mL tube.
4. Add 40 μL of Lipofectamine transfection reagent dropwise onto 1 mL of Opti-MEM media in another 15 mL conical tube (*see* **Note 16**). Vortex the mixture briefly, and let sit at room temperature for 5 min.
5. Add Opti-MEM and transfection reagent mixture to DNA, and let sit for 20 min at room temperature.
6. Change the media on the HEK293T cells, replacing with 9 mL of fresh DMEM media. Then add the 1 mL of Opti-MEM containing DNA to the HEK293T cells, and incubate for 24 h. From this point forward, all equipment should be disposable and bleached to kill viral particles after use.
7. Change to fresh DMEM media the following day, and 48 h later, begin collecting viral supernatant.
8. To collect viral supernatant, pipette the media from the HEK293T cells into a disposable 10 mL syringe with a 0.45 μm filter at the tip. Filter the supernatant to collect only viral particles into an ultracentrifuge tube. Spin the virus in an ultracentrifuge at approximately 20,000–25,000 rpm for 90 min at 4 °C. Remove the supernatant from the pellet, and resuspend the pellet in 100 μL sterile phosphate buffered saline

(PBS). Mix via pipetting and let sit at 4 °C overnight. Aliquot virus and store at −80 °C long-term (*see* **Note 17**).

9. Determine viral titer. Plates 75,000 HEK293T cells into each well of a 6-well culture plate.
10. 24 h later, thaw an aliquot of lentivirus using a 37 °C water bath. Prepare five different dilutions of the virus in DMEM media containing 10 μg/mL polybrene. Recommended dilutions are 1:500, 1:200, 1:100, 1:50, and 1:25 in 1.5 mL total media. Remove the media from 5 wells of the 6-well HEK293T cell plate, and add each viral dilution to an individual well. Count the number of HEK293T cells in the unchanged well. Incubate for 48 h.
11. After 48 h, aspirate media from the HEK293T cells, and replace with 1 mL sterile PBS. Take cells to a fluorescent microscope, and count the number of fluorescent cells (i.e., cells that received the virus). If a well has more than 50% of cells fluorescent, titer concentration can be miscalculated, so only count wells with less than half of the cells transfected by the virus.
12. Calculate the viral titer in transduction units (TU) per mL. Multiply the number of cells that were transduced (the number of HEK293T cells in the untreated well from Subheading 3.2, **step 10**) by the percent of fluorescent cells (Subheading 3.2, **step 11**) and by the dilution factor (e.g., 200, if it was the well receiving the 1:200 viral dilution). Then divide this number by 1.5 mL, i.e., the volume of media containing virus that was added to each well (*see* **Note 18**).

3.3 Delivery of Lentivirus to the Mouse Eye

1. Further details and video of the subretinal injection procedure into the perinatal mouse eye have been previously published [10]. Here, details for both a subretinal and intravitreal injection are described to deliver the lentivirus to the mouse eye at either perinatal or adult stages of development.
2. Using a Sutter P-97 pipette puller or other similar equipment, heat and pull a glass capillary tube to create a 100-μm-diameter needle with a blunt, open tip (*see* **Note 19**). This allows for injection into the intravitreal or subretinal space without damage to the eye while also providing a clear needle for visualization of the lentivirus during the injection procedure. Aspirate silicone through the needle in order to coat the inner walls, and provide passage of the virus through the needle without the virus sticking to the inner walls. After aspiration of silicone through the needle, allow for the needle to dry in a sterile container.
3. Prepare the surgical suite based on institutional animal welfare guidelines and protocols. All procedures should be performed within the barrier facility if animals are to be kept alive after surgery.

4. Obtain a 25 3/4 gauge blood collection set with Luer lock, remove the needle from the tubing by cutting the tubing, and attach the 100 μm needle to the cut end of the plastic tubing. Fill a 1 mL Sub-Q 26 5/8 gauge slip-tip syringe with sterile saline, remove the needle end of the syringe, and attach the syringe to the Luer lock. Eject some of the sterile saline through the tubing and 100 μm needle, ensuring that the setup has been prepared appropriately and there is no leakage. Keep the prepared injection apparatus on a sterile, clean surface during the surgical procedure.
5. Prepare the animal for surgery following the locally approved animal handling protocol, as anesthesia will vary based on the institution and age of the mouse (*see* **Note 20**). When the mouse is fully under anesthesia, with no movement or response to toe pinches, the surgical procedure can begin. The following procedure should be performed under a dissection microscope for close visualization of the eye (*see* **Note 21**).
6. Proptose the eye by pulling apart the eyelids and pinching them underneath the eye to hold it in the proptosed position (*see* **Note 22**). Applying slight pressure on either side underneath the eye with dressing forceps can help to hold the eye in the proptosed position.
7. There are two main routes for delivering lentivirus to the eye. One delivers the virus to the subretinal space, where it is localized between the retinal pigment epithelium (RPE) and the photoreceptor cells. The other delivers the virus to the intravitreal space, where it will mainly affect the bipolar and ganglion cells of the eye (Fig. 4). For a subretinal injection, using a 15-degree microsurgery blade, make a small, scleral incision at or directly posterior to the globe's equator (*see* **Note 23**). The incision should be just large enough for the insertion of the 100 μm needle, and no larger. In pigmented eyes, the darker choroid-RPE will become visible at the incision site. If the animal is albino and the eye is not pigmented, it is recommended to mark the tip of the 15-degree surgical blade with a surgical marking pen, which will leave an ink spot to denote the incision site. For an intravitreal injection, small scrapes can be made with the 15-degree microsurgery blade until a clear fluid (the vitreous) refluxes from the incision site.
8. Aspirate a small, approximately 1 μL bubble of air into the end of the injection needle, in order to separate the lentivirus from the sterile saline. Place the appropriate amount of lentivirus, with a recommended viral titer of 2×10^7 TU/mL (Table 1), onto a sterile, flat surface such as the inner lid of a 10 cm culture plate. For the subretinal injection, aspirate the lentivirus into the tip of the 100 μm needle, and carefully insert the needle

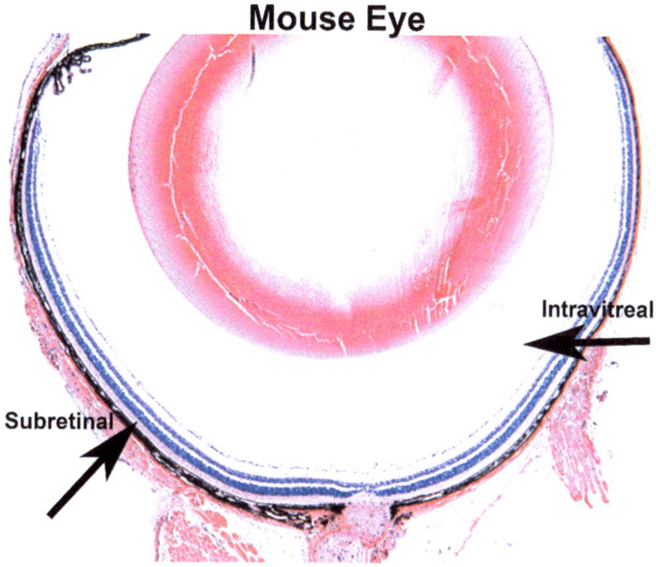

Fig. 4 Subretinal and intravitreal injection sites in the mouse eye. A cross-section of a mouse eye stained with hematoxylin and eosin. Subretinal and intravitreal injection locations are marked by black arrows, where the subretinal injection creates a viral bleb between the photoreceptor cells and the retinal pigment epithelium (RPE) and the intravitreal injection places the virus into the vitreous fluid of the eye

Table 1
Amount of virus based on injection location and age of the mouse. Amounts of virus in microliters that are recommended for injection into the subretinal and intravitreal space of the mouse eye, at perinatal and adult stages of development

Age	Injection location	Amount of virus (μL)
Perinatal	Subretinal	0.5–0.8
Perinatal	Intravitreal	0.8–1.0
Adult	Subretinal	0.8–1.0
Adult	Intravitreal	1.0–1.5

into the incision site and advance it parallel to the outer eye wall in order to enter the subretinal space. For intravitreal injection, insert the needle into the deeper incision site, and advance the needle just slightly parallel to the outer wall to avoid hitting the lens of the eye. It is not necessary to go too deep within the vitreous space for this injection as the virus can more easily spread throughout the vitreous.

9. Apply very light pressure on the syringe plunger to inject the lentivirus. In the subretinal space, moderate back pressure will be felt during injection. The virus creates a subretinal bleb

between the RPE and the photoreceptor cells, causing a temporary retinal detachment at the injection site that will heal within 24 h (*see* **Note 24**). For the intravitreal injection, very little pressure will be felt, and the virus will be easily expelled into the vitreous space. Care must be taken to inject the lentivirus without causing injection of the air bubble or reflux of the virus back out from the injection site. It is also important to remove the needle following the same incision course, to avoid excess damage to the retina and eye (*see* **Note 25**).

10. Place the eye back under the eyelids by applying slight pressure with forceps and pushing down on the eye. Provide pain medication as approved by the institutional animal protocol and welfare guidelines. It is recommended to use a 25 gauge needle to place a single drop of 0.25% bupivacaine (diluted in sterile saline) at the injection site for the comfort of the mouse. Allow the mouse to wake from anesthesia on a heating pad, and then move it to a clean cage after the mouse is ambulatory (*see* **Note 26**). Mice should be monitored after surgery following the institutional guidelines for any signs of distress, and if present, follow the veterinary guidelines for further treatment.

4 Notes

1. TBE can be used instead of TAE.
2. It is best to amplify the mutant calpain from plasmid DNA, provided by companies such as OriGene or Addgene.
3. A leader sequence can be added in the primer design to assist with the restriction enzyme digestion. The leader sequence is approximately 3–6 base pairs (bp) added to the ends of the primer. If used, the PCR product should be digested by the appropriate restriction enzymes prior to ligation.
4. Make sure that the restriction enzymes will not cut within the mutant calpain DNA.
5. Make sure to include the start site and full length of the mutant calpain DNA for correct translation in vivo.
6. This takes approximately 2 min, and stopping and swirling the solution while heating help dissolve the agarose without boiling over the TAE.
7. EtBr is a mutagen and should be handled in a fume hood with nitrile gloves.
8. For cutting DNA bands, it is helpful to leave an empty lane between samples when loading and running the gel.
9. The digestion step can be extended if the restriction enzyme does not have STAR activity, as it will not incorrectly cut outside of its designated sequence.

10. It is recommended to run uncut products on the gel as well as the digested sample to ensure that the correct linearized transfer plasmid and insert are purified from the gel.
11. Avoid too many freeze thaws of ligase buffer containing ATP by making small aliquots of the ligase buffer when first purchasing and only thawing each aliquot when needed.
12. Some plasmids grow more efficiently at 30 °C; check plasmid information before growing overnight.
13. A pipet tip can be used to scrape the remainder of each individual colony. The tip of the pipet can be inserted into a PCR tube containing PCR mix with primers designed to detect the vector backbone and insert sequence. Swirling the pipet tip in the PCR mix provides the plasmid DNA for amplification, and this PCR method can be used to screen the colonies before growing overnight for the presence of the insert and its orientation in the vector backbone.
14. A diagnostic restriction enzyme digest can be performed before sequencing to determine whether or not the insert is present in the plasmid. 500 ng of DNA is typically used for a diagnostic digest, and only 1 h of digestion is necessary for visualizing the insert after running the product on a gel.
15. HEK293T cells can lift easily off of a cell culture plate. Therefore, autoclaved 0.2% gelatin can be added to the culture plate for 20 min. Gelatin is then aspirated, and the culture plate is washed once with sterile PBS and can be used for plating the HEK293T cells on the gelatin coating.
16. Other transfection reagents can be used; it is important to follow the manufacturer's protocols depending on which one is used for the lentiviral production.
17. Virus is good at −80 °C for at least 1 year, but viral titer may slightly decline over time and should be checked periodically after thawing aliquots.
18. Calculating more than one of the wells (the five different dilutions) and averaging the titer calculated from each one are recommended for better accuracy of the TU/mL.
19. Needles smaller than 100 μm can be used; however it is known that 100 μm needles allow for the injection of viruses without risk of damage to the eye or clogging of the needle. Also, when preparing the pulled capillary needles, if glass or debris ends up inside the needle, it can be washed by flushing with 70% ethanol. Allow for the ethanol to fully evaporate or flush afterward with sterile saline before use.
20. Intraperitoneal injection of anesthetic is recommended for this protocol, as the nose cone for isoflurane can block access to the eye.

21. The entire surgical procedure after anesthesia should take under 5 min, and therefore a heating pad during the procedure is not necessary and there is no risk for cataracts. However, when beginning to learn this procedure, more time may be needed. Uninjected adult eyes should be covered with a lubricant, and the procedure may need to be performed on a heated surface to maintain normal body temperature of the mouse.
22. In perinatal mice under 2 weeks of age, Vannas scissors should be used to cut at the eyelid indentation to open the eyelids. Be careful to not make too large of an incision, under 1.5 mm, as the eye will not remain proptosed and will slip back under the eyelids. If needed, a circular punch the size of a mouse eye can be made in a hardened, plastic surface, and this can be placed over the eye after proptosing the eye; the plastic can be pressed down with a small amount of pressure, being careful not to harm the animal or inhibit breathing, and this can help keep the eye accessible during the surgical procedure.
23. To avoid making too deep of an initial cut and damaging the retina, it is recommended to make a couple very short scrapes to open an incision into the scleral space, rather than one deeper cut. A small, acupuncture needle or similar needle can also be used to make the initial incision. Anything with a small, sharp tip that will reduce damage or size of the incision into the eye.
24. For subretinal injection, holding the needle in place for an additional 15 s within the subretinal space after injection of the lentivirus can help to avoid the back pressure flushing the virus back out of the incision site.
25. Although the incision site and/or temporary retinal detachment will heal within 24 h, the eye will remain soft for approximately 1 week, so further injections should be avoided during this timeframe.
26. For perinatal mice, gentle toe pinches can help them fully recover from anesthetic and regain their pink coloring before being placed back with their mother for warmth and care.

Acknowledgment

VBM is supported by NIH grants [R01EY026682, R01EY024665, R01EY025225, R01EY024698, R21AG050437, and P30EY026877], The Doris Duke Charitable Foundation Grant #2013103, and Research to Prevent Blindness (RPB), New York, NY.

References

1. Wert KJ, Bassuk AG, Wu WH, Gakhar L, Coglan D, Mahajan M, Wu S, Yang J, Lin CS, Tsang SH, Mahajan VB (2015) CAPN5 mutation in hereditary uveitis: the R243L mutation increases calpain catalytic activity and triggers intraocular inflammation in a mouse model. Hum Mol Genet 24:4584–4598
2. Bassuk AG, Yeh S, Wu S, Martin DF, Tsang SH, Gakhar L, Mahajan VB (2015) Structural modeling of a novel CAPN5 mutation that causes uveitis and neovascular retinal detachment. PLoS One 10:e0122352
3. Wert KJ, Skeie JM, Bassuk AG, Olivier AK, Tsang SH, Mahajan VB (2014) Functional validation of a human CAPN5 exome variant by lentiviral transduction into mouse retina. Hum Mol Genet 23:2665–2677
4. Mahajan VB, Lin JH (2013) Lymphocyte infiltration in CAPN5 autosomal dominant neovascular inflammatory vitreoretinopathy. Clin Ophthalmol 7:1339–1345
5. Rowell HA, Bassuk AG, Mahajan VB (2012) Monozygotic twins with CAPN5 autosomal dominant neovascular inflammatory vitreoretinopathy. Clin Ophthalmol 6:2037–2044
6. Mahajan VB, Skeie JM, Bassuk AG, Fingert JH, Braun TA, Daggett HT, Folk JC, Sheffield VC, Stone EM (2012) Calpain-5 mutations cause autoimmune uveitis, retinal neovascularization, and photoreceptor degeneration. PLoS Genet 8:e1003001
7. Pastor JC, de la Rua ER, Martin F (2002) Proliferative vitreoretinopathy: risk factors and pathobiology. Prog Retin Eye Res 21:127–144
8. Frank RN (2004) Diabetic retinopathy. N Engl J Med 350:48–58
9. Caspi RR (2010) A look at autoimmunity and inflammation in the eye. J Clin Invest 120:3073–3083
10. Wert KJ, Skeie JM, Davis RJ, Tsang SH, Mahajan VB (2012) Subretinal injection of gene therapy vectors and stem cells in the perinatal mouse eye. J Vis Exp 69:4286

Chapter 19

Methods of Calpain Inhibition to Determine the Role of Calpains in Embryo Development in Amphibians

Ioanna Antoniades, Anna Charalambous, Neophytos Christodoulou, Sara Zanardelli, and Paris A. Skourides

Abstract

Calpains are a family of calcium-dependent intracellular cysteine proteases that regulate important physiological processes by substrate cleavage. Despite the fact that the role of calpains in cell migration and other processes has been extensively studied in vitro, the same does not apply to cell migration and morphogenetic events during embryogenesis, in vivo. Herein, we describe the use of three different methods to selectively block calpain activity in vivo in order to investigate the impact on *Xenopus* gastrulation and neurulation, namely, a calpain inhibitor, a dominant negative, and a morpholino antisense oligonucleotide (MO). We also provide methods to determine the effectiveness of the calpain inhibition and effect on cell fate specification and morphogenetic movements, during embryogenesis in vivo.

Key words *Xenopus*, Calpain, Protein downregulation, CI3 inhibitor, Whole-mount in situ hybridization, Immunofluorescence, Casein zymography, Mesoderm migration, Convergent extension

1 Introduction

Calpains are a family of calcium-dependent intracellular cysteine proteases that regulate important physiological processes (i.e., signal transduction, cell proliferation, apoptosis, and cell motility) by substrate cleavage [1]. Two specific members of this family, Calpains 1 and 2, share 61% of sequence identity and are heterodimers composed of an 80 kDa *catalytic* subunit and a common 30 kDa regulatory subunit, which dissociates upon activation [2, 3].

Despite the fact that the role of Calpain 2 in cell migration and other processes has been extensively studied in vitro, the same does not apply to cell migration and morphogenetic events during embryogenesis in vivo. Herein, we describe the use of three different methods to selectively block Calpain 2 activity in vivo in order

Ioanna Antoniades and Anna Charalambous contributed equally to this work.

Jeannette S. Messer (ed.), *Calpain: Methods and Protocols*, Methods in Molecular Biology, vol. 1915,
https://doi.org/10.1007/978-1-4939-8988-1_19, © Springer Science+Business Media, LLC, part of Springer Nature 2019

to investigate the impact on *Xenopus* gastrulation and neurulation. Specifically we describe the use of a Calpain Inhibitor III (CI3), a dominant-negative mutant (generated by mutating Cys105 to Ser, within the active site of the *Xenopus* Calpain 2 gene, a mutation reported to inactivate the enzyme without affecting overall protein structure and binding capacity) and an antisense morpholino oligonucleotide (MO). The first two approaches efficiently block both Calpains 1 and 2, due to their high sequence similarity. In contrast, the third approach is more specific and selectively blocks the expression of Calpain 2. To determine blocking efficiency, casein zymography analysis is employed and the impact on embryo phenotype characterized. Furthermore, to determine possible effects on cell fate specification, whole-mount in situ hybridization (WISH) experiments are performed. Finally, to address the role of the protease in specific morphogenetic movements, analysis of whole-embryo phenotypes, live video microscopy, and explant-based assays are used [4].

The aforementioned blocking approaches are important as they allow not only the determination but also the regulation of Calpain 2 activity in vivo and thus permit investigation of its role in the various physiological processes, in which it is involved.

2 Materials

Prepare all solutions using analytical/molecular biology-grade reagents.

Diligently follow all waste disposal regulations when disposing waste materials.

2.1 Xenopus laevis Embryos: Collection, Microinjections, and Fixation

1. *Xenopus laevis* eggs collected from induced spawning of adult females, in vitro fertilized using testes from adult males [5, 6].
2. Human chorionic gonadotropin (hCG), 800 units.
3. 10% benzocaine.
4. Testis medium: 90% Leibovitz's L-15 medium, 10% fetal calf serum (FCS), and 50 μg/ml gentamycin.
5. Petri dishes.
6. Microinjector, micromanipulator, forceps, and glass capillary needles.
7. 10× Marc's Modified Ringer's, pH 7.8 (MMR): 0.1 M NaCl, 1.8 mM KCl, 2 mM $CaCl_2$, 1 mM $MgCl_2$, 5 mM HEPES (4-(2-hydroxyethyl)-1-piperazineethanesulfonic acid), 0.1 mM EDTA (ethylenediaminetetraacetic acid), and 0.05 mg/ml gentamycin gel.
8. 4% Ficoll in MMR.

9. 1 × MEMFA (MOPS (3-(*N*-morpholino)-propanesulfonic acid), EGTA (ethylene glycol tetraacetic acid), $MgSO_4$, formaldehyde): 0.1 M MOPS pH 7.4, 2 mM EGTA, 1 mM $MgSO_4$, and 1% formaldehyde.

2.2 Agarose Gel Electrophoresis and Imaging

1. UltraPure 10 × Tris-acetate-EDTA running buffer (TAE): A pH-adjusted blend of UltraPure grade TAE buffer which, when diluted to a 1 × working concentration, contains 40 mM Tris-acetate and 1 mM Na_2EDTA.
2. Agarose: Certified free of RNAses, DNAses, and inhibitors.
3. Ethidium bromide.
4. 100 bp or 1 kb DNA ladder.
5. Nuclease-free H_2O or diethyl pyrocarbonate (DEPC)-H_2O.
6. Agarose gel electrophoresis machine.
7. Gel documentation imaging system.

2.3 Generation of Constructs

1. PCR machine.
2. Primers.

Primer	Sequence
Capn2C105SF	GGAGCCCTGGGTGACAGCTGGCTTC TTGCAG
Capn2C105SR	CTGCAAGAAGCCAGCTGTCACCCAGGGC TCC

3. Plasmids: Capn2R-pCS108.
4. Restriction enzymes: DpnI.
5. Chemically competent bacteria.
6. MiniPrep kit for DNA extraction.

2.4 Whole-Mount In Situ Hybridization (WISH)

1. RNA probe synthesis: SP7 promoter kit.
2. RNA purification kit.
3. 1 × MEMFA.
4. Methanol.
5. Proteinase K (10 mg/ml).
6. PBTw: 1 × PBS, 0.1% Tween 20.
7. 0.1 M triethanolamine (TEO), pH = 7–8.
8. 4% formaldehyde, in DEPC-H_2O.
9. Hybridization solution (Hyb): 50% formamide, 5 × SSC, 1 mg/ml torula RNA, 100 μg/ml heparin, 1 × Denhardt's, 0.1% Tween 20, 0.1% CHAPS, and 10 mM EDTA.

10. 2 × saline-sodium citrate (SSC) buffer: 0.3 M sodium chloride and 30 mM trisodium citrate in DEPC-H_2O, pH 7.
11. RNAse A, 20 mg/ml.
12. Maleic acid buffer (MAB): 100 mM maleic acid and 150 mM NaCl.
13. *Boehringer Mannheim Blocking* (BMB) solution: 2% BMB reagent in MAB.
14. Anti-digoxigenin (DIG) alkaline phosphatase (AP) antibody.
15. AP buffer: 100 mM Tris pH 9.5, 50 mM $MgCl_2$, 100 mM NaCl, 0.1% Tween 20, and 2 mM levamisole.
16. BM purple AP substrate precipitating.
17. Bleaching solution: 0.5 × SSC, 5% formamide, 1.2% peroxide.

2.5 In Vitro Transcription of mRNA

1. Sp6 promoter kit.

2.6 Calpain Inhibitor III (CI3) Preparation

1. Lyophilized CI3 (Calpain Inhibitor III), commercially available from many companies.
2. Dimethyl sulfoxide (DMSO).

2.7 Calpain 2 Antisense Morpholino (MO) Preparation

1. Lyophilized FITC-Capn2 antisense MO and control MO obtained from Gene Tools. The sequence of Calpain 2 MO is GTCTGTCGGCGACCCCGCTC ATGTC (*see* **Note 1**).
2. Nuclease-free H_2O.

2.8 Casein Zymography

1. MK's lysis buffer: 50 mM Tris, pH 8, 150 mM NaCl, 0.5% NP40, 0.5% Triton X-100, 1 mM EGTA, 5 mM NaF.
2. Protease inhibitors.
3. Copper sulfate ($CuSO_4$).
4. Bicinchoninic acid (BCA).
5. Casein zymography gels:
6. Separating gel: 0.2% (w/v) casein salt, 12% (w/v) acrylamide, 0.32% (w/v) bisacrylamide, 375 mM Tris–HCl pH 8.8, 0.04% (w/v) ammonium persulfate (APS), 0.028% (v/v) tetramethylethylenediamine (TEMED).
7. Stacking gel: 4% (w/v) acrylamide, 0.11% (w/v) bisacrylamide, 330 Mm Tris–HCl pH 6.8, 0.04% (w/v) APS, 0.028% (v/v) TEMED.
8. β-Mercaptoethanol.
9. Bromophenol blue.
10. Molecular weight marker.
11. Polyacrylamide gel electrophoresis machine.

12. Pre-running buffer: 25 mM Tris base, 192 mM glycine, 1 mM ethylene glycol-bis(β-aminoethyl ether)-N,N,N′,N′-tetraacetic acid (EGTA), 1 mM 1,4 dithiothreitol (DTT) pH 8.3.
13. Calpain activation buffer: 20 mM Tris–HCl pH 7.4, 10 mM DTT, 1–4 mM $CaCl_2$.
14. Control buffer: 20 mM Tris–HCl pH 7.4, 10 mM DTT), 1 mM ethylenediaminetetraacetic acid (EDTA).
15. Coomassie Blue G-250.

2.9 Mesoderm Migration Assay

1. Hair knives, forceps, coverslips, slides.
2. Calcium- and magnesium-free medium (CMFM): 88 mM NaCl, 1 mM KCl, 2.4 mM $NaHCO_3$, 7.5 mM Tris–HCl pH 7.6.
3. Agarose.
4. Human recombinant activin, 1 U/ml in CMFM.
5. Fibronectin, 0.1 mg/ml in PBS.

2.10 Mesoderm and Neural Convergent Extension Assays (Animal Caps and Explants)

1. Hair knives, forceps, coverslips, slides.
2. 0.5× MMR.
3. Activin (10 ng/ml in 0.5× MMR, for animal cap explants and mesoderm convergent extension).
4. Danilchik's medium (DFA): 53 mM NaCl, 32 mM sodium gluconate, 4.5 mM potassium gluconate, 1 mM $CaCl_2$, 1 mM $MgSO_4$, 5 mM Na_2CO_3 buffered to pH 8.3 with bicine.

2.11 Imaging

1. Forceps, slides, and coverslips.
2. Bright-field imaging: fluorescent stereomicroscope.
3. Fluorescence imaging: bright-field fluorescent microscope equipped with the appropriate filter sets, camera, and the appropriate imaging analysis software.

3 Methods

3.1 Obtaining Embryos

1. Induce ovulation of female frogs (*Xenopus laevis*) by injecting hCG into the dorsal lymph sac. Keep female frogs at room temperature for 9–10 h until egg collection.
2. Sacrifice a male frog by immersion in 0.05% benzocaine for 1 h. Isolate testes and keep in testis medium at 4 °C.
3. The eggs collect in the sac near the cloaca, so simultaneous lateral and vertical pressure should release them. Collect the eggs manually by massaging the belly with one thumb over a clean petri filled with 0.33× MMR.

4. Fertilize eggs immediately after collection. Cut a small piece of testis with scissors, and macerate it with forceps, so it is evenly distributed throughout the media. Leave the embryos for about 20 min. The eggs should rotate within the vitelline membrane, so the animal hemisphere faces upward.
5. About 2 h after fertilization, the first cleavage is visible. De-jelly the embryo for microinjection by adding 1.8% cysteine for 3–5 min, and then wash with 0.33× MMR. Keep the embryos in 0.33× MMR [7].

3.2 Preparation of Calpain Inhibitor III (CI3) and Use in In Vivo Experiments

1. Dissolve the lyophilized Calpain Inhibitor III (CI3) in DMSO to an initial stock concentration of 50 mM (*see* **Note 2**).
2. Add CI3 inhibitor (final working concentration: 100 μM in 0.1× MMR) to embryos at different times of development.
3. For rescue experiments, remove the inhibitor at stage 8 - (pre-gastrula), stages 10.5–12 (gastrula stage), or later stages, by replacing the media in which *Xenopus* embryos are developing with fresh 0.1× MMR.
4. Assess embryo morphology and development. Place *Xenopus* embryos on a slide, and capture bright-field images using a fluorescent stereomicroscope.

3.3 Preparation of mRNA Encoding Calpain 2 Dominant-Negative Mutant and Use in In Vivo Experiments

This procedure consists of two parts: (1) site-directed mutagenesis to generate Calpain 2 dominant-negative DNA (Capn2C105S) and (2) injection of dominant-negative mutant *Xenopus* Calpain 2 mRNA in *Xenopus* embryos for phenotype characterization.

1. Set up a PCR reaction to introduce the mutation cysteine 105 to serine, in the active site of the *Xenopus* Calpain 2 gene. Use 10 ng of pCS108-Capn2R plasmid as template and the primers Capn2C105SF and Capn2C105SR at a final concentration of 0.1 μM (*see* **Note 3**).
2. Use the following PCR program: 95 °C for 1 min, 18 cycles each of 95 °C for 50 s, 65 °C for 30 s, 68 °C for 8 min, followed by 68 °C for 7 min and 4 °C until reaction termination.
3. Add 1 μl of DpnI restriction enzyme, mix well by pipetting up and down, and incubate at 37 °C for 1 h.
4. Transform chemically competent bacteria.
5. Pick colonies, and set up MiniPrep for DNA extraction.
6. Perform in vitro transcription of capped RNA from linearized plasmid encoding the Calpain 2 dominant-negative mutant DNA. Use any commercially available kit, and follow the manufacturer's instructions for in vitro transcription.

7. Use any RNA purification kit, and follow the manufacturer's recommendations to purify the synthesized mRNA. Aliquot and store at −80 °C until injection.
8. Place *Xenopus* embryos in 4% Ficoll, in 0.33× MMR, solution, and inject using a glass capillary pulled needle, forceps, a micromanipulator, and microinjector.
9. Inject embryos, at the two-cell stage or at the tail bud stage, with 1 ng of Capn2C105S mRNA. Allow injected *Xenopus* embryos until stage 8 in 4% Ficoll in 0.33× MMR. Then wash and maintain in 0.1× MMR until the desired stage [8].
10. Assess embryo morphology at different developmental stages. Place *Xenopus* embryos on a slide, and capture bright-field images using a fluorescent stereomicroscope (*see* **Note 4**).

3.4 Preparation of Calpain 2 Antisense MO and Use in In Vivo Experiments

This entails preparation of Calpain 2 antisense MO, injection of Calpain 2 MO in *Xenopus* embryos for phenotype characterization and rescue (*see* **Note 5**), and casein zymography under non-denaturing conditions after MO injection (*see* **Note 6**).

1. Make a 1 mM stock solution by dissolving the lyophilized Calpain 2 MO in nuclease-free H_2O. Aliquot and store at −20 °C until use.
2. Inject *Xenopus* embryos with either 70 ng of Calpain 2 MO alone or together with 500 pg mRNA of HA-Capn2R (for rescue), either at one-cell stage or two out of two blastomeres at two-cell stage, as described above.
3. Allow *Xenopus* embryos to develop in 0.1× MMR, and observe changes in morphology at different developmental stages.
4. Place embryos on a slide, and capture bright-field images using a fluorescent stereomicroscope.
5. Inject *Xenopus* embryos with 50 or 70 ng of Calpain 2 MO either at the one-cell stage or two out of two blastomeres at the two-cell stage.
6. Following injection, allow *Xenopus* embryos to develop in 0.1× MMR up to stage 13 or 19.
7. Prepare embryo lysates by homogenizing injected embryos (stage 13 or 19), by pipetting up and down, in ice cold MK's modified lysis buffer supplemented with protease inhibitors.
8. Centrifuge protein lysates at 15,000 × *g* for 30 min at 4 °C.
9. Quantify total protein concentration by using the BCA assay.
10. Prepare non-denaturing casein polyacrylamide gels and pre-run for 30 min on ice.
11. Run 40 µg of total protein lysate on a non-denaturing casein polyacrylamide gel for 1.5 h at 130 V on ice.

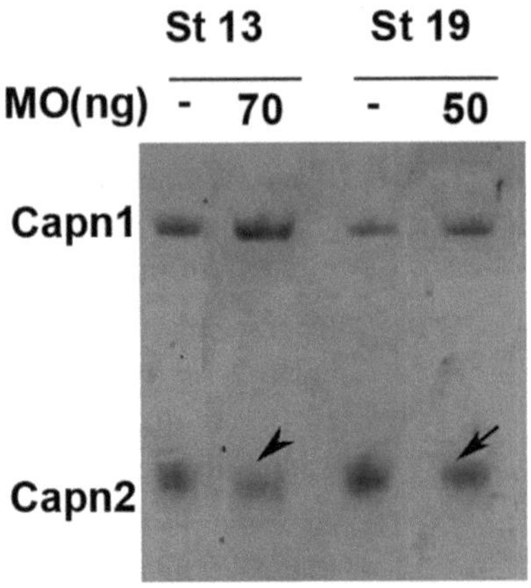

Fig. 1 MO-mediated Calpain 2 downregulation by (**a**) casein zymography analysis showing endogenous Calpain 1 (top) and Calpain 2 (bottom) activity in lysates of MO-injected embryos. Injection of 50 ng MO causes a 40% reduction in Calpain 2 expression at stage 19 (arrow). Injection of 70 ng MO causes a 50% reduction in Calpain 2 expression at stage 13 (arrowhead). Densitometry analysis was performed using ImageJ software, and the samples were corrected for loading in correlation with Capn1. Adapted from Zanardelli et al. (2013) Dev Biol 384(1):83–100 (https:/creativecommons.org/licenses/by/3.0)

12. Incubate the gel at 25 °C in calpain activation buffer containing 1–4 mM calcium and 10 mM DTT overnight (*see* **Note 7**).
13. Stain the gel with Coomassie Blue G-250 (Fig. 1) (*see* **Note 8**).

3.5 Determination of the Role of Calpain 2 in Mesoderm and Neural Cell Fate Specification (See Note 9)

1. Control embryos and Calpain 2-inhibited embryos are used for WISH using antisense RNA probes for the mesodermal markers *X-brachyury* (*Xbra*) and *Chordin* (*Chrd*) or the neural markers *Sox2* and *Pax3* (*see* **Note 10**) [9].
2. Fix embryos at gastrula or neurula stages in 1 × MEMFA, and then dehydrate them in 100% methanol for 1 h (changing methanol every 15 min), in 4 ml glass vials.
3. Rehydrate embryos through a series of methanol solutions (100%, 75%, 50%, and 25% methanol in DEPC-H_2O), 5 min each at room temperature, and then wash three times in PBTw, 5 min each.
4. Permeabilize embryos using 10 μg/ml proteinase K in PBTw for 5 min at room temperature. Wash embryos in 0.1 M TEO twice (5 min each), then in TEO with acetic anhydride (2 × 5 min), and then in PBTw (2 × 5 min) at room temperature.
5. Postfix embryos in 4% formaldehyde in PBTw for 20 min at room temperature, and then wash five times (5 min each) in PBTw.

6. Incubate embryos in hybridization solution for 6 h at 65 °C, and then incubate overnight at 65 °C with the probe diluted in hybridization solution at a final concentration 1 mg/ml.
7. Incubate the embryos with the hybridization solution (without probe) for 3 min at 60 °C, and then wash with 2× SSC at 60 °C (2 × 3 min and 3 × 20 min).
8. Treat embryos with 20 μg/ml RNAse A in 2× SSC for 30 min at 37 °C, wash with 2× SSC at 60 °C for 10 min, and then wash twice with 0.2× SSC for 30 min at 60 °C.
9. Incubate embryos in MAB solution for 30 min at room temperature, block in BMB blocking solution for 1 h at room temperature, and then incubate with anti-DIG AP antibody overnight at 4 °C.
10. Wash embryos in MAB solution (5 × 1 h) at room temperature followed by two washes, 5 min each, in AP buffer.
11. Incubate embryos in BM purple, and check over time for color development. The visualization of color at a specific region determines the presence of mRNA. Stop the chromogenic reaction by washing the embryos in MAB solution at room temperature, and postfix the embryos overnight in 1 × MEMFA at room temperature.
12. Embryos are imaged using a fluorescent stereomicroscope for the acquisition of bright-field images. The expression and distribution of probes in control and Calpain 2-inhibited embryos are compared.

3.6 Examine the Role of Calpain 2 in Mesoderm Migration

1. Dissect animal caps from stage 9 embryos, and dissociate cells in CMFM in agarose (diluted in CMFM)-coated plates. Treat dissociated cells with human recombinant activin (1 U/ml, diluted in CMFM) for 1 h at room temperature (*see* **Note 11**).
2. Plate dissociated cells on coverslips coated with fibronectin (0.1 mg/ml), in 0.5× MMR in the absence (control) or presence of 50 μM CI3 inhibitor, for 2 h at room temperature.
3. Image cells under a bright-field microscope, and acquire images at 5 min intervals for at least 100 min. Use the appropriate software to measure the velocity (μM/min) of single cells, and compare the migration rates between control and CI3-treated mesodermal cells.

3.7 Examine the Role of Calpain 2 in Mesoderm Convergent Extension Using the Animal Cap Elongation Assay

1. Dissect animal caps from stage 8 embryos, and culture on agarose-coated plates in 0.5× MMR containing 10 ng/ml human recombinant activin, in the absence (control) or presence of 50 μM CI3 inhibitor.
2. Image animal caps using a fluorescent stereomicroscope, and compare the activin-induced elongation efficiency between control and treated embryos.

3.8 Examine the Role of Calpain 2 in Neural Convergent Extension Using Neural Plate Explants

1. Inject one-cell stage embryos at the animal pole with 70 ng of the Calpain 2 MO. Allow injected and control (un-injected) embryos to develop up to stages 12–12.5 in 0.1× MMR.
2. Dissect neural plate explants (longitudinal sections) from stage 12 to 12.5 embryos, and culture in DFA medium on agarose-coated plates.
3. Image explants over time under a fluorescent stereomicroscope. Acquire images every 2 h, and compare length changes in explants from control and morphant embryos.
4. For quantification of the elongation efficiency, calculate the elongation rate in control and morphant explants. Use appropriate software to measure changes in length (Δlength), and present mean changes for each condition in plots against time.

4 Notes

1. The MO was designed to anneal to a region of Calpain 2 mRNA that is not conserved among other members of the calpain family.
2. If DMSO solvent is at all wet, it will not freeze at −20 °C, so freeze at −30 °C. DMSO is hygroscopic and will absorb water if subjected to repeated freeze-thaw cycles. To minimize these issues, prepare multiple aliquots.
3. The primers Capn2C105SF and Capn2C105SR were designed starting from nucleotide 297 and including the desired mutation, C105S, in the coding sequence, shown in bold.
4. The morphology of dominant-negative mutant embryos emulated that observed with CI3 treatment.
5. Due to the unavailability of any Calpain 2-specific antibodies, the efficiency of the Calpain 2 MO was evaluated using a GFP-fused Calpain 2 surrogate. The mRNA of the surrogate was co-injected with the Calpain 2 MO, and expression was examined through Western blot experiments and live imaging.
6. Casein zymography is used to determine endogenous Calpain 2 expression levels following MO injections rather than as an activity assay. This is because calcium is added to the reaction buffer, a known inducer of calpain activation in vivo.
7. As a negative control, incubate the gel in the same buffer without calcium and with addition of 1 mM EDTA.
8. After staining of the casein gels with Coomassie blue G-250, both Calpains 1 and 2 showed up as clearing bands. The amount of calpain loaded was proportional to the brightness of the clearing band. Calpain 1 can be easily distinguished from Calpain 2 due to its higher mobility in the gel.

9. Experiments for the examination of mesodermal markers were done using CI3-treated embryos instead of morphants due to insufficient downregulation of Calpain 2 at early stages of development (gastrula stages).
10. Embryos were either treated with 50 μM CI3 inhibitor or injected at four-cell stage with 20 ng of the Capn2 MO at one dorsal blastomere.
11. Activin treatment induces mesodermal tissues in animal cap explants from blastula embryos.

References

1. Wells A, Huttenlocher A, Lauffenburger DA (2005) Calpain proteases in cell adhesion and motility. Int Rev Cytol 245:1–16. https://doi.org/10.1016/S0074-7696(05)45001-9
2. Croall DE, DeMartino GN (1991) Calcium-activated neutral protease (calpain) system: structure, function, and regulation. Physiol Rev 71(3):813–847
3. Sorimachi H, Tsukahara T, Okada-Ban M, Sugita H, Ishiura S, Suzuki K (1995) Identification of a third ubiquitous calpain species--chicken muscle expresses four distinct calpains. Biochim Biophys Acta 1261(3):381–393
4. Zanardelli S, Christodoulou N, Skourides PA (2013) Calpain2 protease: a new member of the Wnt/Ca(2+) pathway modulating convergent extension movements in Xenopus. Dev Biol 384(1):83–100. https://doi.org/10.1016/j.ydbio.2013.09.017
5. Antoniades I, Stylianou P, Christodoulou N, Skourides PA (2017) Addressing the functional determinants of FAK during ciliogenesis in multiciliated cells. J Biol Chem 292(2):488–504. https://doi.org/10.1074/jbc.M116.767111
6. Antoniades I, Stylianou P, Skourides PA (2014) Making the connection: ciliary adhesion complexes anchor basal bodies to the actin cytoskeleton. Dev Cell 28(1):70–80. https://doi.org/10.1016/j.devcel.2013.12.003
7. Charalambous A, Andreou M, Antoniades I, Christodoulou N, Skourides PA (2012) In vivo, site-specific, covalent conjugation of quantum dots to proteins via split-intein splicing. Methods Mol Biol 906:157–169. https://doi.org/10.1007/978-1-61779-953-2_11
8. Charalambous A, Antoniades I, Christodoulou N, Skourides PA (2011) Split-inteins for simultaneous, site-specific conjugation of quantum dots to multiple protein targets in vivo. J Nanobiotechnol 9:37. https://doi.org/10.1186/1477-3155-9-37
9. Charalambous A, Koyioni M, Antoniades I, Pegeioti D, Eleftheriou I, Michaelidou SS, Amelichev SA, Konstantinov LS, Rakitin OA, Koutentis PA, Skourides PA (2015) 1,2,3-Dithiazoles – new reversible melanin synthesis inhibitors: a chemical genomics study. Medchemcomm 6(5):935–946. https://doi.org/10.1039/c5md00052a

Chapter 20

Genetic Models of Calpain Deficiency and Ectopic Expression

Yan Gao, Christine Hall, James MacLeod, and Peter A. Greer

Abstract

Genetic manipulation in cell and animal models can provide important insights into gene function and the relationships between gene mutation and disease. This chapter describes methods to generate models of calpain-1 and calpain-2 deficiency, or their recombinant ectopic expression in cultured cells, and to genotype a conditional transgenic mouse model of calpain-1/calpain-2 deficiency that can be used to explore physiologic roles for these calpains.

Key words shRNA knockdown, CRISPR-Cas9 knockout, Lentiviral vectors, Conditional Cre/loxP *capns1* knockout mice, Calpain, *capn1*, *capn2*

1 Introduction

Experimental methods for manipulating gene expression have matured rapidly in recent years, making it feasible for labs with basic expertise in molecular and cell biology to quickly knock down, knock out, or ectopically express genes of interest in cell culture models. With additional expertise in mouse embryonic stem cell manipulation and mouse embryology, it is possible to manipulate the germline and develop transgenic animal models with targeted gene mutations. Indeed, with the recent application of CRISPR-Cas9 methods, it has become feasible to target genes with insertion or deletion mutations resulting in null (knockout) mutations, as well as subtle missense (knock-in) mutations, or to even correct mutations in the mouse germline [1]. While this chapter will not delve into the methods of gene targeting in mice, a protocol is provided for genotyping an available conditionally targeted (*floxed*) *capns1* mouse strain [2]. This serves as an example for genotyping other gene-targeted mouse strains. This chapter will also provide detailed methods to manipulate the expression of *capn1*, *capn2*, and *capns1* in cultured cells, thereby providing models to study the functions of calpain-1 (CAPN1 + CAPNS1) and

Jeannette S. Messer (ed.), *Calpain: Methods and Protocols*, Methods in Molecular Biology, vol. 1915,
https://doi.org/10.1007/978-1-4939-8988-1_20,

calpain-2 (CAPN2 + CAPNS1). These methods can be readily applied to other genes.

2 Materials

2.1 PCR Genotyping

1. Taq DNA polymerase for PCR genotyping can be obtained from multiple commercial sources (e.g., NEB, FroggaBio, Thermo Fisher Scientific) and will typically be supplied with 10× reaction buffer to achieve 1× standard reaction of 10 mM Tris–HCl (pH 8.3 at 25 °C), 50 mM KCl, and 1.5 mM $MgCl_2$. Troubleshooting of PCR reactions may include increasing final $MgCl_2$ concentrations in 0.5 mM increments using a 50 mM $MgCl_2$ stock solution to achieve best results. This adjustment may be necessary depending on EDTA content in solutions used to purify DNA.
2. Deoxyribonucleotides (dATP, dCTP, dGTP, dTTP) are often sold by suppliers of Taq DNA polymerase as stock solutions at 10 mM concentrations of each dNTP. A more economical option is to purchase PCR-grade dNTPs, make stocks at 200 mM in 10 mM Tris–HCl (pH 7.5), 1 mM EDTA (stored at −80 °C), and make a working solution with 10 mM of each dNTP (stored at −20 °C).
3. Oligonucleotide primers for distinguishing the wild-type (*WT*), floxed (*flox*), or Cre-deleted knockout (*KO*) *capns1* alleles are the following (also *see* Fig. 1):

   Ex8FOR: 5′-ATGAACTTCCAGGGGCCTTTGAGGC-3′.
   Ex9REV: 5′-GCCGGATGATCATGCTGTAGAGATG-3′.
   Ex11REV: 5′-CTCAGCACTCAGAGTCCGAAGGAAG-3′.

4. Custom-synthesized oligonucleotides can be obtained from any reputable commercial or academic supplier. Dissolve oligonucleotides at 1000 pM/μL in 10 mM Tris–HCl (pH 7.5), 1 mM EDTA (stored at −80 °C). Dilute working stocks with dH_2O to 100 pM/μL (stored at −20 °C).
5. DNA extraction buffer for tail biopsies or cell pellets consists of:
   (a) 50 mM Tris (3.0275 g).
   (b) 100 mM EDTA (18.612 g).
   (c) 100 mM NaCl (2.922 g).
   (d) 1% SDS (25 mL 20% SDS).
   (e) Adjust to pH 8.0.

   Make up to 500 mL with dH_2O.
   Store at room temperature.

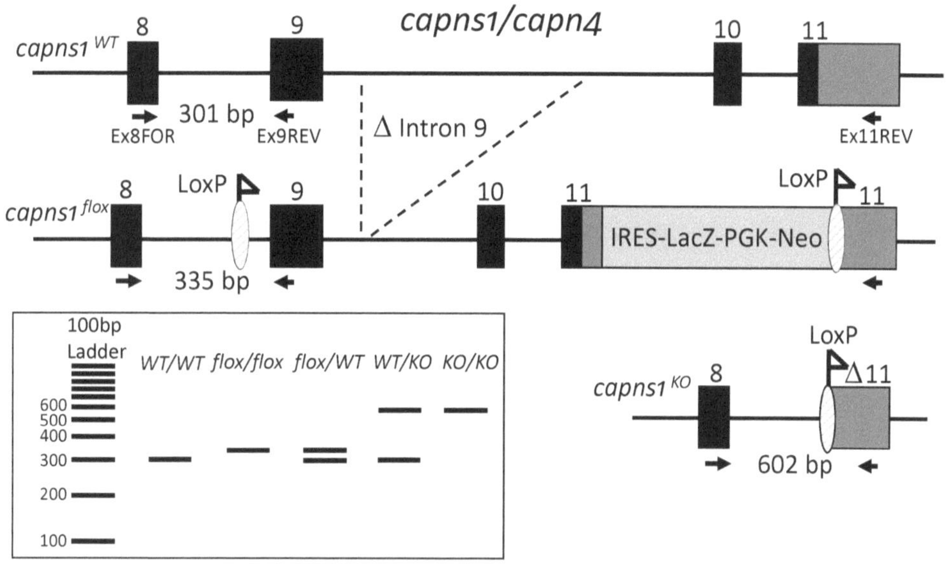

Fig. 1 PCR genotyping floxed capns1 mice. Wild-type (*capns1*^{WT}), floxed (*capns1*^{flox}), and Cre recombinase-deleted knockout (*capns1*^{KO}) alleles of the mouse *capns1/capn4* gene from the Tan et al. publication [2] are shown. Black boxes are coding and dark gray boxes are noncoding regions of the indicated exons. The internal ribosome entry site (IRES)-LacZ-PGK-Neo cassette (light gray) was inserted in the 3′ untranslated region of exon 11. LoxP sites recognized by Cre recombinase are shown as hatched white ovals with flags. The targeting strategy also involved deleting a nonessential portion of intron 9, indicated with dashed lines. Annealing positions for genotyping PCR primers are indicated (Ex8FOR, Ex9REV, and Ex11REV), and the sizes of the corresponding PCR products are shown between these primers. A representation of a genotyping gel analysis is shown with the interpreted genotypes indicated for each lane

6. Proteinase K (Sigma) is dissolved in water at 10 mg/mL and stored as 1 mL aliquots at −20 °C.
7. Prepare 5 M NaCl by adding 14.6 g NaCl to a 50 mL screw cap centrifuge tube and dissolving in dH_2O up to 50 mL.
8. TE-RNase A DNA buffer consists of 10 mM Tris–HCl (pH 8), 1 mM EDTA, and 100 μg/mL RNase A. Prepare 10 mL aliquots in snap-cap tubes, heat in a 65 °C water, bath for 30 min to inactivate any contaminating DNase activity, and store at −20 °C.
9. 6× DNA sample buffer prepared in a 15 mL graduated tube.
   (a) 25 mg bromophenol blue (migrates with ~300 bp DNA).
   (b) 4 g sucrose.
   (c) Add dH_2O up to 10 mL.

   Nutate at room temperature to dissolve components. Store at 4 °C to avoid microbial growth.
10. 10× TBE buffer for DNA agarose gel electrophoresis consists of:

(a) 432 g Tris (0.89 M).
(b) 220 g boric acid (0.89 M).
(c) 29.8 g EDTA (0.02 M).
(d) Adjust volume to 4 L with dH_2O.
(e) Adjust to pH 8.3.

Dilute 10× with distilled water to make up a 1× TBE working solution for casting agarose gels and running in electrophoresis unit.

11. UltraPure™ agarose powder (Invitrogen) for both analytical and cloning applications.
12. Ethidium bromide (Sigma). Prepare 10 mg/mL stock, and use at 100 ng/mL to stain DNA for UV imaging. Wear gloves when handling and eye protection when visualizing with UV light (*see* **Note 1**).

2.2 Lentiviral Production for Knockdown, Knockout, and Overexpression

1. Polybrene (hexadimethrine bromide—Sigma) to enhance lentiviral infection is prepared as a 1000 × stock (5 mg/mL) in dH_2O, sterilized using a 0.22 μM syringe top filter, and stored in a 15 mL screw-top centrifuge tube at 4 °C.
2. HEK293T (ATCC CRL-3216) and MDA-MB-231 (ATCC HTB-26) cells are grown in Dulbecco's Modified Eagle Medium (DMEM, Gibco) with 10% FBS and added antibiotics and antimycotics (Gibco) in a 37 °C incubator with 5% CO_2. Cells are maintained in a logarithmic growth state by splitting 1 to 4 upon reaching confluence using Trypsin/EDTA (Gibco). Cells that grow to super-confluent states are discarded and replaced (*see* **Note 2**).
3. Plasmid construction is performed by standard molecular biology and microbiology methods. Plasmid DNA for transfections is produced using Geneaid™ Plasmid Maxi Kit (FroggaBio).

3 Methods

3.1 Genotyping Conditionally Targeted Capns1/Capn4 Mice

Conditional targeting of the mouse *capns1* gene (previously known as *capn4*) is described in detail elsewhere [2]. The targeting strategy deleted nonessential sequences in intron 9 and introduced loxP sites into intron 8 and the 3′ noncoding region of exon 11, downstream from an inserted IRES-LacZ-PGK-Neo reporter/selection cassette (Fig. 1). The floxed allele ($capns1^{flox}$) still expresses functional CAPNS1, but Cre-mediated deletion of the loxP-flanked exons 9 and 10 and the coding region of exon 11 results in a null allele ($capns1^{KO}$). PCR primers in exon 8 and exon 9 are used to distinguish between the *capns1*^{flox} and *capns1*^{WT} alleles. PCR between exon 8 and exon 11 becomes possible after Cre-mediated deletion of sequences between the loxP sites. This section provides a protocol for preparing genomic DNA from mouse biopsies and genotyping these *capns1* alleles.

3.1.1 Extraction of Genomic DNA from Mouse Tissues

1. Cut approximately 5 mm from mouse tails at weaning, and transfer to labeled Eppendorf tubes. If not extracting DNA immediately, biopsy tissues can be stored at −20 °C.
2. Add 700 μL of DNA extraction buffer/proteinase K mixture (630 μL extraction buffer plus 70 μL 10 mg/mL proteinase K). Mix by inversion, and incubate in a 55 °C water bath overnight.
3. Add 250 μL 5 M NaCl. Nutate at room temperature for 30 min.
4. Centrifuge at room temperature at 13,500 rpm (15,689 × *g*) for 10 min.
5. Transfer all of the aqueous phase (~850 μL top layer) to a new tube using a P1000 tip with 3 mm cut off the end with scissors (this reduces DNA shearing during pipetting) (*see* **Note 3**).
6. Add 540 μL isopropanol to aqueous phase. Mix by repeated inversion until DNA precipitate forms. If DNA clump is large enough, capture it with a P200 tip and transfer to a new tube. If you do not see a large clump of DNA, mix on a nutator for 30 min, centrifuge for 5 min at 13,000 rpm (14,549 × *g*) at room temperature, decant supernatant (being careful not to dislodge pellet), quick spin, and remove residual supernatant with P200 tip.
7. Air-dry DNA for 20 min.
8. Add 100 μL TE-RNase A and let DNA dissolve at 37 °C for 30 min or overnight at 4 °C.

3.1.2 PCR Genotyping Genomic DNA from Mouse Tissues

1. Add 1 μL DNA sample to each PCR tube.
2. Prepare a PCR cocktail of all components except Taq DNA polymerase, and template DNA (make enough for all planned reactions, plus one).
3. Cool on ice, then add Taq DNA polymerase, and mix immediately before adding complete PCR cocktail to tubes with DNA sample.

Components	25 μL reaction	Cocktail for 10 reactions
10× PCR buffer	2.5 μL	25 μL
50 mM $MgCl_2$	1 μL	10 μL
10 mM dNTP	0.5 μL	5 μL
100 pmol/μL Ex8FOR	0.1 μL	1 μL

(continued)

Components	25 μL reaction	Cocktail for 10 reactions
100 pmol/μL Ex9REV	0.1 μL	1 μL
100 pmol/μL Ex11REV	0.1 μL	1 μL
dH_2O	19.45 μL	194.5 μL
Taq DNA polymerase	0.25 μL	2.5 μL

Ex8FOR: 5′-ATGAACTTCCAGGGGCCTTTGAGGC-3′.
Ex9REV: 5′-GCCGGATGATCATGCTGTAGAGATG-3′.
Ex11REV: 5′-CTCAGCACTCAGAGTCCGAAGGAAG-3′

4. Add 24 μL PCR cocktail to each DNA sample and mix by pipetting up and down three times. Discard used tip. Change to new tip to add and mix PCR cocktail with each DNA sample.
5. Overlay each reaction with one drop of PCR oil and close lids. Oil overlay is not required if using a PCR machine with a heated lid.
6. Place tubes in preprogramed PCR machine, and begin PCR cycle.

Cycling conditions	
a. 94 °C	5 min (preheat)
b. 94 °C	30 s
c. 61 °C	30 s
d. 72 °C	45 s (38 cycles of steps b–d)
e. 72 °C	10 min
f. 4 °C	hold (until analysis)

7. Add 5 μL 6× DNA gel-loading buffer to each completed PCR reaction.
8. Use P20 to mix and transfer 20 μL to wells of 2% agarose gel. Including EtBr in the gel and anode chamber will allow the progress of electrophoresis to be checked at intervals using a UV lamp.
9. Load 100 bp DNA ladder into last lane. Electrophorese until resolution of 301 and 335 bp products is apparent (*see* Fig. 1).

3.2 CAPNS1 Knockdown Using Lentiviral shRNA

Dharmacon markets lentiviral vectors expressing shRNAs directed against over 15,000 human and 15,000 mouse annotated genes with multiple shRNAs per gene. We recommend purchasing a panel of 3–5 shRNA vectors for a given gene and empirically testing them for target knockdown by immunoblotting analysis of transduced cells. The following is a description of stable CAPNS1 knockdown in human MDA-MB-231 breast cancer cells using the pLKO.1 vector system (*see* **Note 4**) expressing a *CAPNS1*-directed shRNA (TRCN0000003548) targeting nucleotides 506–526 (5′-CGCCACAGAACTCATGAACAT-3′) of the human *CAPNS1* transcript (X04106.1).

3.2.1 Generation of Lentivirus from Transfected HEK293T Cells

Addgene is a nonprofit entity that provides plasmids which are ideally suited for lentiviral production (https://www.addgene.org/viral-vectors/lentivirus/). There are many excellent resources on the Addgene website, including protocols and virus biosafety information that new users of viral vectors should familiarize themselves with. All virus work needs to be done in a Biosafety Level 2 (BL2) facility with protocols approved by the institutional biosafety committee (*see* **Note 5**).

HEK293T cells are an SV40 virus-transformed human embryonic kidney cell line that is well suited for lentivirus production by transfection with appropriate lentiviral provirus and packaging plasmids [3]. The following lentivirus production protocol is for transfections in 10 cm dishes. For other culture formats, scale up or down according to the culture dish surface area.

1. Day 1: Trypsinize log-growing HEK293T cells 18 to 24 h prior to transfection so that the cell density reaches the optimal ~75% confluency at the time of transfection.
2. Day 2: 30 to 60 min before transfection, the media on HEK293T plates is changed to 5 mL fresh complete culture medium (DMEM +10% FBS + antibiotics).
3. For each dish, dilute a total of 15 μg of DNA (8 μg pLKO.1 transfer vector + 6 μg packaging plasmid pCMVΔR8.74 + 2 μg envelope plasmid pMD2G) into 500 μL of serum-free DMEM, pipetting to mix.
4. In another tube, dilute 45 μL of PolyJet reagent (Cat# SL100688, SignaGen Laboratories) into 455 μL of serum-free DMEM, pipetting to mix.
5. Add the diluted PolyJet reagent immediately to the diluted DNA solution all at once (do not mix the solutions in the reverse order). Immediately pipette up and down three to four times to mix.
6. Incubate for 15 min at room temperature to allow DNA-PolyJet transfection complexes to form.
7. Add the 1000 μL transfection complex dropwise onto the 5 mL of medium in each dish, and move the plate gently back and forth to distribute transfection complexes evenly.
8. DNA-PolyJet complex-containing medium can be replaced with 6 mL fresh complete medium 6–8 h post-transfection.

9. Days 3 and 4: Harvest the virus-containing media, and replace with 6 mL fresh complete medium. Filter the virus-containing media through a 0.45 μm syringe top filter.
10. Use virus immediately to infect target cells (*see* below), or aliquot in cryovials, and store at −80 °C.
11. Harvest second batch of virus-containing media on day 4. Filter and use or store as described above.

3.2.2 Infection of MDA-MB-231 Cells for CAPNS1 Knockdown

MDA-MB-231 cells are a human triple-negative basal-like breast cancer cell line. The pLKO.1 vector system expressing a *CAPNS1*-directed shRNA (TRCN0000003548) has been proven effective in our hands at knocking down CAPNS1 to nearly undetectable levels [4].

1. MDA-MB-231 cells are grown in DMEM +10% FBS + antibiotics/antimycotics and split 1 to 4 the day before infection to achieve approximately 50% confluence in 6 cm plates.
2. Replace media with 2 mL fresh complete media, 1 mL virus-containing media (fresh or thawed), and polybrene to achieve a final concentration of 5 μg/mL. Fresh virus may be added after 24 h to achieve higher rates of transduction. Puromycin selection at 1 μg/mL is begun 48 h after infection. This selection should be applied in parallel to uninfected control cells to confirm puromycin toxicity (*see* **Note 6**).
3. Split puromycin-selected (lentivirus transduced) cells when they achieve confluence, and continue culturing in the presence of puromycin. Use one plate to generate a cell lysate for immunoblotting analysis to determine the degree of CAPNS1 knockdown (human-specific antibody to CAPNS1 from Santa Cruz Biotechnology, Inc.; sc-32325).
4. Once knockdown has been established, freeze down multiple vials of these early passage cells as knockdown may not be maintained over many passages.

3.3 Calpain Knockout Using CRISPR-Cas9

CRISPR-Cas9 methods have revolutionized gene manipulation, making it relatively routine for labs with skills in cell and molecular biology to knock out genes of interest in cultured cells. This section describes application of this method to the disruption of *CAPNS1*, *CAPN1*, and *CAPN2* genes in human cells to produce cells lacking calpain-1 (*CAPN1 KO*), calpain-2 (*CAPN2 KO*), or both calpain-1 and calpain-2 (*CAPNS1 KO*). A detailed description of the CRISPR-Cas9 method to knock out genes in human cells has been published by Feng Zhang and colleagues [5]. That publication includes an appendix with suggested sgRNA target sequences for 18,080 human genes (*see* **Note 7**). We selected two target sequences from that appendix for *CAPNS*1 and independently designed two sgRNA target sequences for each of the *CAPN1*

Table 1
CRISPR-Cas9 targeting of *CAPNS1*, *CAPN1*, and *CAPN2*

Gene	Target sequence	cDNA nucleotides	Gene ref.#
CAPNS1#1	GGCGGCTGCGCAGTACAACC	371–390	X04106.1
CAPNS1#2	CCAACGAGAGTGAGGAGGTC	433–452	X04106.1
CAPN1#1	CGAGTGGTACGAGTTGCGCA	448–467	NM_005186.3
CAPN1#2	CCCAATTCCTCCAAGACCTA	405–424	NM_005186.3
CAPN2#1	CAGCCGAAGGGAGCGAGTTC	627–646	BC021303.2
CAPN2#2	TGGGGAGAAGTGGAGTGGAC	980–999	BC021303.2
hU6 F	GAGGGCCTATTTCCCATGATT	2608–2628	LentiCRISPRv2
LKO.1	GACTATCATATGCTTACCGT	2779–2798	LentiCRISPRv2

and *CAPN2* genes (Table 1). Detailed instructions for designing sgRNA encoding oligonucleotides and cloning them into LentiCRISPRv2 [6] (Addgene, https://www.addgene.org/52961/) are provided by the GeCKO Lentiviral CRISPR Toolbox resources from the Zhang Lab (http://genome-engineering.org/gecko/). LentiCRISPRv2 vectors direct expression of both the gene-directed sgRNA and the Cas9 endonuclease in transduced puromycin-selected cells. Other vectors can also be found at this resource.

3.3.1 Generation of CRISPR-Cas9 Calpain Knockout Vectors

1. Order sense and antisense oligonucleotides corresponding to the two gene target sequences shown in Table 1, including 5′ and 3′, extended sequences to enable annealing with BsmBI-digested LentiCRISPRv2 (*see* below). Consult the GeCKO Lentiviral CRISPR Toolbox resources from the Zhang Lab for additional information (http://genome-engineering.org/gecko/). Oligos to target the human *CAPNS1*#1 target sequence shown in Table 1 are illustrated below, along with the genomic target sequence, including the adjacent PAM trinucleotide (NGG) and the encoded amino acid sequence (*see* **Note 8**).

```
                                             PAM
Genomic            5'-AGCGAGGCGGCTGCGCAGTACAACCCGGAGCCC-3'
Sequence           3'-TCGCTCCGCCGACGCGTCATGTTGGGCCTCGGG-5'
                        S  E  A  A  A  Q  Y  N  P  E  P
Sense oligo        5'-CACCGGGCGGCTGCGCAGTACAACC
Antisense oligo        3'-CCCGCCGACGCGTCATGTTGGCAAA-5'
```

2. Digest LentiCRISPRv2 with BsmBI and gel purify 13 kbp vector away from 2 kbp filler fragment.

3. Phosphorylate oligo pairs and ligate with gel-purified vector.
4. Transform Stbl3 bacteria with ligation mixture. Alternatively, DH5α or any recombination-deficient bacterial strain can be used.
5. Pick colonies and prepare miniprep DNA for sequencing with the hU6 F or LKO.1 sequencing primers shown in Table 1. Confirm correct insertion of 20mer targeting sequence.
6. Prepare maxiprep DNA from sequence-validated gene-specific LentiCRISPRv2 knockout vectors.
7. Prepare lentivirus by co-transfection of HEK293T cells with gene-specific LentiCRISPRv2 shuttle vector, packaging plasmid pCMVΔR8.74, and envelope plasmid pMD2G, as described above in Subheading 3.2.1—Generation of Lentivirus from Transfected HEK293T Cells.
8. Harvest virus from HEK293T cells and infect MDA-MB-231 cells as described above in Subheading 3.2.2—Infection of MDA-MB-231 Cells for CAPNS1 Knockdown.
9. Perform immunoblotting analysis of the polyclonal population, and compare CAPNS1 protein band intensity to parental MDA-MB-231 cell lysates. Expect to see a significant reduction of the targeted gene product in lysates of the polyclonal population. Then proceed to clone isolation.
10. Trypsinize the polyclonal population to single cells, count and plate in 96-well dishes to achieve wells with single cells.
11. Expand clones into 12-well dishes and then to 6-well dishes, and reseed the 12-well dishes to prepare lysates for immunoblotting analysis.
12. Identify null clones by immunoblotting. Expand and freeze down multiple vials of cells for long-term liquid nitrogen storage.

3.4 Ectopic Expression of Calpain

Phenotypes observed in CRISPR-Cas9 knockout cells can be validated as being directly due to loss of the targeted gene or further explored by structure-function analysis by ectopically reexpressing recombinant wild-type or mutant versions of the targeted gene (*see* **Note 9** and **10**). Here we describe making retroviral vectors to reexpress *CAPNS1*, *CAPN1*, or *CAPN2* in their respective CRISPR-Cas9 knockout cells or ectopically overexpress these genes in other cells of interest.

3.4.1 Cloning Murine Capns1, Capn1, and Capn2 into Retroviral Vectors

1. Generate oligonucleotides to PCR amplify the complete open reading frames of mouse *capns1* (AF058298) (*see* **Note 11**), *capn1* (NM_001110504), or *capn2* (NM_009794) (Table 2). Incorporate unique restriction sites into these oligonucleotides

Table 2
Gene-specific PCR primers

Gene	Oligonucleotide
CAPNS1 FOR#1	5′-GCA**GGATCC**GACTCAGCTGCGATCATGTTC-3′
CAPNS1 FOR#2	5′-GGT**GGATCC**GCCATGGGCATCCTGGGCGGAGTCA-3′
CAPNS1 REV	5′-CAGTC**GTTAAC**TCCCATTCAGGAATACATAG-3′
CAPN1 FOR	5′-AGC**AGATCT**AGCCACCAGAGGATGACAGAGG-3′
CAPN1 REV	5′-GGCG**GTTAAC**GGCGGGGAGGCGTGGCTTG-3′
CAPN2 FOR	5′-GATC**CTCGAG**ATGGCGGGCATCGCGATAAAGC-3′
CAPN2 REV	5′-GCAT**GTTAAC**TCAGAGTACTGAAAAACTCAGCC-3′

to facilitate directional cloning into an expression vector of choice. In this case we have incorporated 5′ BamHI (*capns1*), BglII (*capn1*), or XhoI (*capn2*), respectively, and 3′ HpaI sites for all three to facilitate directional cloning between those sites in the retroviral vector pMSCVhyg (Clontech). Similar approaches can be used for other vectors (*see* **Note 12**).

2. Use a high-fidelity thermostable DNA polymerase such as KOD Hot Start from Novagen or PfuUltra from Agilent Genomics to amplify the murine *CAPNS1*, *CAPN1*, and *CAPN2* open reading frames from a convenient source of cDNA using the above gene-specific PCR primers.
3. Check an aliquot by agarose gel electrophoresis to confirm the size of the PCR products: *capns1* (850 or 650 bp, using CAPNS1 FOR#1 or #2, respectively); *capn1* (2243 bp); and *capn2* (2122 bp).
4. Purify the PCR products with PureLink Quick PCR Purification Kit (Invitrogen, catalog no. K310001) to remove primer dimers or short spurious PCR products.
5. Digest PCR products, and vector with appropriate restriction enzymes, resolve on agarose gels to locate bands of interest using long wavelength UV lamp, and then excise and purify DNA using PureLink Quick Gel Extraction Kit (Invitrogen, catalog no. K210012).
6. Ligate vectors with inserts and transform bacteria to obtain clones.
7. Select six colonies for each cloning and prepare miniprep DNA for sequencing validation.
8. Prepare maxiprep DNA from clones with correct sequence.
9. Co-transfect *CAPNS1*, *CAPN1*, and *CAPN2* retroviral vectors with packaging plasmids into HEK293T cells to produce

amphotropic retrovirus (as described in Subheading 3.2.1 above), and infect MDA-MB-231 gene-specific KO cells.

10. Validate rescue of gene expression by immunoblotting (CAPNS1, Abnova M01, clone 3C4; CAPN1, Santa Cruz, sc-32,325; CAPN2, Cell Signaling, #2539) (*see* **Note 13**).

4 Notes

1. Always use UV protective eyewear when looking at EtBr-stained DNA using UV lamp.
2. HEK293T cells are grown in DMEM +10% FBS + antibiotics in a 37 °C incubator with 5% CO_2. They should be split with trypsin/EDTA upon reaching confluence. Cells that have been maintained at confluency for any length of time should be discarded and replaced.
3. Use scissors to cut 2 mm off tips used to pipette genomic DNA.
4. The pLKO.1 vector system is the original lentiviral system used by The RNAi Consortium (TRC), a public-private initiative from the Broad Institute, to provide genome-wide knockdown reagents for human and mouse genes. pLKO.1 vectors include a puromycin selection feature. Later generation vectors have added a GFP reporter (pGIPZ) or RPF and doxycycline-inducible features (pTRIPZ). Plasmids encoding packaging components required to produce lentiviral particles (pCMVΔR8.74 and pMD2G) are available from Addgene or Dharmacon.
5. Lentivirus produced with this protocol is capable of infecting humans. All work should be carried out in a BL2 facility. Gloves should be worn at all times, and all used materials (plastic, glass) should be sterilized before disposal according to institutionally approved biohazard procedures. Infected and retrovirally transduced cells should be considered potential sources of infectious virus until absence of packaging component genes or proteins can be verified as absent using PCR or ELISA methods, respectively. When using third-generation lentivirus production protocols, the probability of infectious virus beyond 72 h is very low.
6. Lentivirus titers produced with these methods can often achieve multiplicity of infection (MOI) rates giving near complete survival of transduced cells after puromycin challenge. For this reason, it is important to include non-transduced cells as a negative control during puromycin selection. Virus titers can be determined by performing dilution series of virus, infecting cells, and counting colonies after puromycin selection.

7. Two target sequences are provided in Table 1 for each gene. We recommend trying two or more sgRNA targets per gene to increase the probability of achieving a good knockout frequency.
8. If selecting sgRNA target sequences manually, note that the 20mer must be followed immediately by the NGG "PAM" sequence; where N is any nucleotide, but the G residues are invariant. This PAM sequence is absolutely required for recognition by Cas9 nuclease.
9. If possible, take advantage of the PAM requirement by designing a rescue construct with a silent mutation of one of the invariant G residues of the PAM sequence. Otherwise, rescue constructs should have two or more silent mutations introduced into the 20 nucleotide sgRNA target sequence. These changes will render your rescue construct resistant to the continuing expression of the gene-specific sgRNA and Cas9. Note that in the *CAPNS1* example given above, the second nucleotide of the PAM is the third nucleotide of a proline codon. Mutating this G to any other base destroys the PAM but retains a proline codon.
10. The QuikChange II XL kit (Agilent Technologies, Cat # 200521-5) provides a convenient approach to generating mutations in the sgRNA target sequence or associated PAM trinucleotide.
11. We have found it is difficult to PCR amplify the 5′ region of *CAPNS1*, which is highly GC rich and very repetitive. A functional CAPNS1 subunit can be produced lacking the glycine-rich N-terminus by starting at the third in-frame ATG methionine-60 codon using the CAPNS1 FOR#2 primer shown above. This primer also introduces a C > G change immediately after the ATG to make it a better Kozak consensus sequence to improve expression.
12. cDNAs corresponding to mouse, rat, or human *CAPNS1* and mouse *CAPN1* and *CAPN2* can be obtained from the authors. These cDNAs can also be obtained from Addgene or commercial sources including Dharmacon. Alternatively, reverse transcription and high-fidelity PCR can be used to clone them from most sources of mRNA, as these calpains are ubiquitously expressed. Regardless of source, the sequence should be validated.
13. Knockout of *CAPNS1* will not only result in loss of the CAPNS1 protein but also a substantial reduction in CAPN1 and CAPN2 levels. This is likely due to instability of the catalytic subunits in the absence of the regulatory subunit.

Acknowledgment

This work was supported by a grant from the Canadian Institutes of Health Research (81189) to PAG.

References

1. Wang H et al (2013) One-step generation of mice carrying mutations in multiple genes by CRISPR/Cas-mediated genome engineering. Cell 153(4):910–918
2. Tan Y et al (2006) Conditional disruption of ubiquitous calpains in the mouse. Genesis 44 (6):297–303
3. Zufferey R et al (1998) Self-inactivating lentivirus vector for safe and efficient in vivo gene delivery. J Virol 72(12):9873–9880
4. Grieve S et al (2016) Calpain genetic disruption and HSP90 inhibition combine to attenuate mammary tumorigenesis. Mol Cell Biol 36 (15):2078–2088
5. Shalem O et al (2014) Genome-scale CRISPR-Cas9 knockout screening in human cells. Science 343(6166):84–87
6. Sanjana NE, Shalem O, Zhang F (2014) Improved vectors and genome-wide libraries for CRISPR screening. Nat Methods 11(8):783–784

INDEX

Jeannette S. Messer (ed.), *Calpain: Methods and Protocols*, Methods in Molecular Biology, vol. 1915, https://doi.org/10.1007/978-1-4939-8988-1, © Springer Science+Business Media, LLC, part of Springer Nature 2019

F

G

H

I

K

L

M

P

R

S

U

MIX
Papier aus verantwortungsvollen Quellen
Paper from responsible sources
FSC® C105338

If you have any concerns about our products,
you can contact us on
ProductSafety@springernature.com

In case Publisher is established outside the EU,
the EU authorized representative is:
Springer Nature Customer Service Center GmbH
Europaplatz 3, 69115 Heidelberg, Germany

Printed by Libri Plureos GmbH
in Hamburg, Germany